The Notochord

The Notochord
Development, Evolution and Contributions to the Vertebral Column

P. Eckhard Witten and Brian K. Hall

CRC Press
Taylor & Francis Group
Boca Raton London New York

CRC Press is an imprint of the
Taylor & Francis Group, an **informa** business

Cover Image: Cross section of the notochord of a three-month-old (5.2 cm total length) embryo of the lesser spotted dogfish, *Syliorhinus canicula*. The notochord is surrounded by a cartilaginous notochord sheath and by hyaline cartilage. Image width = 660 μm. Agnes Boutet (Roscoff) is thanked for providing the embryo. Processing and microphotography by PEW.

First edition published 2022
by CRC Press
6000 Broken Sound Parkway NW, Suite 300, Boca Raton, FL 33487-2742

and by CRC Press
2 Park Square, Milton Park, Abingdon, Oxon, OX14 4RN

© 2022 Taylor & Francis Group, LLC

CRC Press is an imprint of Taylor & Francis Group, LLC

ISBN: 9781498787918 (hbk)
ISBN: 9781032162683 (pbk)
ISBN: 9781315155975 (ebk)

DOI: 10.1201/9781315155975

Typeset in Times
by codeMantra

Contents

SECTION I Development and Evolutionary Origin of the Notochord

SECTION II Function of the Notochord and Notochord Sheath

SECTION III Nature and Fate of the Notochord and Vertebrae across the Vertebrates

SECTION IV Relationships: Notochord-Cartilage, Notochord-Vertebrae, Notochord-Vertebral Column

Series Preface

In recent decades, evolutionary principles have been integrated into biological disciplines such as developmental biology, ecology and genetics. As a result, major new fields emerged, chief among which are Evolutionary Developmental Biology (or Evo-Devo) and Ecological Developmental Biology (or Eco-Devo). Inspired by the integration of knowledge of change over single life spans (ontogenetic history) and change over evolutionary time (phylogenetic history), Evo-Devo produced a unification of developmental and evolutionary biology that generated unanticipated synergies: Molecular biologists employ computational and conceptual tools generated by developmental biologists and by systematists, while evolutionary biologists use detailed analysis of molecules in their studies. These integrations have shifted paradigms and enabled us to answer questions once thought intractable.

Major highlights in the development of modern Evo-Devo are a comparison of the evolutionary behavior of cells, evidenced in Stephen J. Gould's 1977 proposal of changes in the timing of the activity of cells during development – heterochrony – as a major force in evolutionary change, and numerous studies demonstrating how conserved gene families across numerous cell types "explain" development and evolution. Advances in technology and in instrumentation now allow cell biologists to make ever more detailed observations of the structure of cells and the processes by which cells arise, divide, differentiate and die. In recent years, cell biologists have increasingly asked questions whose answers require insights from evolutionary history. As just one example: How many cell types are there, and how are they related? Given this conceptual basis, cell biology – a rich field in biology with a history going back centuries – is poised to be reintegrated with evolution to provide a means of organizing and explaining diverse empirical observations and testing fundamental hypotheses about the cellular basis of life. Integrating evolutionary and cellular biology has the potential to generate new theories of cellular function and to create a new field, *Evolutionary Cell Biology.*

Mechanistically, cells provide the link between the genotype and the phenotype, both during development and in evolution. Hence, the proposal for a series of books under the general theme of *Evolutionary Cell Biology: Translating Genotypes into Phenotypes* to document, demonstrate and establish the central role played by cellular mechanisms in the evolution of all forms of life.

Brian K. Hall
Sally A. Moody

Preface

To the best of our knowledge, we believe that this is the first book to be devoted to the notochord. How can that possibly be? The notochord, a tubular organ ventral to the dorsal neural tube in all chordates and the structure that gives the name to the phylum (Chordata), was the object of intensive research in the decades following the publication of *On the Origin of Species* by Charles Darwin (Darwin, 1859). With excitement, as proof for the descent of vertebrates from invertebrate chordate ancestors, the existence of the notochord and members of the Chordata were identified, described and classified (Chapters 1 and 2)

Uniformly across the chordates (including vertebrates) early in development the notochord is composed of large vacuolated cells known as chordocytes that secrete an extracellular sheath known as the notochord sheath. Although a defining organ (synapomorphy) of the chordate phylum, the notochord in jawed vertebrates is often regarded as a transient embryonic structure with limited or no function in adults. Many textbooks imply or state that the limited function during early development is the end of the story about notochord cells, structure, function, development and evolution. Not so. Despite its persistence as a central chordate feature, the notochord has undergone an enormous amount of evolutionary change. The notochord retains its basic functions as a signaling center during development required for the patterning of the dorsal nerve cord and the vertebral centra, and retains the function of providing a flexible skeleton for vertebrate embryos. These conserved roles only touch the surface of the function, maintenance and fate of the notochord and notochord cells.

The limited view of notochord function that prevails in textbooks is likely based on the erroneous beliefs that our own, mammalian, notochord is short-lived and has a function only during early embryonic development. Not so. In all vertebrates, the notochord is the signaling center that initiates vertebral development. Depending on taxonomic group, early steps in vertebral body development either involve mineralization of the notochord sheath or chondrogenic somite-derived cells that surround the notochord. Another important role of the notochord is patterning and development of the intervertebral joints. Most vertebrate species retain a notochord lifelong as functional and metabolically active component of the intervertebral disks.

The aim of this book is no less than to evaluate all the evidence available to understand the development and evolution of the notochord across the vertebrates and the role the notochord plays in vertebral and vertebral column development. This journey will take us back to studies published over 160 years ago and forward to studies published recently. Section I begins with how the notochord develops and produces the notochord sheath, and how discovery of the vertebrate notochord and the same structure in non-vertebrate chordates transformed our ideas on the evolutionary origin of the vertebrates and led to the recognition of a phylum Chordata. We review the ongoing debate about the germ layer origin of the notochord, whether it arises from embryonic mesoderm or ectoderm, whether its origin has been conserved throughout vertebral evolution, and whether all sections of the notochord arise from the same germ layer.

We continue, in Section II, to discuss how the notochord specifies ectoderm as neural and establishes both the anterior-posterior (A-P) and dorsal-ventral (D-V) axes of the embryonic neural tube, using signaling pathways based on the genes *Shh*, *Pax*, and Bmps and the Bmp inhibitors noggin and gremlin. The early development of notochord cells, genes expressed by the cells and the cells' products evoke two important topics that are further addressed in Sections III and IV: the relationship between notochord and cartilage, that both have collagen type II-based extracellular matrices, and the basic patterning function of the notochord for the vertebral column.

The two chapters in Section III evaluate the evidence for the role and fate of the notochord across the vertebrates. Whether the notochord is segmented has been a long-standing issue, relating as it does to how segmentation of the vertebrae and vertebral column is (are) initiated and controlled. Four modes of initiation of vertebral body centra emerge: (i) mineralization or chondrification of the notochord sheath, (ii) deposition of bone around the notochord, (iii) deposition of cartilage around the notochord, and (iv) formation of cartilage continuous with the cartilaginous vertebral arches. Next we provide an overview about the fate and the function of the notochord in adult vertebrates, other than the better-known central part of the mammalian intervertebral disk. Notochords may remain in adult vertebrates to provide axial support as a hydroskeleton, transform into a functional component of the intervertebral joints to form the intervertebral disks or transform into cartilage (later also into bone) in amphibians and reptiles, including birds.

The two chapters in Section IV draw conclusions on the two major themes: the relationship between notochord and cartilage (Chapter 8) and the evolution of the vertebral column from a notochord perspective (Chapter 9). Both topics involve assessments of homology of cells (chordocytes and chondrocytes), developmental modules, tissues (notochord and cartilage), organs (vertebrae and vertebral columns) and developmental mechanisms (deposition of a notochord sheath and of cartilage extracellular matrix, initiation of vertebral development through mineralization of the notochord sheath or by chondrogenesis or osteogenesis of sclerotomal mesenchymal cells), and, finally, decisions of levels of homology.

Some readers may remark that more emphasis could have been placed on the somites. Naturally, the authors are fully aware of the important function of the somites. Many excellent textbooks and scholarly articles address the function of somites but sideline the notochord. Indeed, this book focuses on the notochord. Development and evolution are viewed from a notochord perspective which we hope will complete the literature and our knowledge about the organ that defines our own phylum.

P. Eckhard Witten
Brian K. Hall

Acknowledgments

The authors are particularly grateful to John G. Maisey (New York) who contributed expertise, insight and text about the chondrichthyan notochord. The authors are very grateful to all colleagues who granted permission to use figures from their scientific works, as acknowledged in the figure legends. This book has a history. It would not have been written if Brian Hall had not accepted PEW as a postdoc in his laboratory, without everything PEW learned from Brian and for the fact that PEW's research on notochord-related topics started at Dalhousie University (Halifax, Canada).

PEW gratefully acknowledges colleagues with whom he had and has the privilege of collaborating on notochord and vertebral column developmental-related topics. Foremost Brian Hall (Halifax) and Ann Huysseune (Ghent). Further, and in no particular order, PEW is very grateful to Agnes Boutet (Roscoff), Antonella Forlino (Pavia), Christoph Winkler (Singapore), Clara Boglione (Rome), Paul Coucke (Ghent), the late Doris W. Au (Hongkong), Erika Kague (Bristol), Giorgos Koumoundouros (Heraklion), Björn Busse (Hamburg), M. Leonor Cancela (Faro), Per Gunnar Fjelldal (Matredal), Daria Larionova (Ghent) and Santosh P. Lall (Halifax). Harald Rosenthal (Neu Wulmstorf) is thanked for supporting the authors' joint research in Halifax. PEW is grateful for contributions made by his former PhD students (now grown scientists): Anabela Bensimon Brito (Marseille), Adelbert De Clercq (Ostend), Arianna Martini (Rome). Current PhD students (soon grown scientists) who contributed to the subject are Lucia Drábiková, Silvia Cotti and Claudia Di Biagio. Their contributions are possible through the initiative of Marc Muller (Liège) and funding from the *European Union's Horizon 2020 research and innovation programme* under the Marie Sklodowska-Curie grant agreement No. 766347. PEW thanks the Skretting Aquaculture Research Centre (Stavanger) and the Vehice Histopathology Laboratory (Puerto Montt) for supporting aquaculture-related research on the notochord.

BKH is grateful to PEW for the invitation to join him in this project and for the many interactions initiated by the research and writing. Thank you Eckhard. Comments on portions of the text by Gloria Arratia (Kansas), Olivier Pourquié (Boston), Claudio Stern (London) and Matt Vickaryous (Guelph) are acknowledged. Research in the BKH laboratory, including developmental studies of the notochord, somites and vertebrae, was supported from 1968 to 2017 by grant A5056 from the National Research Council (NRC) of Canada and then by the Natural Sciences and Engineering Research Council of Canada (NSERC). PEW and BKH are thankful to Ann Huysseune and June Hall for commenting on the manuscript and for careful proof-reading.

Authors

P. Eckhard Witten (Dr. rer. nat.) was educated as a Zoologist at the Zoological Institute and Zoological Museum of the University of Hamburg (Germany). His first postdoctoral position brought him to the Centre for Biomechanics, Department Osteopathology at Hamburg's University Hospital. For a second postdoc he went to the laboratory of Brian K. Hall (Dalhousie University, Halifax, Canada) to study jaw and tooth alterations in wild Atlantic salmon. It was in Brian Hall's laboratory and in collaboration with Santosh Lall (NRC, Halifax) that PEW became engaged in research on the vertebral column and thus on the notochord. PEW continued research on the subjects in Norway as head of the Fish Health Department at the Skretting Aquaculture Research Centre (Stavanger). Today PEW is a professor at Ghent University in Belgium. Together with Ann Huysseune, he co-supervises the Evolutionary Developmental Biology laboratory, which specializes in the development and evolution of skeletal tissues, ranging from the notochord to the vertebral column, dermal skeleton and dentition. Several figures in this book derive from research on zebrafish and Japanese medaka in the context of the laboratories' biomedical research projects. PEW was co-founder of the European Society for Evolutionary Developmental Biology. He and his colleague Leonor Cancela are founders and organizers of the international conference series Interdisciplinary Approaches in Fish Skeletal Biology (IAFSB.org).

Brian K. Hall, Ph.D., D.Sc., LL.D. (hc), F.R.S.C., University Research Professor Emeritus at Dalhousie University, was trained in Australia as an experimental embryologist. His research on the differentiation of skeletal tissues led him to earlier stages of embryonic development and the origin and function of skeletogenic neural crest cells. Comparative studies using embryos from all five classes of vertebrates provided a strong evolutionary component to his research. These studies, along with analyses of the developmental basis of homology, played significant roles in establishing evolutionary developmental biology. A Fellow of the Royal Society of Canada, Foreign Fellow of the American Academy of Arts and Sciences and recipient of the $100,000 2005 Killam Prize in Natural Sciences from the Canada Council for the Arts, he was one of eight individuals awarded the first Kovalevsky Medals in 2001 to recognize distinguished scientists of the 20th century in comparative zoology and evolutionary embryology.

Section I

Development and Evolutionary Origin of the Notochord

Each of the three chapters in this section has superficially simple aims: Discuss the nature of the notochord and notochord cell and examine how both develop in vertebrate embryos (and to a lesser extent in non-vertebrate chordate embryos [Chapter 1]); provide a history of the discovery of the notochord in chordates and vertebrates and discuss our understanding of notochord evolution (Chapter 2); and examine the evidence for a mesodermal and/or endodermal origin of the notochord in vertebrate embryos.

The first task (Chapter 1) is not as straightforward as it seems at first glance. In all vertebrates, the notochord comprises large vacuolated cells (chordocytes) connected by desmosomes and surrounded by an expanded and multilayered basement membrane known as the notochord sheath. In all vertebrates and non-vertebrate chordates examined, the notochord is the signaling center that triggers the differentiation of neural ectoderm, neural plate, and neural tube using antagonists of Bmp and expression of *Shh*. Teleost and amniote notochords stand apart in demonstrating divergent functions in later embryonic stages. Understanding the nature of the notochord sheath turns out to be surprisingly complicated; the sheath has undergone a considerable evolutionary change in different groups of vertebrates. That said, the notochord shares strong similarities with cartilage in structure and early morphogenesis and in biochemical composition of the sheath when compared with the extracellular matrix of cartilage. In adult vertebrates, the notochord may be retained as the primary axial skeleton, replaced by a cartilaginous or ossified vertebral column, form intervertebral disks or form intravertebral cartilage.

DOI: 10.1201/9781315155975-1

The second task (Chapter 2)—the evolutionary origin of the notochord—should also be straightforward; look at the fossil record. However, notochords fossilize poorly, especially in the basal chordates, where we would expect to find them. The question of who qualifies as a chordate, on the other hand, has a long and exciting history discussed in this chapter. Discovery of the notochord in embryos previously assigned to a plethora of animal groups either led those animals to be classified as chordates and discussed as potential chordate ancestors, or removed them from the chordates when their 'notochord' was discovered to be a different structure, albeit lying above (dorsal to) the neural tube. Non-chordates, such as annelids, became potential chordate ancestors. Why? Because they have a ventral nerve cord and associated axochord. To have been chordate ancestors, they would have had to invert their dorso-ventral axis by ninety degrees, which sounds preposterous. The annelid hypothesis is supported by the patterns of gene expression shared by axochord and notochord but must be refuted based on the phylogentic position of annelids. This illustrates how much the nature of evidence and the evidence of nature have changed since the 19th century.

The third task (Chapter 3)—the germ layer of origin of the notochord—has been debated even longer than has the evolutionary origin of the notochord a debate usually traced back to Karl von Baer who, in 1828, identified the notochord (his chorda dorsalis) as the precursor of the vertebral column. One hundred years later, Edwin Goodrich concluded that the notochord is endodermal in origin in all vertebrates (Goodrich 1930). However, the location of the endoderm—the dorsal roof of the developing gut—contains both endodermal and mesodermal cells, the anterior portion of the notochord can arise from different cells than does the posterior portion, and early in development notochord precursor cells are admixed with future mesodermal cells in what is known as the chordamesoderm.

1 Introduction to the Notochord and to Notochord Cells

CONTENTS

1.1 INTRODUCTION

The notochord is a closed tube that consists of large vacuolated cells surrounded by an extracellular matrix as a sheath (known as the *notochord sheath*[1]) that it secretes, positioned immediately ventral to the dorsal nervous system. The notochord has for centuries been recognized as one of the few organs that unite chordates. The Phylum Chordata includes the vertebrates. The presence of a vertebral column that forms *around and from* the notochord separates vertebrates from all other animal phyla, then and still referred to as the "invertebrate phyla", as if absence of a vertebral column was a unifying feature, which it is not.

The notochord is a *synapomorphy* for the chordates. A synapomorphy or shared-derived character is a feature (a character) present in an ancestral species and in its evolutionary descendants to the exclusion of all other organisms. Four synapomorphies separate chordates from all other animals: (i) a dorsal notochord, (ii) a hollow dorsal nerve cord, (iii) pharyngeal pouches and pharyngeal slits and (iv) a postanal

---

[1] The notochord sheath is also referred to as the notochordal sheath, perinotochordal sheath, chordal sheath or expanded basement membrane. We use notochord sheath.

DOI: 10.1201/9781315155975-2

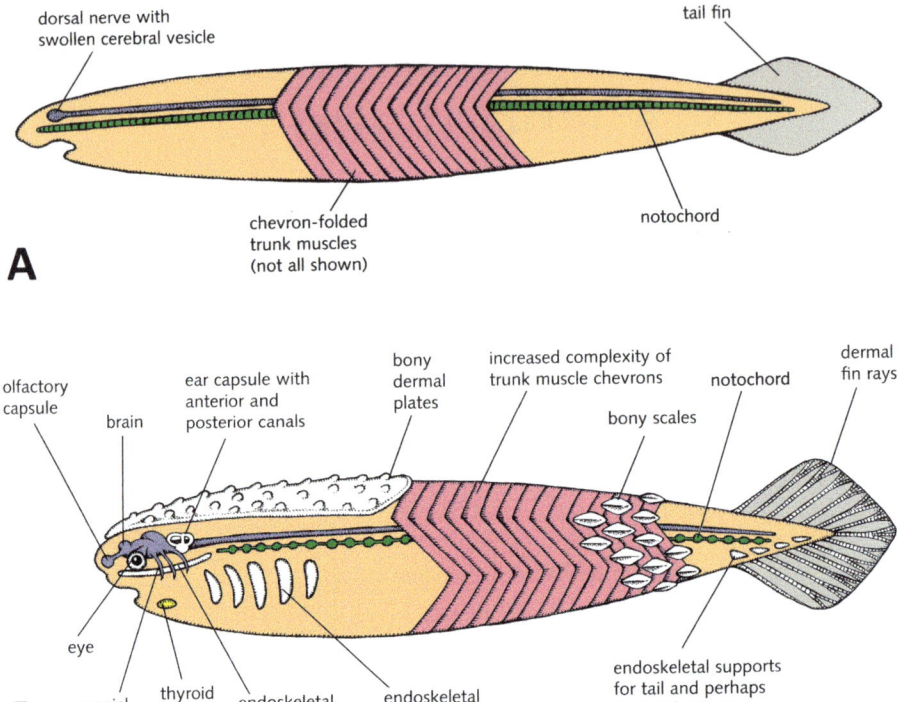

FIGURE 1.1   (A) Generalized and simplified characters of a chordate. The continuous noto-chord (green) is located below the dorsal nerve cord (blue). Chevron-shaped muscles are found in aquatic vertebrates. (B) Generalized and simplified characters of a vertebrate. The notochord remains as a continuous rod with intersegmental dilatations. The concept of a chordate, respectively vertebrate, archetype represented by these two figures was introduced by Richard Owen (Owen 1848). (Figures modified from Maisey, J. G., 2000. *Discovering Fossil Fishes.* Westview Press, Boulder Colorado.)

tail. A fifth character, (v) the endostyle, is a filter-feeding organ and precursor of the thyroid gland, found in a subset of chordates — urochordates, cephalochordates and larval lampreys, and as thyroid gland in all vertebrates (Figure 1.1).

A *novelty* or novel structure is the appearance of a feature in a lineage that was not present in the ancestors of the lineage. To define a character as novelty requires that the feature was not present in the ancestor and implies that a precursor of the structure was not present in the ancestor (Hall 2013). The notochord would have been a novel structure when it or its precursor originated in the most primitive chordates that are members of the deuterostome clade. But, what do we mean by a precursor? Obviously not something we would recognize as a notochord. What about cells/tissues that could have the cellular basis for the origin of the notochord? What about pre-existing genes or signaling pathways that could have been modified at the origination of notochord cells (Stone and Hall 2004; Hall and Kerney 2012)? Would such cells or genes represent latent homologues of the notochord? Could cells in pre-chordates that served a similar structural function — structural support, say — be considered as potential precursors? If so, where in the organism would we look for such structures; along the dorsal body or elsewhere? The fossil record alone cannot answer the question about the origin of the notochord. Consequently, over the past 200 years, almost every invertebrate phylum has been proposed as a starting point for giving rise to chordates and further to vertebrates (Holland et al. 2015). Questions about the origin of the notochord are taken up in this book, especially in Chapters 2 and 8. These important questions pertain to any novel structure such as the neural crest and neural crest cells (Stone and Hall 2004; Hall 2005, 2018). Therefore, understanding the notochord and its origination may well illuminate the origin of other features, including the origin of chordate synapomorphies.

1.2   STRUCTURE OF THE NOTOCHORD

The notochord or chorda dorsalis is the defining organ of the chordate phylum, which comprises cephalochordates, urochordates and vertebrates.

In cephalochordates, the notochord extends along the entire antero-posterior length of the animal; members of the genus *Branchiostoma* (amphioxus) are prominent and well-studied representatives. Urochordates have a notochord in the tail of neotenic adults (larvaceans) or in the tail of planktonic larvae prior to metamorphosis (tunicates). The vertebrate notochord is a closed tube that extends from the hypophyseal (pituitary) region of the brain to the tip of the tail. The appearance of the notochord precedes the origin of the vertebral column in vertebrates both in development and in evolution (Chapters 2 and 3).

Despite its importance as a defining feature, the notochord is one of the most enigmatic vertebrate structures.

A. Notochord Cells (Chordocytes)

A differentiated notochord comprises large vacuolated cells connected by desmosomes (and some gap junctions) and surrounded by an expanded and elaborated basement membrane as a *notochord sheath*. The notochord is a flexible rod with a constant length, enabling it to function as an axial hydroskeleton (Wood and Thorogood 1994; Nordvik 2007). Osmotic pressure generated by the large, vacuolated inner cells is counteracted by a strong notochord sheath that encircles the notochord cells. Such a sheath was described as long ago as 1876 in dogfish and other cartilaginous fish (*Scyllium canicula*, *S. stellare*, *Pristiurus* and *Torpedo*) by Balfour (1876, 1877). Even earlier, Kölliker (1859) provided a detailed analysis and classification scheme for comparing the sheath across vertebrates. His analysis is discussed in Chapter 2 in the context of notochord evolution (Figure 1.2).

Experimental studies using urodeles enabled Mookerjee (1953) to conclude that the sheath compresses the notochord cells, a constriction that explains the longitudinal growth of the notochord. Notochords isolated from embryos of the alpine newt *Triturus alpestris*, the newt *T. palmatus* and the Mexican axolotl *Ambystoma mexicanum* when maintained in organ culture failed to form a sheath, while notochords transplanted to a site *in vivo*, where they maintained contact with somitic cells, formed a sheath. Mookerjee (1953) interpreted his results as an evidence for the somitic origin

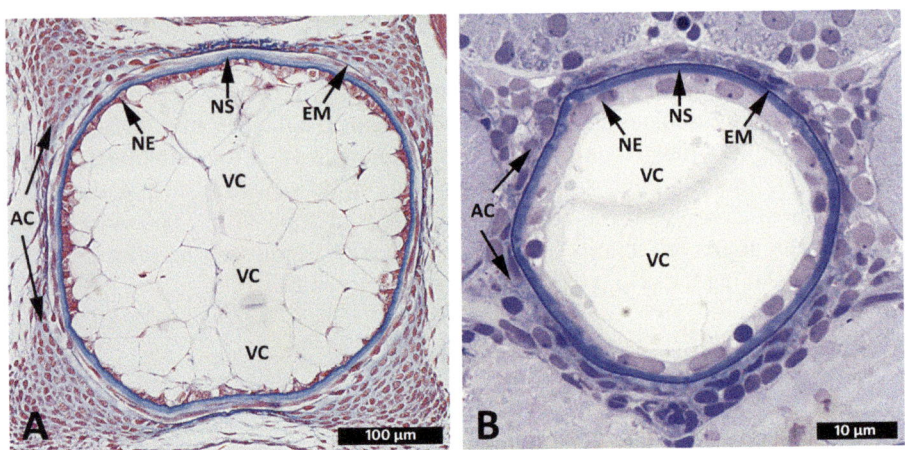

FIGURE 1.2 Notochord cross section from (A) a dogfish embryo (*Scyliorhinus canicula*) and (B) a zebrafish embryo (*Danio rerio*). The notochord comprised of vacuolated cells (VCs) and the basic components of the notochord sheath (NS) are visible. The two species show that the principal components of the notochord have been preserved for more than 400 million years of gnathostome evolution. VC, vacuolated notochord cells; NE, notochord epithelium; NS, notochord sheath; EM, elastic membrane, which is the outer limit of the notochord sheath; AC, autocentrum formed from sclerotome-derived cells that form a tube around the notochord. The cells of the autocentrum eventually provide most of the material for the vertebral centra. In dogfish and zebrafish (representing elasmobranchs and teleosts, respectively) the very first anlagen of vertebral centra arise within the notochord sheath. (A) Azan staining, (B) Toluidine Blue staining. (Image provided by PEW.)

of the sheath after somites received a signal(s) from the notochord. We know today that Mookerjee (1953) was right about the cross talk between notochord and somites but wrong about the somitic origin of the notochord sheath. The notochord sheath is primarily a product of the notochord, as discussed in Chapters 6 and 7.

The large cells in the center of the notochord are surrounded by the notochord epithelium, which is an epithelial-like layer of non-vacuolated cells known as *chordoblasts*. Before differentiation into notochord epithelium and vacuolated cells, all notochord cells are referred to as chordoblasts. After differentiation into the two cell types, we propose that the name chordoblasts be limited to the cells of the notochord epithelium that continue to give rise to vacuolated cells and continue to produce the notochord sheath. We propose that the vacuolated notochord cells be named *chordocytes*, although some literature refers to these cells as chordoblasts, which is confusing once the two cell types have fully differentiated. The parallel we use here is with cartilage where the term *chondroblast* refers to cells of the early chondrogenic anlagen and, after differentiation, to the cells in the perichondrium, while the name *chondrocytes* is used for the differentiated inner cells produced by chondroblasts. Indeed, a major conclusion of our analysis, discussed fully in Chapter 8, is that this is more than a parallel but represents a fundamental homology of cell types (chondro- and chordo-) and of organs (notochord and cartilage).

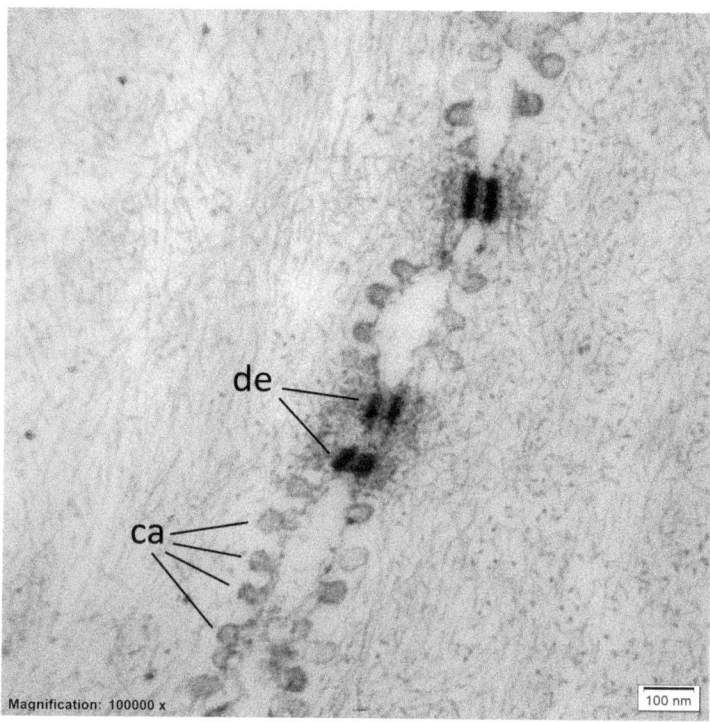

FIGURE 1.3 Typical of vertebrate notochords, cell membranes of two adjacent notochord cells from a zebrafish (*Danio rerio*) are joined by desmosomes (de). Caveolae (ca) are abundant. (Transmission electron microscopical image provided by PEW.)

Notochord cells are joined by desmosomes and possess caveolae (invaginations of the cell membrane; Figure 1.3). Caveolae are highly enriched with cholesterol, with their main constituents being the scaffolding proteins caveolin-1 and caveolin-2. Caveolae are implicated in a membrane-mediated mechanical response of notochord cells (Lotz and Hsieh 2014). The fate of cells as notochord in zebrafish embryos is regulated by Jagged-1–Notch signaling, which allows the development of non-vacuolated cells at the expense of vacuolated cells (Yamamoto et al. 2010).

Development of desmosomal connections between individual chordoblasts and the linking of these cells into the notochord epithelium was examined in some detail in a stage-by-stage analysis of the Japanese medaka (*Oryzias latipes*). Desmosomes appear at stage 28 and by stage 32 (one day before hatching) have matured with cytoplasmic plaques and tonofilaments (Ekanayake and Hall 1991).

B. THE NOTOCHORD SHEATH

The notochord epithelium is surrounded by the multilayered notochord sheath. Key early references on the structure of the developing notochord are Kölliker (1859), Klaatsch (1895), von Ebner (1896), Schauinsland (1903) and other authors, summarized by Goodrich (1930).

The primary notochord sheath is secreted by the notochord at an early stage when the cells are still flat and arranged as a stack of coins. This outer part of the sheath is an elastic membrane that later becomes the outer elastic layer of the notochord. Inside the outer elastic membrane, notochord cells secrete a laminin and collagen type-II rich matrix, which is the notochord sheath proper. Once the notochord cells have differentiated in vacuolated cells (chordocytes and notochord epithelium (chordoblasts)), a basal lamina appears inside the notochord sheath. The notochord epithelium also secretes a very thin inner elastin layer, but this layer is not always detected. The ultrastructure of the sheath was studied in urodeles by Waddington and Perry (1962), in chick embryos by Bancroft and Bellairs (1976), in mouse embryos by Jurand (1974) and in basal osteichthyans by Schmitz (1998b). As shown in Figure 1.4 and discussed in detail in Chapter 6, the sheath in all vertebrates consists of:

- an inner epithelial basement membrane secreted by notochord epithelial cells and, sometimes detectable, a thin inner elastin layer;
- a "thick" fibrous sheath composed of collagen fibers; and
- an elastica externa composed of elastic fibers and forming the definitive outer limit of the notochord sheath.

Ekanayake and Hall (1991), Schmitz (1995), Bensimon-Brito et al. (2012a), Lim et al. (2017) and other researchers have studied the ultrastructure of the notochord in teleost fish. Three layers also comprise the sheath of the zebrafish notochord

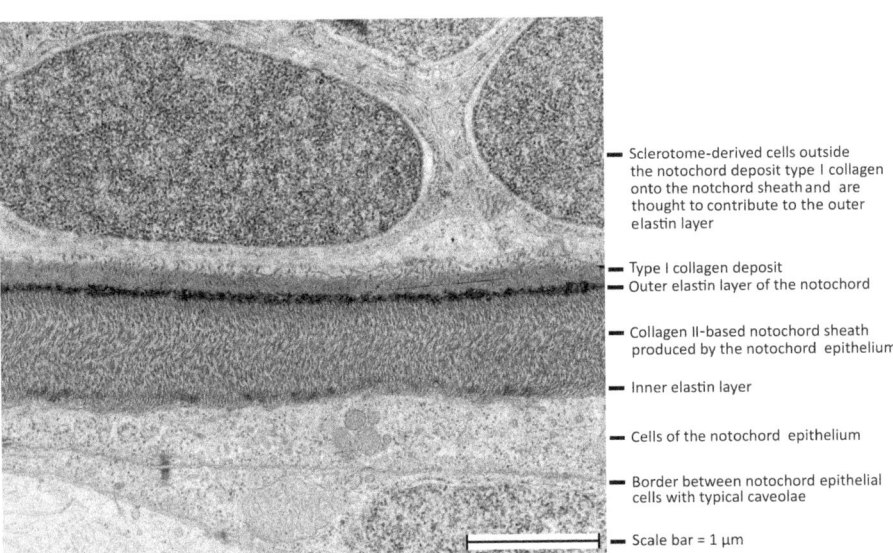

Sclerotome-derived cells outside the notochord deposit type I collagen onto the notchord sheath and are thought to contribute to the outer elastin layer

Type I collagen deposit
Outer elastin layer of the notochord

Collagen II-based notochord sheath produced by the notochord epithelium

Inner elastin layer

Cells of the notochord epithelium

Border between notochord epithelial cells with typical caveolae

Scale bar = 1 μm

FIGURE 1.4 The notochord sheath of a Japanese medaka (*Oryzias latipes*) with the major components identified from the notochord cells outward. Desmosomes and caveolae that connect the notochord cells are visible in the left lower corner. (Transmission electron microscopical image provided by PEW.)

(Parsons et al. 2002). Laminin is a major component of the sheath; three isoforms are essential for the early differentiation of the notochord (Pollard et al. 2006).

The fate of these layers varies in different vertebrate groups and relates to the degree of development of the vertebral axial skeleton. Cyclostomes, which lack vertebral centra, retain a thick notochord sheath. Tetrapods, which have a well-developed vertebral skeleton, may retain only a trace of the sheath within the vertebral centra but extend the sheath and its components into the spaces between the vertebrae (Goodrich 1930, 1958; Gardiner 1983, and discussed in Chapters 6 and 9).

The inner layer of the notochord sheath shares collagen type II and proteoglycans with cartilage matrix, and in some vertebrates is cellular, as discussed in Section 5.3B. The outer layer of the notochord sheath (elastica externa) contains high amounts of elastin in living jawless vertebrates (lampreys and hagfish; Schinko et al. 1992) and in jawed vertebrates. The outer elastin layer of the notochord sheath can become fenestrated (for example, in chondrichthyans) or can disappear later in life (Goodrich 1930).

The origin and function of the notochord sheath has been a matter of discussion as far back as Goodrich (1930) and Remane (1936; see references therein and see Chapter 6 for a discussion of the available evidence). The notochord is not invaded by nerve fibers, blood vessels or lymphatic vessels, all of which are features that the notochord shares with cartilage (Hall 2015).

1.3   FUNCTION OF THE NOTOCHORD AND NOTOCHORD CELLS

Although a defining organ of the chordate phylum, the notochord is often regarded as a transient structure with limited function. Such a limited view of notochord function is likely based on the erroneous belief that our own, mammalian, notochord is short-lived and has only a limited skeletal function during early development. Today, we know that mammals, including humans, retain the notochord in the intervertebral disks and that the notochord has a life-long role in maintaining the function of the intervertebral disks (Risbud and Shapiro 2011).

A. The Primary Chordate Axial Skeleton

The classic textbook view is of the notochord as the primitive axial skeletal tissue of vertebrates, and indeed, the notochord is both a skeletal tissue and an organ. With its highly vacuolated chordocytes and extracellular sheath(s) the notochord serves as the skeletal organ that prevents compression of the embryo along the A-P axis, a function that continues into the adult in many species (Wainwright et al. 1976). The vertebrate notochord has important signaling functions even earlier in development, functions in vertebral development, and has one of more life-long functions in almost all vertebrates including mammals. This book elaborates this broader view of the function of the notochord across vertebrates.

In introducing the structure of the notochord above, it was noted that in a mature notochord, osmotic pressure generated by the large chordocytes is counteracted and directed along the longitudinal axis by the constraining notochord sheath (Koehl et al. 2000). Notochord cells, however, exercise a mechanical (and morphogenetic) role before the sheath forms; i.e., chordoblasts have extrinsic force-generating and growth-directing potential. In studies of notochord developing in urodele amphibians Mookerjee et al. (1953) and Waddington and Perry (1962) explained this morphogenetic property as a function of the increasing closeness of association between notochord cell membranes, which "might provide the motive force which brings about the morphogenesis of the organ" (Waddington and Perry 1962, p. 459). This property was further demonstrated in studies on the coral reef pearlfish *Carapus homei*, which undergoes a 50%–60% decrease in body length during metamorphosis after settlement on the reef. Reduction in the size of the notochord cells increases the compressive force of these cells along the longitudinal axis that both shortens the notochord and is sufficient to displace the already ossified vertebrae (Parmentier 2016).

B. SIGNALS FOR NEURAL AND ECTODERMAL INDUCTION

Early in development the notochord is the signaling center that provides antagonists of bone morphogenetic proteins (BMP) that trigger the differentiation of neural ectoderm, the neural plate and the neural tube. The signaling role of the notochord continues through the production of sonic hedgehog (SHH), a polypeptide that is produced by the notochord and secreted as an extracellular signal that specifies the ventral region (floor plate) of the neural tube early in embryonic development (Chiang et al. 1996; McMahon et al. 1998; Teillet et al. 1998a). Subsequently, the ventral neural tube is itself a source of SHH. The receptor for SHH, Patched-1 (PTCH1), is in highest concentration ventrally in the floor plate and lowest within the dorsal neural tube. The notochord and ventral tube therefore collectively establish a concentration gradient of SHH that specifies and maintains the dorso-ventral (D-V) axis of the developing neural tube (Placzek 1995; Ribes and Briscoe 2009). Inactivate the *Shh* gene in the mouse and both notochord and a floor plate fail to form and the neural tube lacks D-V patterning (Chiang et al. 1996; Balmer et al. 2016). In zebrafish (*Danio rerio*) *Wnt* signaling induces notochord cell fate in a population of notochord/floor plate bipotential cells by downregulating *sox2*. *Notch* signaling specifies hypochord cells from a notochord/hypochord[2] bipotential cell population. Despite these differences these two populations are developmentally equivalent in the zebrafish tail-bud (Row et al. 2016).

C. SIGNALS FOR SOMITIC MESODERMAL INDUCTION

Early in tetrapod embryonic development, the notochord is the signaling center that triggers the differentiation and segmentation of mesoderm into somites that subsequently give rise to sclerotomes and then to the vertebrae. Note that in teleost fish the anlagen of the vertebrae arise by segmented mineralization of the notochord sheath and not from sclerotomal cells as they do in tetrapods (Ward et al. 2018). Thus in teleost fish, but possibly also in other vertebrates, mineralization of the notochord sheath defines the identity of each vertebra (vertebral centrum) *prior to cartilage and bone formation*, topics discussed in Chapters 7 and 8. In the wake of Lauder's concept according to which segmentation of the notochord and vertebral

---

[2] The hypochord is a transient rod of cells ventral to the notochord. The hypochord is thought to aid in positioning the developing dorsal aorta.

arches are independent (modular) events (Lauder 1980), in Chapter 9 we discuss how notochord segmentation that is independent of the somitic segmentation defines vertebral centra.

D. Integration of Axial Structures at the Phylotypic Stage

These important signaling and integrative functions of the notochord early in development and in the adult have been used to explain, in large part, why the notochord persists when its axial skeletal function is taken over by vertebrae, and indeed, why the notochord has persisted throughout vertebrate evolution. Evolutionary biologists have used essential inductive interactions early in development to develop the concept of *epigenetic burden* (Riedl 1978) and *epigenetic traps* (Wagner 1989a, b) where epigenetic reflects the cell-to-cell interactions and burden or trap signifies an evolutionary constraint ensuring maintenance of the notochord even when it does not persist into the adult. A recent analysis of, admittedly, a restricted number of vertebrates (the Chinese soft shell turtle, *Pelodiscus sinensis* and the domestic chicken, *Gallus domesticus* — a non-avian and an avian reptile) uses the relatively high conservation of inductive signaling cascades — such as the one initiated by sonic hedgehog (SHH) — between turtle and chick notochords as evidence for developmental burden. In contrast, turtle and chicken somites and neural tube show lower levels of gene conservation (Fujimoto et al. 2021).

Such conservation across the larger taxonomic scales of phyla is at the basis of the concept of *phylotypic stages,* a time or period early in development when all embryos show the greatest extent of morphological and molecular similarity (Hall 1997; Irie and Kuratani 2014). Interestingly, although it was once believed that the most conserved evolutionary stages would be the earliest stages in development, the phylotypic stage is *not* at the very onset of development — zygote or gastrula stage — but at neurulation in what is known as the pharyngula stage in vertebrates,[3] which is the earliest stage when the four features that define vertebrates — notochord, dorsal nerve cord, tail bud (Section 3.5), and branchial (pharyngeal) slits — are present. Many developmental biologists had a poster in their laboratories in which the developmental biologist Lewis Wolpert asserted (1991) that it is

"not birth, marriage or death, but gastrulation which is truly the most important time in your life."

As evolutionary-developmental biologists 35 years later we might want to amend this famous saying to:

"it is not birth, marriage, death or gastrulation but neurulation and the pharyngula stage that is truly the most important time in your life and in the life of your phylum."

The phylotypic stage is often illustrated using an hourglass model, in which the narrow "waist" at the middle of the hourglass represents the stage with the least variation and the most highly conserved features. The phylogenetically oldest genes

---

[3] Other phylotypic stages are the planula larva in hydrozoans, the segmented larva in polychaetes, the extended germ-band stage in insects, and the veliger larva in gastropods (Slack 2003). Note that all are larval and not early embryonic stages.

that share similar patterns of expression and that are under the strongest purifying selection characterize the phylotypic stage/period (Elinson and Kezmoh 2010; Drost et al. 2017).

A further explanation for the persistence of the notochord is its important contributions to the tissues, development and function of the intervertebral joints, contributions that reflect the active evolutionary history of the notochord, a topic taken up in Chapters 6 and 7. A vertebral column without intervertebral joints would be non-functional. Birds are an interesting exception, in which the notochord plays an inductive role rather than providing cells for the intervertebral joints. Birds replace the notochord with intervertebral cartilage but attracting the cartilage precursor cells to the joints is a function of the notochord. Eventually, birds fuse lumbar (notarium) and sacral vertebrae (synsacrum) to gain stiffness for flying (Romer 1951, p. 159; James 2009).

The teleost notochord began to receive increased attention after zebrafish and medaka became established as model organisms. Teleost fish hatch early as embryos. The notochord is present at hatching, prior to the development of the vertebral column, when it functions as the axial skeleton of the free-living embryos. Thus, the notochord is a prominent and well-developed structure and the subsequent formation of vertebral anlagen in the notochord sheath can be observed in embryos outside the egg. The notochord, the mineralized notochord sheath, vertebral bone, tendons and axial muscles develop as an integrated functional unit.

E. Fate in Later Ontogeny

Later in development, many vertebrates replace parts of the notochord with cartilaginous or bony vertebrae but the notochord usually remains in spaces within or between the vertebrae (Wake and Lawson 1973; Witten et al. 2005; Risbud et al. 2010). In squamate reptiles, for example, the notochord becomes constricted by the vertebrae but may persist throughout the length of the vertebral column (McLean and Vickaryous 2011). In mammals, notochord cells are found inside the intervertebral disks and, as in other vertebrates, notochord cells acquire a cartilage-like phenotype with aging (Trout et al. 1982a, b). There is ample evidence that notochord cells are required to maintain a healthy, functional, intervertebral disk throughout life (Shapiro and Risbud 2014).

Several vertebrates possess, as adults, a non-constricted, continuous notochord as a functional axial skeleton (Witten and Hall 2021). This is the basic condition for extant (hagfish and lampreys) and extinct (ostracoderms) jawless vertebrates. Hagfish have a strong notochord but only rudimentary vertebral arches confined in the posterior part of the body and no centra. Hagfish deposit a specialized chondroitin sulfate into the notochord that reinforces its skeletal function (Ueoka et al. 1999). A persistent and non-constricted notochord is also present in various chondrichthyans and several basal osteichthyans such as coelacanths, lungfish and sturgeons. In these vertebrates the notochord is anything but transient, continuing to grow throughout the life of the animal, often attaining an impressive size (Remane 1936; Leprévost et al. 2017). In the absence of complete vertebral centra the notochord is the fully functional main axial skeletal support.

1.4   DIFFERENTIATION OF NOTOCHORD CELLS

A. STACK OF COINS

The notochord first becomes visible as a structure called a "stack of coins" (Figure 1.5). This early notochord has exactly the diameter of a single cell. One row of flat cells and all cells are arranged perpendicular to the length of the notochord, thus "stack of coins" (Figure 1.5). The stack of coins arrangement is ancient; it is present in amphioxus and tunicates (Remane 1936; Stach 1999; Kovalevsky 1866a, b, 1867, 1871a). The same arrangement is found in lampreys (Pasteels 1958), chondrichthyans (Boeke 1908), in teleosts such as salmon, zebrafish and tilapia (Melby et al. 1996; Peters 1963; Fleming et al. 2004) and in urodele amphibians (Mookerjee et al. 1953). Kovalevsky (*ibid*) emphasized the similarity of early notochord development in amphioxus, ascidians and vertebrates (Chapter 2).

A very similar "stack of coins" arrangement is also found in early-differentiated rod-like cartilage elements (Huysseune 1989, 2000; Bird and Mabee 2003; Zhang and Cohn 2006; Zhang et al. 2006). Cole (2011) compared rod-like cartilage elements in chordates and protostome invertebrates and discusses the possible common origin of cartilage-like tissues in protostomes and in deuterostomes.

B. NOTOCHORD EPITHELIUM

Notochord cells divide and differentiate into an outer layer of cells (chordoblasts) designated as the notochord epithelium (following Jurand 1962) and an inner population of vacuolated cells (chordocytes) that initially fill the entire lumen of the notochord

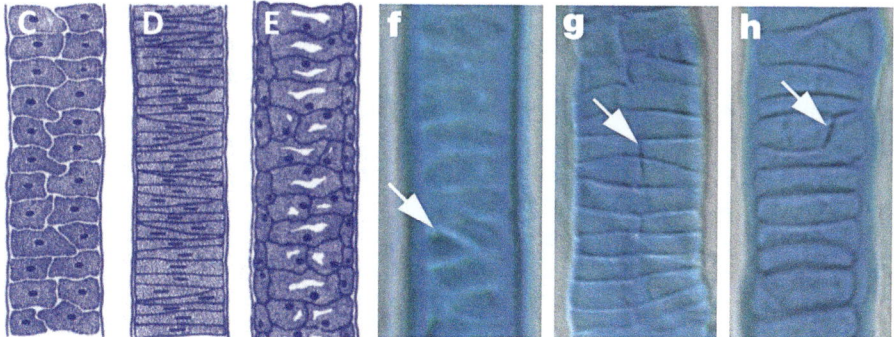

FIGURE 1.5   The arrangement of cells as a stack of coins in early embryonic notochord and cartilage elements. Labeling of the original figures has been kept. The most typical arrangement is portrayed in D. (C–E) represent early stages of urodele notochord development as seen in *Ambystoma* and in *Triturus*. (F) chondrocytes from the mouth cirrus of a larval Amphioxus (lancelet). (G) chondrocytes of a larval gill bar of *Petromyzon* (sea lamprey). (H) chondrocytes of a larval zebrafish gill bar. F × 1000, G × 700, H × 500, arrows point to the acidic extracellular matrix. (C–E) Adapted from Mookerjee, S., et al., 1953. *J Embryol Exp Morph* 1: 399–409. (F–H) From Jandzik, D., et al., 2015. *Nature* 518: 534–538. With permission.

(Schauinsland 1906; Schaffer 1930; Melby et al. 1996). Some literature refers to the cells of the notochord epithelium as notochordoblasts (Schmitz 1995; Grotmol et al. 2003) or as notochord basal cells (Schmitz 1998a, b). As introduced earlier, we use the name chordoblasts for these cells. Grotmol et al. (2006) provide an excellent description of the differentiation process of cells in the notochord sheath of Atlantic salmon and demonstrate when during embryonic development the cells in the notochord center become vacuolated.

The large chordocyte vacuoles are Golgi-derived vesicles that share properties with giant lysosomes (Ellis et al. 2013). The large fluid-filled vacuoles exert hydrostatic pressure against the notochord sheath providing the mechanical properties of the notochord responsible for its elongation and stiffness during growth (Adams et al. 1990; Koehl and Long 2000). Chordocytes can be arranged in different patterns that often appear to be irregular (Mookerjee et al. 1953 for *Triturus tigrinum, T. alpestris* and *T. palmatus*; Barteczko and Jacob 2002 for the caecilian *Ichthyophis kohtaoensis*). A central condensation and flatting of vacuolated notochord cells seen in actinopterygians is called a funiculus (Schauinsland 1906) or notochord strand (Schaffer 1930). The notochord strand becomes surrounded by large extracellular vacuoles that develop in adults in addition to the chordocytes (Schmitz 1995). Anteriorly and posteriorly the notochord can blend into cartilage (Chapter 8). Otherwise, the notochord is a completely closed system.

C. Development of the Notochord Sheath

There is agreement on the cellular origin of the notochord sheath proper.

As early as 1894, Claus showed in detailed developmental studies that the external elastic layer of the notochord in chondrichthyans is a product of the notochord epithelium (Claus 1894). In a range of developmental stages of the shark *Acanthias* examined 120 years ago, Claus studied the successive formation of the elastica externa from notochord cells, which is the membrane that forms before sclerotome differentiation or the formation of any other skeletogenic cells. Thus, Claus concludes that the elastic membrane must be a product of the notochord. Later, Ridewood (1921) also argued that early developmental stages show that the elastic membrane is secreted by the notochord before the presence of sclerotome-derived cells around the notochord and thus it is a product of the notochord.

The notochord sheath, with or without cells, is a strong connective tissue wall made from collagen type II and elastic fibers. It consists of an inner fibrous layer and an outer elastic layer (Schmitz 1998a, b; Welsch and Storch 1971). The cells of the notochord epithelium produce the notochord sheath that contains no cells in most vertebrates. Possibly, the notochord epithelium may also give rise to cells *inside* the notochord sheath, present in the bichir *Polypterus* in which the sheath is cellular (Schmitz 1998a, b). In chondrichthyans, however, cells inside the notochord sheath are definitely sclerotome-derived and invade the sheath from outside (Goodrich 1930).

Whether the cells that populate the notochord sheath in all vertebrates with a cellular notochord sheath derive from the notochord epithelium is a long lasting and still ongoing debate. Gegenbaur (1862) argued that cells in the notochord sheath derive from the notochord epithelium. He viewed the sheath as a notochord-derived cartilage that is covered by an external and an internal elastin layer (Gegenbaur 1862). Claus (1894), Gadow and Abbott (1895) and Ridewood (1921) agree with Goodrich's view that the cells found in the notochord sheath of chondrichthyans invade the notochord sheath from outside, a process in which the outer elastic membrane is disrupted (Goodrich 1930).

1.5 NOTOCHORD AS CARTILAGE

With or usually without cells inside the notochord sheath, the notochord sheath has a prominent outer layer made of elastin. Sometimes an inner elastic layer also is recognized. Below the outer elastin layer is a thick layer chiefly composed from collagen type II and proteoglycans. As discussed in detail in Chapter 8, the composition of the notochord sheath is essentially identical to cartilage matrix.

In either case, a discussion that remains is whether the cells of the notochord epithelium, that without doubt produce the cartilage-like matrix of the notochord sheath, can also produce cartilage cells, chondrocytes. Since the notochord is a completely closed system, cartilage that develops inside the notochord (Wake and Lawson 1973; Jonasson et al. 2012) must derive from notochord cells. Moreover, when injured, the notochord acquires a cartilaginous fate (Loizides et al. 2014; Lopez–Baez et al. 2018). In this regard, and in the context of the evolutionary origin of notochord and type II-containing collagen, it may be that type II was first expressed in the notochord, co-opted by cartilage and elaborated into an ECM early in chordate evolution; invertebrate cartilages lack type II collagen (Chapter 8).

While acknowledged as the initial supporting skeletal structure of the embryo, the notochord usually is discussed as a skeletal structure that differs from cartilage and bone. Cartilage is rarely regarded as a tissue that shares features of the notochord; the vacuolated chordocytes and elaborated basement membrane (an epithelial structure) of the notochord are contrasted with the ECM and mesenchymal organization of cartilage. Notochord is considered able to become cartilage on the basis of evidence such as the existence of notochord cartilage discussed in Chapter 8. Although notochord has been regarded as a subtype of cartilage (Chapter 8), notochord is not usually identified explicitly as a member of a cartilage "superfamily." Many studies, however, show that the notochord can produce cartilage. We discuss the relationship(s) between notochord and cartilage in depth in Chapter 8.

1.6 SUMMARY

The differentiated notochord, one of four synapomorphies that separate chordates from all other animals, consists of large vacuolated cells (chordocytes) connected by desmosomes and surrounded by an expanded and multilayered basement membrane known as the notochord sheath. The history of the development of the sheath has been studied since the 1860s and is examined in some detail in this chapter. The inner layer of the sheath contains, in common with cartilage, collagen type II, proteoglycans and in some vertebrates, cells. Morphogenesis of the notochord is based on the "stack of coins" arrangement, a mechanism that the notochord shares with developing rods of cartilage in vertebrates and in invertebrates.

Early in development and continuing to the pharyngula or phylotypic stage, the vertebrate notochord is the signaling center that triggers the differentiation of neural ectoderm, neural plate and neural tube with antagonists of Bmp and production of SHH as the key signaling factors. In tetrapods the notochord is the signaling center that triggers the differentiation and segmentation of sclerotomal mesoderm into somites and then into vertebrae. In teleost fish, however, mineralization of the notochord sheath provides the anlagen of the vertebral centra. The fate of the notochord varies in different vertebrates, either (i) being retained as the primary axial skeleton, (ii) replaced, (iii) forming intervertebral disks between the vertebrae or forming (iv) intravertebral cartilage within the vertebrae. The chapter concludes with an introduction of the topic of notochord as cartilage and of the challenges such a concept raises.

2 Discovery and Evolutionary Origin of the Notochord

CONTENTS

2.1 INTRODUCTION

This chapter treats the discovery of the vertebrate notochord and the recognition that all chordates are united by the possession of a notochord. Animals with a notochord belong to the chordate phylum Chordata, named from the Latin *chorda* or cord. Chordates are a major group of *deuterostomes*, i.e., animals in which the primary

DOI: 10.1201/9781315155975-3

opening of the embryo, the blastopore, becomes the anus. In the other major group, *protostomes*, the blastopore becomes the mouth. The two other major groups of deuterostomes — echinoderms and hemichordates — have featured prominently in the discussion of the evolutionary origin of the notochord.

2.2 CHORDATES

Some 45,000 extant species of chordates possess a notochord, a single dorsal nerve cord, pharyngeal clefts and extensive sensory and other organs. Extant chordates consist of three groups (Classes):

PHYLUM CHORDATA

Class Urochordata or Tunicata. Three groups, Ascidiacea (sea squirts)., Appendicularia (Larvacea), and Thaliacea (salps.)
Class Cephalochordata or Acrania. The class contains the genera *Branchiostoma* (amphioxus) and *Epigonichthys,* formerly *Asymmetron*
Class Vertebrata or Craniata. All jawless and gnathostome vertebrates.

A fourth class Hemichordata (hemichordates), aligned with chordates in the late 19th century, is now recognized as a non-chordate phylum, primarily on the basis of the absence of a notochord and larval stages that resemble those of echinoderms.

A brief summary of the notochord in each chordate class and of what was regarded as a homologous feature in hemichordates (now known as the *stomochord*) is provided below.

A. CEPHALOCHORDATES

Cephalochordates are literally chordates in which a notochord extends to the anterior-most tip of the head (hence *cephalo*). Extant cephalochordates are represented by the genera *Branchiostoma* and *Epigonichthys* (Poss and Boschung 1996). The most studied of the 30–35 species is the European lancelet, *Branchiostoma lanceolatum*. The literature also refers to this species as amphioxus or as Mediterranean amphioxus. Amphioxus, however, is a common name and not a generic name. Consequently, the name amphioxus (lower case a) can be used for other lancelets that belong to the genus *Branchiostoma*.

The elongate lancelet body is bilaterally flattened and a tail is present. All are marine and free-living. The notochord extends along the entire length of the animals and is retained throughout life. Notochord development follows what is now known to be a conserved chordate pattern with notochord cells initially arranged in the typical "stack of coins" pattern (Stach 1999, and see Chapters 1 and 5).

The differentiated notochord in amphioxus is, however, unusual among chordates in consisting of muscle cells, both laterally compressed cells containing microfiber bundles, and cells lacking microfibrils but with intracellular canals and known as Müller's tissue (Eakin and Westfall 1962). Welsch (1968) concluded from the available evidence that the structure of the muscle cells in the notochord of *Branchiostoma lanceolatum* does not resemble the notochord cells (chordoblasts) in vertebrates but resembles the fine structure of the oblique-striated muscles in annelids and molluscs.

B. Tunicates

Urochordates or tunicates consist of some 2,000 species of marine, free-living or sessile, solitary or colonial animals that lack a coelom. As the name urochordate suggests, a notochord is restricted to the tail of the larvae, which are either planktonic and neotenic (larvaceans) or planktonic and immature (ascidians). Virtually all that we know about tunicate notochords come from studies of ascidians, starting with Alexander Kovalevsky's discoveries of a notochord in ascidians (Figure 2.1 and see below).

Interestingly, not all ascidians have a notochord. Members of two families—Molgulidae and Styelidae — develop as larvae that lack a tail and so are known as tailless or anural larvae. In at least two species of sea grapes (*Molgula provisionalis* and *M. pacifica*), the larval stage has been lost completely. This represents an evolutionary switch from indirect development (having a larva in the life history) to direct development (development without a larval stage). Tailless larvae fail to develop a notochord; recall that the notochord is restricted to the tail in tunicates. Cells that form the notochord may be redirected to contribute to the endoderm in the tailless larvae of these two species, indicating a cellular plasticity and mechanism of respecification important for the evolution of novel variation. Absence of zygotic

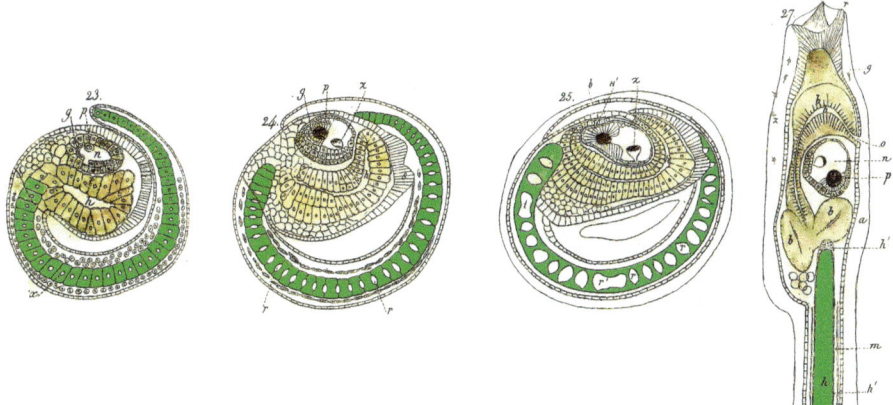

FIGURE 2.1 Kovalevsky's description of the notochord in the larva of a tunicate, the sea squirt *Ciona intestinalis* (*Ascidia intestinalis* in the original publication; Kovalevsky 1866b). The notochord (green) was described as *"Achsenstrang"*. The stack of coin arrangement of the notochord cells is clearly visible in the left-hand image (23). Forty-five notochord cells are clearly shown in image 24. Kovalevsky noted the emergence of vacuoles (25, r), described as *"ovale Körper zwischen den Zellen,"* and the fusion of the vacuoles (25, r'), which leads to a hollow notochord (27, h) with a peripheral notochord epithelium (27, h'). Discovery of the notochord in amphioxus and in tunicates was strong additional evidence for the evolutionary theory as it revealed the origin and descent of vertebrates. Kovalevsky's notochord discoveries were immediately picked up by Darwin and by Haeckel and incorporated into their works (Darwin 1871; Haeckel 1868). Kovalevsky, went on to become the leading Russian embryologist of the late 19th century and a strong Darwinian. (Modified from Kovalevsky, A. 1866b. *Entwicklungsgeschichte der einfachen Ascidien. Mém Acad Imp Sci St-Petersbourg* 10: 1–19.)

transcripts of the genes *Uro2* (*Lynx*) and *Uro11* (*Manx*) was implicated by Swalla et al. (1993) in the failure of tail muscle to develop in tailless larvae.

Tails do form in hybrids that develop when eggs from a species with tailless larvae are fertilized with sperm from a species with tailed larvae. *Manx* that encodes a DNA-binding zinc finger protein is downregulated in tailless larvae but upregulated in hybrid larvae (Swalla and Jeffery 1996). Similarly, *Manx* is inhibited and neither notochord nor tails develop in hybrid embryos exposed to antisense *Manx* DNA, implicating *Manx* as a major signaling gene for tail and/or notochord development.

2.3 HEMICHORDATES: A PHYLUM OUTSIDE THE CHORDATES

The 100 species of hemichordates ("half a cord"), which are marine benthic acorn worms, lack a tail as adults and have a specialized larva known as a tornaria, a type of larva very similar to the echinoderm starfish bipinnaria larva. In fact, before the adult worm was discovered, the larvae were believed to be starfish larvae (Romer 1951).[1] *Balanoglossus* and *Saccoglossus* are two commonly studied genera.

Acorn worms lack a notochord but have a structure in the anterior proboscis that is known as a *stomochord* or buccal diverticulum. The stomochord is a short anterior extension from the buccal cavity that may facilitate locomotion. Discovered by William Bateson in the genus *Balanoglossus*, Bateson (1884a, b) identified this dorsal structure as a notochord (Figure 2.2). As a consequence, Bateson:

   i. erected the class Hemichordata within the chordate phylum;
  ii. proposed a hemichordate-like organism as the chordate ancestor; and
 iii. proposed that the non-chordate ancestor had a notochord, i.e., origination of the notochord heralded the origin of the chordates.

Not taken up with much enthusiasm until recently (although see Garstang 1928), the current phylogenetic position of hemichordates as a separate phylum outside the chordates, leaving tunicates and vertebrates as sister chordate taxa, renders a hemichordate chordate ancestor unlikely. That said, and as discussed in Section 2.5, the last 25 years has seen proposals for a hemichordate-like intermediate in the transformation of an annelid-like ancestor to a chordate. This research also led to modifications of Bateson's original proposal (in which the invertebrate ancestral chordate possessed features of chordates such as a notochord), leading to the conclusion that the notochord arose in primitive chordates from the roof of the archenteron (Lowe et al. 2003, 2015; Lowe 2008). Discovery of a central nervous system (CNS) in hemichordates and of *Shh* signaling from the stomochord—recall notochord *Shh* in neural induction in vertebrates introduced in Chapter 1—led to the proposal that the hemichordate ancestor may have possessed more cellular or genetic precursors of the primitive chordate than had been recognized previously; for the evidence, see Satoh et al. (2014) and the discussion in Section 2.5C.

---

[1] Whether echinoderms that are phylogenetically more deeply rooted than cephalochordates have a notochord, and whether hemichordates have a notochord has long been debated (see, for example, Bowler 1996 and Rychel et al. 2006).

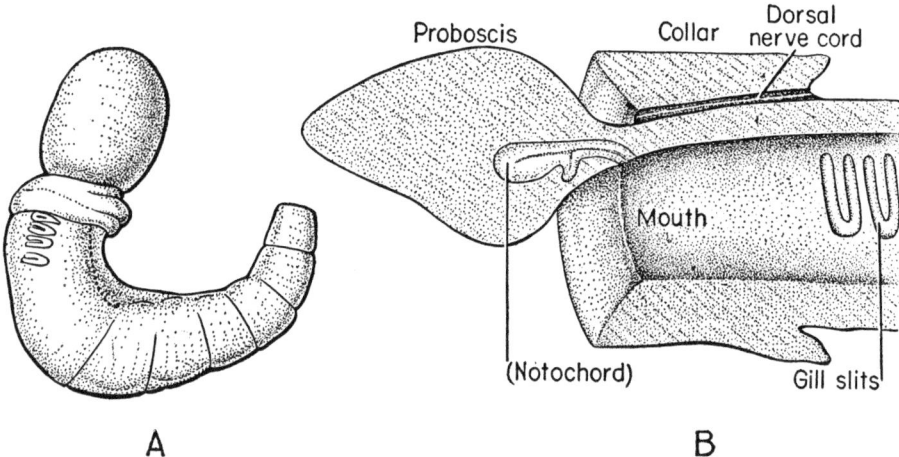

FIGURE 2.2 An acorn worm from the genus *Balanoglossus* (Phylum Hemichordata, Order Enteropneusta). These benthic marine animals received much attention for possibly being similar to (closely related to) chordate ancestors. Two conserved chordate characters are an anterior digestive tract with pharyngeal slits (gill slits) and a dorsal nerve cord. (A) A young acorn worm. The rugby ball-like anterior structure (top) is the proboscis, below which is the collar region and pharyngeal slits, four of which are shown. (B) Internal anatomy of the anterior region. Because of the presence of other chordate characters, a hollow tube like structure inside the proboscis was interpreted as notochord. This interpretation is no longer supported by most investigators (see text for details). (Images adapted from A. S. Romer, 1951. *The Vertebrate Body*. W.B. Saunders Company, Philadelphia.)

2.4 DISCOVERY THAT AMPHIOXUS AND ASCIDIANS POSSESS A NOTOCHORD

The emergence in the 1860s and 1870s of the classification and recognition of vertebrates as one of several groups of chordates was based on embryological evidence and on comparative adult morphology (Bowler 1996; Hall 1998a, b). As we will see, in the search for the likely nature of the chordate ancestor, some researchers emphasized embryological evidence, others evidence from features of adult chordates, resulting in several major divergent camps of opinion.

The question of the origin of chordates is intimately linked to questions about the origin of the notochord. We concentrate on the identification by a now famous Russian embryologist, Alexander Kovalevsky (1840–1901)[2] of a notochord in amphioxus (a cephalochordate) and in the vase tunicate *Ascidia* (now *Ciona*) *intestinalis* based upon research he undertook in the mid-1860s (described in Kovalevsky 1866b). Interestingly, the evolution of the notochord has been taken up again this century as molecular, developmental genetic, and phylogenetic data have provided totally new ways of viewing ancestor–descendant relationships and the origin of new features; below and see Annona et al. (2015).

---

[2] We follow the U.S. Library of Congress transcription for Ковалевский (Kovalevsky). Other Romanizations of his name include Kovalevskii, Kovalevsky and Kowalewsky.

A. Amphioxus

Kovalevsky (1866a, 1867, 1877) showed that amphioxus possesses the major features of early embryonic development found in vertebrates (Figure 2.3):

    i. a dorsal notochord;
    ii. a dorsal neural tube that arises from neural folds that fuse to form a hollow neural canal;
    iii. gill slits, and
    iv. a simple gut (archenteron) that arises by invagination at the gastrula stage of development.

Initially, Kovalevsky thought that the notochord arose from blocks (either segments or somites) of muscle cells. Further studies convinced him that the notochord arose from the roof of the developing gut (the archenteron; Kovalevsky 1877).

    The features of the amphioxus notochord and its mode of embryonic development were the most simple structure and mechanism of embryogenesis known for any chordate/vertebrate. Although Kovalevsky concluded that amphioxus displayed features of invertebrate *and* of vertebrate development, he did not discuss any evolutionary implications of his findings in his first paper[3]. Others did, inter-

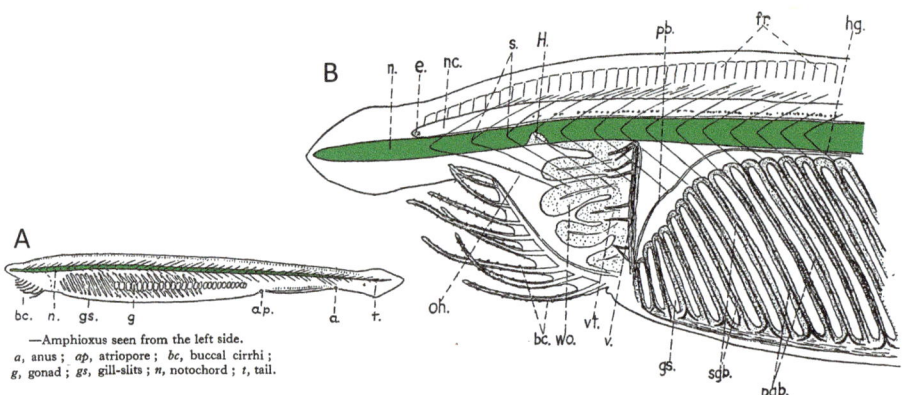

—Amphioxus seen from the left side.
a, anus ; *ap*, atriopore ; *bc*, buccal cirrhi ; *g*, gonad ; *gs*, gill-slits ; *n*, notochord ; *t*, tail.

—Amphioxus, anterior end seen from the left side by transparency.
bc, buccal cirrhi ; *e*, eye-spot ; *fr*, fin-ray boxes ; *gs*, gill-slit ; H, Hatschek's pit ; *hg*, hyperpharyngeal ciliated groove ; *n*, notochord ; *nc*, nerve-cord ; *oh*, oral hood ; *pb*, peripharyngeal ciliated band ; *pgb*, primary gill-bar ; *s*, septa between myotomes ; *sgb*, secondary gill-bar ; *v*, velum ; *vt*, velar tentacle ; *wo*, wheel-organ or ciliated organ of Müller.

FIGURE 2.3 (A) de Beer's scheme of the characters of a lancelet, commonly referred to as amphioxus (anterior to the left). The notochord (green) extends along the entire length of the animal which led to the scientific (and common) names Cephalochordata (cephalochordates) for the group. No other group of chordates has a notochord that extends to the anterior tip of the animal. (B) Details of the anterior region showing typical chordate characters, a dorsal nerve cord (nc) above a notochord (n, green) that extends further anteriorly than the nerve cord, and a digestive tract with pharyngeal (gill) slits (gs). (Modified from de Beer, G. R., 1928. *Vertebrate Zoology*. Macmillan, New York, pp. 6, 7.)

---

[3] A student of Ernst Haeckel's, Kovalevsky, went on to become the leading Russian embryologist of the late 19th century and a strong Darwinian.

preting the embryology of amphioxus as an intermediate between vertebrates and invertebrates (Metchnikoff 1866). Soon amphioxus was being proposed as representing the likely condition of the ancestral chordate, and chordates were proposed to have evolved from an amphioxus-like organism. Coming contemporaneously with Haeckel's grand synthesis of development and evolution summed up in the phrase "ontogeny recapitulates phylogeny" (Haeckel 1866), these conclusions found a ready audience. Amphioxus remains a strong candidate for an invertebrate model of the vertebrate ancestor but the lack of any plausible amphioxus-like fossils earlier than the late Paleozoic constitutes a problem for the "Amphioxus hypothesis" (Holland 2005) until the discovery of a Lower Cambrian craniate *Haikouichthys ercaicunensis* (Shu et al. 1999; and see Section 2.7). As we will see in Section 2.5 below, origins of the notochord may be from an even less obviously likely source.

B. Ascidians (Tunicates)

Complicating the evolutionary interpretation, Kovalevsky discovered a notochord and other chordate embryonic features in pre-metamorphic larval tunicates. He called the notochord *Achsenstrang* or *Achsencylinder* (Kovalevsky 1866b, 1871a). It consisted of a solid cord of cells, surrounded by a cellular sheath and with a fluid-filled lumen, the latter having the appearance of the fluid-filled body cavity or coelom. Previously grouped with molluscs, Kovalevsky proposed a close relationship between ascidians and vertebrates, a relationship that led to the hypothesis that vertebrates could have evolved from an ancestral free-swimming tunicate-like larva, a hypothesis that is still actively discussed today (Annona et al. 2015; Holland et al. 2015). Leading 19th century morphologists such as Ernst Haeckel, Karl Gegenbaur and Charles Darwin adopted the tunicate larva (ascidian tadpole) as the likely form of the ancestral chordate.

> Ascidians are related to the Vertebrata, in their manner of development, in the relative position of the nervous system, and in possessing a structure closely like the *chorda dorsalis* of vertebrate animals. …We should thus be justified in believing that at an extremely remote period a group of animals existed, resembling in many respects the larvae of our present Ascidians, which diverged into two great branches — the one retrograding in development and producing the present class of Ascidians, the other rising to the crown and summit of the animal kingdom by giving birth to the Vertebrata.
>
> *(Darwin, 1871, p. 163)*

Less than 20 years after Kovalevsky's discovery and the recognition of urochordates (tunicates) as a group within the chordates, a third group of deuterostomes, the *hemichordates* was considered to possess an anteriorly located vestige notochord (Metchnikoff 1881; Bateson 1886, and see above). The hypothesis that hemichordates could be the most ancestral chordates, much discussed and often refuted, has been revived and is still discussed today (Section 2.3, and see discussions in Hall 2015 and in Lowe et al. 2015).

2.5   ANNELIDS AS CHORDATE ANCESTORS[4]

Consideration of transformations between animals as introduced above did not start in the 1870s. Jean-Baptiste Lamarck proposed molluscs as vertebrate ancestors 60 years earlier (Lamarck 1809) while the publication of *The Origin of Species* by Charles Darwin in 1859 firmly set considerations of relationships in an evolutionary context. Over time, almost every invertebrate phylum has been proposed to contain possible vertebrate ancestors. The two groups that received most attention were acorn worms (hemichordates) and annelids (Holland et al. 2015).

Molluscs or annelids as vertebrate ancestors seems a stretch. Background to such notions is found in a famous debate that occurred over 2 months in 1830 before the French Academy of Sciences (Académie des Sciences) between two French naturalists, Georges Cuvier (1769–1832) and Étienne Geoffroy Saint-Hilaire (1772–1844). The issue could not have been more polarized:

- All animals are modifications of a single unified plan (Geoffroy) versus;
- All animals fall into one of four separate and non-overlapping groups (*embranchements*)[5] (Cuvier).

Geoffroy sought evidence for the single plan in features—that we now call homologies—shared between vertebrates and molluscs (in particular, the cuttlefish *Sepia officinalis*), especially in their segmented organs (Hall 1998a; McBrine et al. 2009). Cuvier placed molluscs and vertebrates into two separate *embranchements*. To interpret the organs of molluscs and vertebrates as similar required that the vertebrate body be bent back, placing the mouth near the anus, with the legs attached to the head, essentially turning the cuttlefish inside out and upside down! Such a major transformation, discussed in the following section, has been proposed as allowing an annelid-like vertebrate ancestor.

Although neither Geoffrey nor Cuvier carried out their debate in the context of evolution, they used the concepts of a unified plan (*Bauplan*) and homology, both of which became two of the most important propositions of the evolutionary theory. Like Geoffroy, Kovalevsky was not limited to thinking within Cuvier's four embranchements—vertebrates, molluscs, arthropods and annelids, and zoophytes (sponges and other phyla)—in searching for ancestors of the chordates (Kovalevsky 1871b, 1877). Rather, he followed the morphology espoused by Geoffroy for whom dorso-ventral body axes could be inverted, bridging arthropods and annelids on the one hand and vertebrates on the other (Bowler 1996; Hall 1998a, 2015).

A. THE 1870S: MIDLINE STRUCTURES, SEGMENTATION AND AN ANNELID ORIGIN FOR CHORDATES

Annelids were known to have a fibrous midline structure associated with the ventral nerve cord, a feature that was interpreted as a potential annelid equivalent to the

---

[4] Parts of this discussion are adapted from Chapter 4 in Hall (1998a) and from Witten and Hall (2021).

[5] Cuvier's four *embranchements* are Vertebrata, Articulata (arthropods and segmented worms), Mollusca (all other soft bilaterally symmetrical invertebrates), and Radiata (cnidarians and echinoderms).

notochord (provided that you allowed transformations between Cuvier's *embranche-ments*). Dohrn (1875) concluded that these midline structures were muscles that migrated from the nerve cord, and proposed them as a potential source of vertebral cartilage. Later others, proposed that annelids had a cartilaginous notochord (Lwoff 1893). The two important topics of (i) notochord–cartilage relationships, and (ii) the presence of cartilage in invertebrates are taken up in Chapters 8 and 9.

Independently in 1875, Anton Dohrn and Carl Semper proposed an annelid or proto-annelid as the chordate ancestor using, not embryological evidence, but evidence from segmentation of adult morphology (Dohrn 1875[6]; Semper 1875, 1876–1877). Despite the apparent substantial obstacles of a ventral nervous system and a mouth situated above the brain in annelids that contrasted with a dorsal nervous system and mouth below the brain in vertebrates (Figure 2.4), Semper and Dohrn argued that the fundamental feature of body segmentation overrode these differences; segmentation of adults was the driving evolutionary process.

Did segmentation evolve once or many times? If once, then segments in all animals are equivalent (homologous). If more than once then segments (and segmentation) in different animals are not homologous unless those animals share a recent common ancestor. Both Semper and Dohrn saw body segmentation as an ancestral, primitive feature shared by invertebrates and vertebrates: interestingly, the notochord was not regarded as being segmented; but see Chapter 7. If segmentation is shared and primitive, then annelids and vertebrates must have arisen

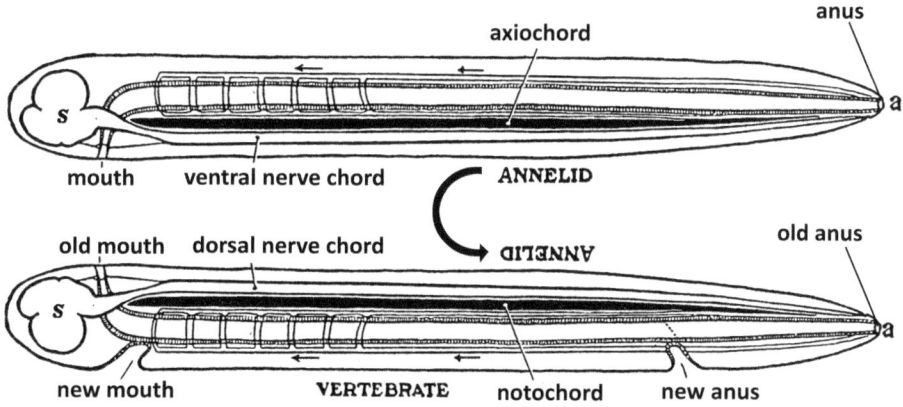

FIGURE 2.4   Dorso-ventral inversion of an annelid or annelid-like ancestor to a vertebrate chordate would have required the evolution of a new ventral mouth and anus and loss of the old, transformation of the ventral axiocord (axochord) into a dorsal notochord, and transformation of the ventral nerve cord to a dorsal nerve cord lying ventral to the notochord. See Section 2.5B for further details. (Figure modified from Romer, A. S., 1951. *The Vertebrate Body.* W.B. Saunders Company, Philadelphia and Hall, B. K., 1998a. *Evolutionary Developmental Biology.* 2nd Edition. Chapman and Hall, London/Kluwer Academic Publishers, Netherlands.)

---

[6] See Maienschein (1994) for an analysis of Dohrn's research.

from a common segmental ancestor and amphioxus and ascidians are degenerate, not basal. A common ancestor of annelids and vertebrates would have had a noto-chord and segmentally arranged organs and gill slits. Dorsal rather than ventral fusion of the nerve cord and the development of a new mouth gave this ancestor basic vertebrate characters. In his monograph on *Amphioxus and the Ancestry of the Vertebrates*, Willey (1894) summed up his discussion of the Dohrn-Semper hypothesis with:

> We conclude, therefore, that the ventral mouth of the craniate Vertebrates is the homo-logue of the primordial dorsal mouth as we find it in the Protochordates...

(1894, p. 282)

B. THE 20TH C: DORSO-VENTRAL INVERSION

As just discussed, the theory of vertebrate origin from annelids was warmly advo-cated during later decades of the 19th century. In the first half of the 20th century, the likelihood of the dorso-ventral inversion required for chordates to have had an annelid or annelid-like ancestor came under substantial criticism to the point that it was regarded as ludicrous. A leading vertebrate zoologist and palaeontologist Alfred Romer did not favor the idea that vertebrates are inverted annelids. Romer expressed his grave doubts with the remark "even a worm may have some idea as to which way is up" and further commented on the annelid hypothesis (Romer 1951, pp. 27, 28). Figure 2.4 illustrates the structures that would have had to change fundamentally for such an inversion to occur.

The lowly angleworm (earthworm, Oligochaeta) is none too prepossessing as an ancestor, but there are numerous marine annelids of a more progressive and attractive nature. Annelids have a typical bilateral symmetry, as do vertebrates, and, in correla-tion with this, are, like typical vertebrates, active animals, in contrast with the sessile organisms common in many invertebrate phyla. Then too, as discussed in the previous section, annelids are segmented, as are vertebrates (although to a less obvious extent).

As in vertebrates, the annelid CNS is composed of a brain-like mass at the ante-rior end of the body and a longitudinal nerve cord (Figure 2.5). So far, so good. But beyond this point, the comparison breaks down. Even segmentation is neither a game-changing character nor a perfect argument; annelids are segmented in every respect, from skin to gut lining, vertebrate segmentation is primarily confined to part of the middle body layer. Annelids have, it is true, a longitudinal nerve cord but it is solid, not hollow, and it is ventral not dorsal (Figure 2.5). This last point was especially troublesome to advocates of the dorso-ventral inversion theory. They "resolved" the difficulties by assuming that a vertebrate is a worm upside down. This is hard to swallow (even a worm may have some idea as to which way is up) and involves further perplexities.

The worm mouth is on the underside of the head, and so is that of a vertebrate. A reversal of surfaces implies that the old mouth of the worm has closed and been replaced, historically, by a new one (Figure 2.4). Attempts have been made to find traces of the theoretical old mouth opening in vertebrate embryos — it should pass upward and forward through the brain to the top of the head! — but without convincing results.

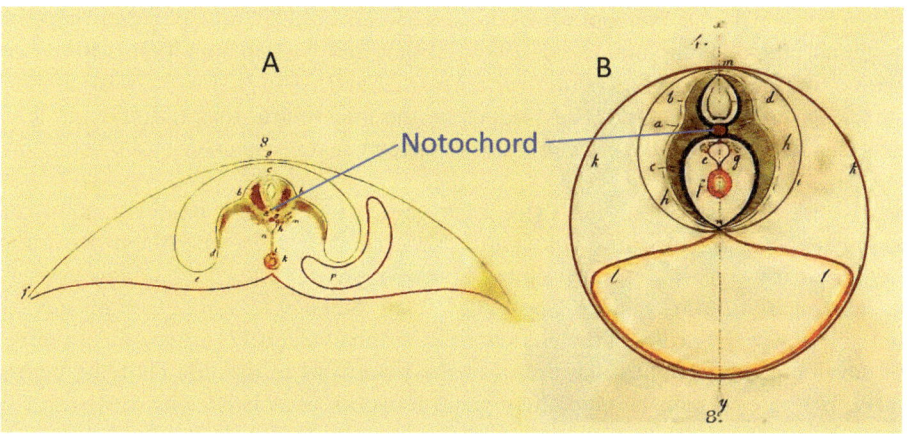

FIGURE 2.5   (A) Karl Ernst von Baer's first description of the notochord in vertebrates was in a 5-day-old chicken embryo (von Baer 1828, 1837), illustrated here in a cross-section. The notochord is indeed tiny in relation to the size of the entire embryo. von Baer (1828) not only named the notochord "*Rückensaite*," but also introduced the term *Chorda Dorsalis* and observed the presence of a notochord sheath. According to von Baer, previous investigators mistook the notochord for the anlage of the neural tube. (B) A cross-section of a generalized vertebrate embryo with a prominent notochord as precursor of the vertebral column (von Baer 1837). In the wake of von Baer's discovery of the chicken notochord, Richard Owen not only described the notochord in a fossil marine reptile (which he named *Plesiosaurus rostratus*) but also recognized that the notochord was preserved as part of the ossified vertebral column (Owen 1847). (Modified from von Baer, K. E., 1828. *Über Entwicklungsgeschichte der Thiere. Beobachtung und Reflexion*. Erster Theil. Gebrüder, Bornträger, Königsberg and von Baer, K. E., 1837. *Über Entwicklungsgeschichte der Thiere. Beobachtung und Reflexion. Zweiter Theil*. Gebrüder Bornträger, Königsberg.)

Even if this difficulty in orientation can be solved, there are more ahead. There is no trace in an annelid of a notochord, although see the structure labeled axiocord (axochord) in Figure 2.4 and the discussion of an axochord in Section 2.5B. There is no trace of the seemingly crucial chordate feature — internal gills. And the type of mesoderm formation is in contrast; in chordates, a pouch type of mesoderm formation is the basic pattern, in annelids the mesoderm splits into layers. There is, thus, almost no positive reason to believe in a descent of vertebrates from annelids; and there are so many difficulties in the way that there is today little reason to take stock in an annelid theory. All that said, molecular studies over the past 30 years have resurrected the possibility of dorso-ventral inversion and so brought annelids back onto the stage of chordate ancestry.

C. THE 1990s: THE STOMOCHORD, AXOCHORD AND GENE-BASED HOMOLOGY HYPOTHESES

Revival of an annelid-like ancestor for chordates in the mid-1990s was based on several lines of evidence that included (i) adopting dorso-ventral inversion as possible, (ii) non-homology of the hemichordate stomochord with the notochord, and

(iii) anatomical and genetic similarities between annelid and chordate nervous systems (Nübler-Jung and Arendt 1994; Arendt and Nübler-Jung 1996). Based on shared transcription factors (*Brachyury*, *Fox-A*, *Fox-D*, *Sox-D*, *Sox-E*, *Twist*) and the signaling molecules Noggin and Hedgehog, the mid ventral longitudinal muscle of annelids was renamed the *axochord* and proposed as the precursor of the notochord (Lauri et al. 2014, and see below).

Two caveats are important: these signaling factors are not uniquely expressed in either the axochord or the notochord. Furthermore, the axochord appears not to have the early signaling genes used by the notochord in early chordate development (Chapter 1, and see Hejnol and Lowe 2014, 2015; Satoh et al. 2012). Complicated as the question of the origin of the notochord is, Satoh et al. (2012) argued that in terms of developmental genetics, the origin of the notochord in the first chordate comes down to the evolution of regulatory gene networks associated with *Brachyury*— the key notochord regulatory gene (Chapter 8)—and their role in altering the role of *Brachyury* from initiating ventral mesoderm to form muscle to initiating dorsal mesoderm to form notochord. Satoh et al. (2014) used such evidence to argue for an evolutionary relationship of the stomochord to the chordate anterior pharynx.

All that being said, increasing numbers of similarities between annelid axochord and chordate notochord are emerging from ongoing research, indicating that reliance on gene networks associated with a single gene, even such a key gene as *Brachyury*,[7] might be simplistic.

Using a wider approach to understand notochord origins, Lauri et al. (2014) investigated the development of the marine polychaete annelid *Platynereis dumerilii*, which they show possesses a population of mesodermal cells (the future axochord) that converges and extends toward the ventral midline (as notochord precursors do toward the dorsal midline) and expresses a combination of genes that are specific to the notochord in chordates. Using the three criteria of specificity, conservation and function, the ventral midline cells express the genes *Brachyury*, *Fox-A*, *Fox-D*, *Sox-D*, *Sox-E*, *Twist*, *Noggin* and *Shh* (as indicated above); the hemichordate stomochord does not express *Brachyury* (although other regions of the embryo do), *Fox-A* or *Noggin*.

Furthermore, like the notochord, the axochord possesses a thick collagen-rich extracellular matrix and an axochord is found in a basal annelid, *Owenia fusiformis*. The phylogenetic analysis presented by Hejnol and Lowe (2014) shows that contractile, midline mesodermal cells could have been present in the bilaterian ancestor and so have been positioned for modification during the long history of chordate evolution. At the same time, the phylogenetic analysis by Hejnol and Lowe (2014) refutes any homology between the annelid axochord and the chordate notochord at the level of characters; Lowe et al. (2015) point out that the muscle cells that form the annelid axochord have no notochord-like signaling functions.

---

[7] As a gene and transcription factor, *Brachyury* (from the Greek βραχύς, ουρά for short tail) creates the usual problems associated with names for the same gene in different vertebrates. In humans, the protein *Brachyury* is encoded by the *T-box transcription factor T* (*TBXT*) gene. Until 2018, *TBXT* was known as *T*. *No tail* (*ntl*) has been proposed as the name for *Brachyury* in zebrafish, but No Tail is more usually used for a zebrafish mutant of *Brachyury* (Schulte Merker et al. 1992, and see Chapter 8); we use *Brachyury* for the gene across the chordates, in the hope of avoiding confusion.

Hejnol and Lowe (2014) emphasize the significance of the annelid debate, a debate that should not be underestimated. We must assume that early bilaterians were small animals with modest body plans and that bilaterian evolution was characterized by convergent evolution of complex body plans in different phyla. Contrasting to this scenario is the assumed homology of axochord and notochord. Homology implies that early bilaterians already had a very complex Bauplan and requires that complex characters such as notochord, axochord or CNS have been lost in most bilaterian lineages. Obviously, this is not the most parsimonious assumption.

Speculations about a possible dorsal ventral inversion and relocation of the CNS should consider the organization of the CNS, which is solid in invertebrates and hollow in vertebrates. In invertebrates, neurons are located in the periphery of the CNS. Cell connections fill the lumen and thus are limited in numbers. The structure of the vertebrate CNS is the opposite. Neurons are located in the center and neuronal connections are peripheral and essentially unlimited in number (Meinertzhagen 2010). Any hypothesis that assumes axochord-notochord homology must not only explain the dorsal ventral inversion of the body and inversion of mouth and anus but also the complete reorganization of the structure of the CNS.

2.6   VERTEBRATES

Vertebrates have a notochord that runs the full length of the body from the base of the brain to the tip of the tail (see Chapter 1).

Karl Ernst von Baer (1792–1876) studying chicken development, was the first to discover the notochord and to recognize it as the embryonic axial skeleton and precursor of the vertebral column (Figure 2.5). von Baer pointed out that earlier investigators mistook the anlage of the notochord for the neural tube. von Baer (1828) coined the term dorsal strand or chorda dorsalis (*Rückensaite*) for this feature. Later he used the name chorda vertebralis (*Wirbelsaite*) to reinforce his conclusion that the notochord is the precursor of the vertebral column (von Baer 1837). Discovery of a notochord in other vertebrates, including cyclostomes (lampreys) led to the recognition of the *homology of the notochord across the vertebrates.*

A. Albert Kölliker's 1859 Paper

Given two broad and potentially separate evolutionary issues — notochord development and vertebral development — it is amazing to find that both were tackled in depth in a paper written by Albert Kölliker over 160 years ago (Kölliker 1859). The title of Kölliker's paper was "On the structure of the Chorda Dorsalis of the Plagiostomus[8] and some other fishes, and on the relation of its proper Sheath to the development of the Vertebrae." Kölliker recognizes four layers to the notochord:

- an outer elastic acellular membrane, often perforated;
- the sheath proper, a cellular fibrous connective tissue;

---

[8] The Plagiostomi were then recognized as an order of fishes that contained sharks and rays.

- an inner elastic membrane of reticulated elastic fibers; and
- "gelatinous substance of the chorda itself, made up of soft cartilage cells" (1859, p. 215).

Kölliker concluded that: (i) the notochord remained intact in species that lacked vertebral centra, such as cyclostomes, sturgeons (*Acipenser* spp.) and the South American lungfish *Lepidosiren paradoxa*; and (ii) that the notochord was limited to the middle of each vertebral centrum in species with ossified vertebrae, but could be separated into "as many parts as there are interstices between the vertebrae" in "most of the higher animals" [i.e., tetrapods] (Kölliker 1859, p. 216). Kölliker also categorized relationships between the notochord and the cartilage of the base of the cranium, and relationships between the notochord and vertebral development (*ibid.* pp. 217–222), the latter forgotten until very recently and the topics of Chapters 7 and 8.

B. Notochord Structure and Function in Vertebrates

As discussed herein, especially in Chapters 7 and 8, we have recently learned much about the nature of the relationship between the notochord and vertebral development, especially the important primary role that the notochord plays in vertebral skeletogenesis in many organisms. Few individuals, however, have tackled the issue of the comparative structure of the notochord at the cellular level across the vertebrates. Such studies, often requiring electron microscopic levels of analysis, should illuminate notochord and notochord cell evolution.

2.7   SEARCHING FOR CHORDATES AND EARLY VERTEBRATES IN THE FOSSIL RECORD

Alexander Kovalevsky's discovery of the presence of a notochord in tunicates (Kovalevsky 1866b) triggered the search for vertebrate ancestors within the new phylum chordates, although the search for invertebrate ancestors did not cease. An obvious place to look to resolve chordate and vertebrate ancestry is the fossil record. Indeed, several scenarios for the origin of chordates are based on the fossil record, for example, the calcichordate hypothesis (independent origins of cephalochordates, urochordates and vertebrates from calcichordates, a group of echinoderms) and the vetulicolians hypothesis (Cambrian organisms, now regarded as a sister group to tunicates) concisely reviewed by Holland (2005).

Of course, identification of a hemichordate or cephalochordate in the early fossil record (which means lower or middle Cambrian) is not evidence for the *ancestor of chordates* but is evidence for when chordates may have arisen and for the morphological organization of early chordates. A major difficulty arises in identifying chordates in the fossil record if the notochord is poorly preserved. Although the notochord withstands decay better than other soft tissues in taphonomy experiments (Briggs 1995; Raff et al. 2006), normally only mineralized skeletal tissues leave traces in the fossil record (Johanson et al. 2012; Trueman 2013). As the notochord, like cartilage and connective tissue, preserves poorly, considerable lengths have to be gone to confirm the presence of a notochord. Philippe Janvier emphasized that for fossils that are preserved as either soft-tissue imprints or minute skeletal fragment it is sometimes difficult for palaeontologists to tell which 'structures' are reliable vertebrate remains and which merely reflect our idea of the features an ancestral vertebrate should possess (Janvier 2015).

A. CAMBRIAN FOSSILS: CANDIDATE CHORDATES

As we will see below, if a putative notochord is preserved as a midline structure, there is room for interpretation and debate as to whether the structure is really a notochord. Debates about the presence of a notochord influence our ideas on the evolution of early chordates and early vertebrates. Especially in the oldest fossils, such as those in the Early and Middle Cambrian, identification of the notochord is difficult.

Analysis of 114 specimens from the Middle Cambrian Burgess Shale led Conway Morris and Caron (2012) to interpret *Pikaia gracilens* as a stem chordate, not because of unequivocal identification of a notochord but on the basis of some 100 segmented sigmoidal myomeres (muscle blocks) in each individual. Identification of a notochord in these specimens was extremely tentative. The preserved structure may have been (i) a notochord (their interpretation, made in no small part because of the presence of other chordate characters), (ii) a protonotochord, or (iii) a non-homologous structure, the result of independent evolution and convergence with chordates.

Preservation of the notochord is better in two Early Cambrian species—Early Cambrian or perhaps Lower Cambrian *Cathaymyrus diadexus* and *C. haikoensis*—from the Chengjiang formation in Yunnan Province, China (530 MYA). They were classified as cephalochordates (and so not stem chordates) on the basis of the identification of a notochord and the presence of segmented myomeres (Shu et al. 1996, 1999, 2003; Luo et al. 2001). So, given that these are Early Cambrian in age, a notochord in *Pikaia* would not be inconsistent with when the notochord and chordates originated. But these taxa postdate the origin of the notochord.

As noted in Section 2.4, lower or early Cambrian craniates (possible hemichordates) have been identified. *Haikouichthys ercaicunensis* from the Chengjian fauna of China is one (Shu et al. 1999). Another, the middle Cambrian filter-feeding *Oesia disjuncta* from the Burgess Shale in Canada was interpreted as a primitive hemichordate by Nanglu et al. (2016). Geographically, these fossils indicate independent origins of hemichordates, although it should be noted that classification as hemichordates is not based on evidence of the presence of a notochord but on cladistic analysis of a number of other characters.

B. CANDIDATE VERTEBRATE: *JAMOYTIUS KERWOODI,* CONODONTS AND THE "TULLY MONSTER"

a. Jamoytius kerwoodi

On the basis of two specimens, White (1946) named an Early Silurian fossil *Jamoytius kerwoodi* (in honor of the English palaeontologist, James Alan Moy-Thomas [1908–1944]) and described it as the most primitive vertebrate then known. White came to this conclusion despite the absence of most vertebrate or even chordate characters in these two specimens. Now assumed to represent its own order of jawless vertebrates — Jamoytiiformes — *Jamoytius* was considered a vertebrate because of a dorsal midline structure that was interpreted as a notochord, but it could have been intestine or part of a branchial basket (Sansom et al. 2010). Early vertebrates, like early chordates, can be enigmatic.

b. Conodonts

Conodonts are another enigmatic group of fossils that have fueled the debate about the origin of vertebrates. Conodonts first appeared in the Cambrian 485 MYA and disappeared in association with the Triassic-Jurassic extinction event 200 MYA. Conodonts have been identified as plants, annelids, molluscs, primitive chordates, vertebrates or even stem gnathostomes. Derek Briggs, on the basis of the identification of segmented muscles, fins with fin rays, and a notochord, classified conodonts as chordates (Briggs 1992). Almost 20 years later, Susan Turner together with seven other colleagues (Turner et al. 2010) concluded that conodonts possess insufficient characters to classify them as vertebrates, craniates or jawed vertebrates, and that they may not even be chordates.

The "conodont debate" exemplifies the difficulties in recognizing a notochord, for example, to decide if a midline structure is a dorsal notochord or a ventral annelid gut (Blieck et al. 2010). This is especially problematic when preserved mineralized elements bear no obvious resemblance to any known vertebrate skeletal elements. In the absence of a notochord — and in the absence of virtually any vertebrate characters (Turner et al. 2010)— the conodont discussion focused on tooth-shaped mineralized microfossils known as conodont elements. Scott (1934) considered these to be essentially identical to the jaw apparatus of annelid worms. Recent detailed analyses have led to the conclusion that conodont elements are unrelated to vertebrate teeth (absence of dentine and enamel) or to jaws (absence of bone) (Turner et al. 2010; Murdock et al. 2014; Hall 2015; Donoghue and Rücklin 2016). Conodonts remain enigmatic.

c. The "Tully Monster"

Recognition and identification of the notochord are also central in the debate about another candidate vertebrate, the "Tully Monster" *Tullimonstrum gregarium*, a soft-bodied organism from the late Carboniferous Mazon Creek biota, 309–307 MYA. Previously identified as a nemertean worm, polychaete worm, gastropod, arthropod or conodont, an analysis of 1200 specimens by McCoy et al. (2016) claimed evidence for the presence of a notochord, cartilaginous vertebrae, gill pouches, and multiple tooth rows around the mouth. These characters coupled with a detailed phylogenetic analysis led McCoy and colleagues to classify *Tullimonstrum* as a vertebrate and stem lamprey.

This assignment of the "Tully Monster" to the vertebrate clade is rejected by Lauren Sallan, Philippe Janvier and their colleagues (Sallan et al. 2017). They argue that the light colored stain (the hypothetical notochord) extends anterior to the supposed eye bars (never the case in any vertebrate), and further, that by removing the notochord from the equation, other traits of the fossil are not easily interpreted as vertebrate characters.

Similarly, the earliest craniates (named ostracoderms[9] by Edward Cope in 1889) are known only from their mineralized dermal skeletons, and again, we have minimal data about their internal anatomy, skeletal or otherwise (Maisey 2000).

2.8 DEVELOPMENTAL BIOLOGY AND PALAEONTOLOGY

The problems that palaeontologists can have in identifying the notochord in early chordates and the discussion of molecular comparisons between chordates and annelids raises the question of how much developmental biology can contribute to resolve chordate ancestry. In this realm, evolutionary biology faces the problem that developmental biology normally focuses on a few highly derived vertebrates such as mice, domestic chickens and zebrafish. For practical purposes, ontogenetic data are essentially unavailable for most extant species and of course largely unavailable from the fossil record (Maisey 1988). Two papers, the second written at the invitation[10] of the editor of *Palaeontology* to complement the first (Conway Morris 1994; Hall 2002) demonstrate the mutual benefit to be obtained when palaeontology uses developmental and molecular biology and *vice versa*. Indeed, the history of the skeletal systems in vertebrates can best be understood through the reciprocal illumination that comes through a combined palaeontological and developmental biological approach (Maisey 1986).

2.9 SUMMARY

This chapter treats the discovery of the notochord as a feature uniting all chordates that brought some groups into the chordate phylum through comparative embryological studies in the 19th century. Essential features defining a chordate are discussed, followed by a discussion of the classes of chordates, which are cephalochordates, ascidians (tunicates) and vertebrates, and the nature of the notochord in each group. How hemichordates came to be included and then excluded from the chordate family on the basis of the possession of a stomochord also is discussed, as is the discovery that annelids with their ventral nerve cord and axochord were, and still are by some researchers, regarded as potential chordate ancestors. The use of classes of genes to establish homology played a major role in the "annelids as chordate ancestors" discussion. The chapter concludes with a discussion of the search for chordates and early vertebrates in the fossil record when the notochord is so poorly preserved, and of how palaeontology and developmental biology have reinforced one another in the search to understand the nature and evolutionary origin of the notochord.

---

[9] Ostracoderms are not a monophyletic group but a polyphyletic assemblage of six groups of jawless vertebrates. Even the two well-recognized groups — heterostracans and cephalaspids — are not natural groups. The six recognized groups are galeaspids, pituriaspids, osteostracans, pteraspidomorphs, thelodonts and anaspids (Janvier 2015).

[10] The request was "to write an article on 'why palaeontology needs (or should be interested in) evo-devo', along the lines of Conway Morris' 'why molecular biology needs palaeontology' but from the opposing view point" (Hall 2002, p. 647).

3 Germ-Layer Origin of the Notochord
Endoderm or Mesoderm

CONTENTS

3.1 INTRODUCTION

Today, textbooks and journal papers describe the notochord as a *mesodermally* derived midline structure but this has not always been the case. The classic literature from the 19th century clearly identified the notochord as an *endodermally* derived organ. The debate over whether the notochord is mesodermal or endodermal therefore is an ancient one. Is it possible, of course, that the embryonic origin of the notochord — from mesoderm or from endoderm — is not the same in all chordates or differs along the anterior-posterior embryonic axis in individual species, or even in individual embryos? Just how conserved are the developmental and evolutionary origins of such an essential chordate tissue? This chapter addresses these questions.

3.2 THE NOTOCHORD, VERTEBRAL COLUMN AND NEURAL TUBE

Based on his research into chicken embryos, Karl von Baer (1792–1876), the founder of modern comparative embryology, recognized the notochord as the embryonic precursor of the vertebral column and distinguished the notochord from the neural tube (Figure 2.5A). Von Baer (1828) introduced the term *chorda dorsalis* (*Rückensaite*) for what we now know as the notochord. Later (von Baer 1837) he changed the name to *chorda vertebralis* (*Wirbelsaite*) to emphasize that the notochord is the precursor of the vertebral column (see Figure 2.5B). Today, chorda dorsalis remains as the

DOI: 10.1201/9781315155975-4

established scientific term, in no small part because of the discovery by Kovalevsky (1866a, b) of a notochord in tunicates (see Chapter 2). Indeed, not all chordates possess a vertebral column.

Before the identification of the notochord as a feature of all chordate embryos, it was often mistaken for the neural tube. von Baer (1828) put an end to this confusion. Ever since von Baer recognized the notochord, scientists have investigated its embryonic origin and its contribution to the development of the vertebral column in craniates (chordates with a skull) and gnathostomes (chordates with a skull and jaws). Craniates are synonymous to vertebrates (Janvier 2015). Craniates are subdivided into jawless vertebrates (*agnathans*) and jawed vertebrates (*gnathostomes*). Jawless vertebrates carry the name vertebrates although most of them do not develop a true vertebral column with vertebral bodies around the notochord. Some jawless vertebrates such as extant hagfish possess small cartilaginous modules located ventral to the notochord that have been interpreted as vertebral column elements (Ota et al. 2011). A vertebral column has been identified in a jawless vertebrate *Palaeospondylus gunni* of uncertain affinities from the Middle Devonian (390 Ma) of Scotland (Janvier 2015). As this species demonstrates, the fossil record of jawless vertebrates provides minimal data about their internal anatomy, skeletal or otherwise (Maisey 2000) so that, in most cases, one must assume that a true vertebral column was absent. In contrast, jawed vertebrates have a vertebral column comprising paired arcualia, paired vertebral (hemal) arches and complete vertebral centra, or a combination of these elements.

Vertebral centra can be made from cartilage, mineralized cartilage, mineralized notochord sheath and/or bone. Arches can be made from cartilage, mineralized cartilage and/or bone but not from the notochord sheath (Gardiner 1983; Arratia et al. 2001).

About 100 years after von Baer's publication, Goodrich (1930) emphasized that the *evolutionary* origin of the notochord remains to be clarified (as it still does today; Chapter 2) but, for Goodrich, there was no question about the *developmental* origin of the notochord:

> The notochord invariably develops from the dorsal wall of the archenteron as a thickening or upfolding which becomes nipped off from before backwards, and continues to grow at its posterior end as the embryo lengthens.
>
> *(Goodrich 1930, pp. 7).*

Goodrich's view about notochord development was based on several comprehensive studies dating back to the 17th century (Figure 3.1).

3.3   DEVELOPMENTAL ORIGIN OF THE NOTOCHORD

Developmental and molecular studies provide a wealth of knowledge about early notochord development and genes expressed by the notochord and critical for notochord initiation/development (see Chapter 4). Concerning the germ-layer origin of the notochord, all three germ layers (ectoderm, endoderm and mesoderm) have at various times been proposed as giving rise to the notochord (Remane 1936). Similar discussions exist for other vertebrate key characters, for example the origin of teeth (Huysseune et al. 2022a). We discuss evidence for an endodermal, mesodermal or ectodermal origin of the notochord below.

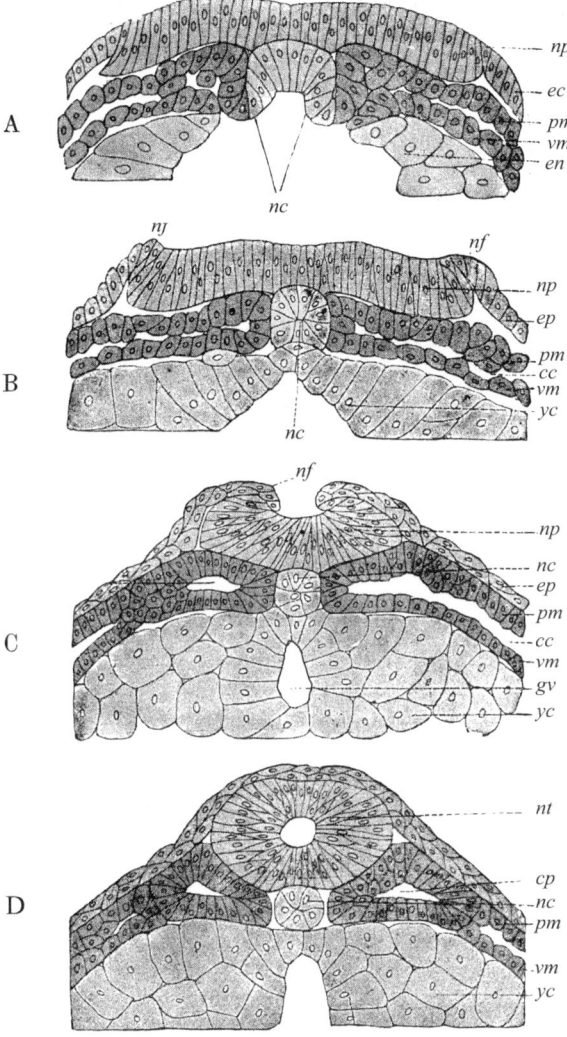

FIGURE 3.1 The classic early 20th century view of the development of the notochord shown as cross sectioned embryonic stages of a urodele (*Triton sp.*) with dorsal to the top. (A) Embryo prior to neural fold formation. (B) Onset of neural fold (nf) formation. (C) Beginning of closure of the neural folds around the medullary canal. (D) The neural tube (nt) is closed and covered dorsally with epithelial ectoderm. Left and right pairs of somites (also seen in C) are differentiating from the parietal mesoderm (pm) creating a coelomic cavity (cc). The notochord (nc) in this scheme is assigned to endoderm in the roof of the archenteron (B, C) from which it has separated as a cylindrical rod (D). The cells that form the notochord are clearly distinguished from the dark grey shaded mesoderm cells that give rise to the somites. Notice that in (A) the notochord anlage (nc) has no connection to endoderm but connects laterally to the parietal. mesoderm. Labels (translated into English by PEW): cc, coelomic cavity; cp, cavity of the primitive segment; ec, ectoderm; en, endoderm; ep, epidermis; gv, gut volume; nc, notochord; nf, neural fold; np, neural plate; nt, neural tube; pm, parietal mesoderm; vm, visceral mesoderm; yc, yolk cells. (Modified after Hertwig, O., 1915. *Lehrbuch der Entwicklungsgeschichte des Menschen und der Wirbeltiere.* 10th Edition, Verlag von Gustav Fischer, Jena, p. 782.)

A. Notochord as Endodermal

When Goodrich (1930) pointed out that the development of the notochord is largely settled, many investigators had reached a consensus that the notochord is of endodermal origin (Figure 3.1). Observations had been consolidated around the interpretation that the notochord folds off from the roof of the archenteron, which is the anlage of the gut (Moore et al. 2016). By assuming that both the development and the origin of the notochord are highly conserved, Kovalevsky (1866b) reviewed the knowledge about notochord development in different groups of chordates and compared this with his own observations on ascidians. Kovalevsky concluded that early notochord development always follows the same trail. Indeed, only later stages of notochord development deviate, reflecting the fate and function of the structure, which include loss, constriction, persistence, transformation into cartilage, or growth and transformation into an intervertebral disk (Musgrove 1891).

Development of the notochord starts after gastrulation. In amphioxus at the early neurula stage, the mesoderm is represented by a single cell layer located in the dorsal roof of the archenteron (endodermal/mesodermal structure; Figure 3.2A1–D1). At the late neurula stage, all the rostral somites pinch off from the archenteron roof simultaneously and together with the notochord (Onai et al. 2015).

Goodrich (1930) and Remane (1936) emphasized that, not only in amphioxus, but in all vertebrates, the notochord invariably develops as a thickening of or as an unfolding from the dorsal wall of the archenteron. The notochord anlage in mammals is initially connected to the anterior part of the gut, located close to Seessel's pouch (Hertwig 1915), an evagination of the future endodermal pharynx, an anatomical association that fits with an endodermal origin of the notochord (Figure 3.3). During gastrulation and early neurulation stages in human embryonic development, the notochord temporarily fuses with the endodermal roof of the archenteron from which it separates during the third week of development (de Bree et al. 2018). The connection between the notochord and the gut is only lost after disruption of the pharyngeal membrane (Hertwig 1915, p. 691).

If the anterior notochord does not completely separate from the endoderm roof, an irregular flask-shaped depression of the mucous membrane is generated from permanent notochord remains. This embryological notochord residual, known as a pharyngeal bursa, occurs as Thornwaldt cysts in about 4% of healthy adult humans (Babic 1990, 1991; Cetinkaya 2018). After separation of the notochord from the endoderm, growth of the notochord occurs at its posterior end. After the primitive streak has been formed, the growth point of the notochord is situated in front of the neurenteric canal, the dorsal lip of the blastopore (Goodrich 1930; Remane 1936; see Section 3.5).

An important point to note at this stage is that the process discussed above represents one of the two mechanisms of neurulation. *Primary neurulation* begins cranially *after* the three germ layers have segregated from one another. All vertebrates, however, have a mechanism of *secondary neurulation*, which begins caudally *without* caudal cells separating into germ layers. Consequently, the caudal region of vertebrate embryos develops from a blastema of cells known as the *tail bud*. Primary/secondary neurulation and tail buds are discussed in Section 3.5.

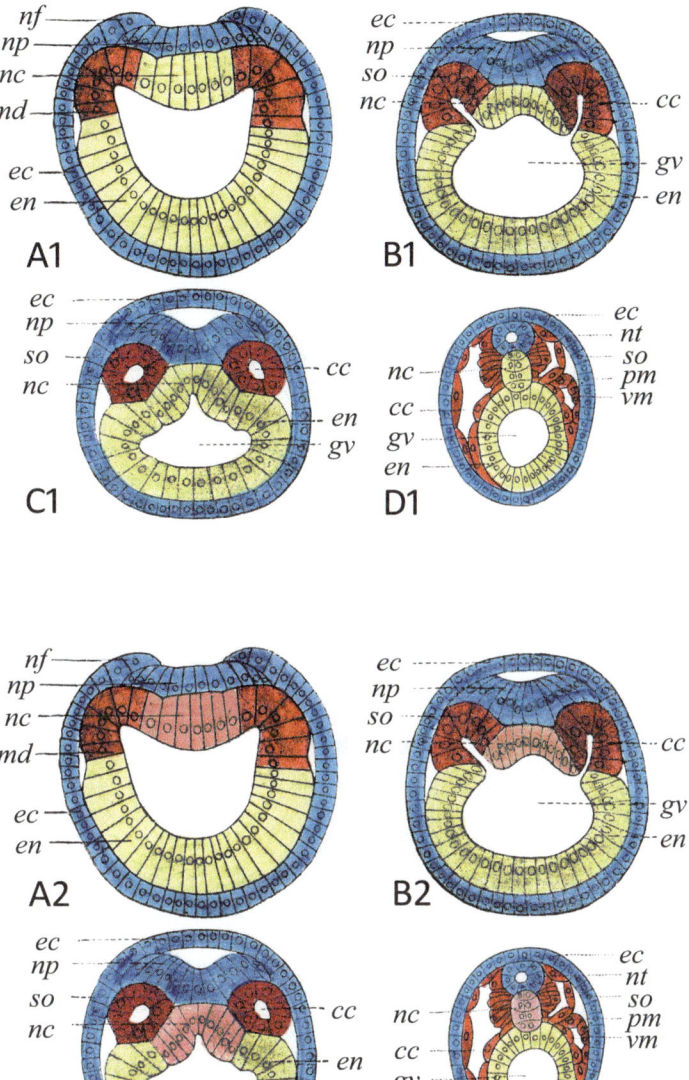

FIGURE 3.2 Development of the notochord in early to late neurula stage amphioxus embryos shown as cross (transverse) sections, according to Hatschek (1893). Germ layers are colored following a long tradition as ectoderm (blue), endoderm (yellow) and mesoderm (red). A1–D1 Represent the classical view and Hatschek's interpretation according to which the notochord derives from the roof of the archenteron and is thus of endodermal origin (yellow). (A1) A section through the middle of the body of an embryo with the first mesoblastic pairs of somites (md). (B1) Section of an embryo with the fifth pair of somites (so) and the coelomic cavity (cc) in formation. (C1) Transverse section through the anterior part of the first somite of an embryo with five well-formed pairs of somites. (D1) Transverse section through

(Continued)

FIGURE 3.2 (*CONTINUED*)

the middle of the body of an embryo with 11 pairs of somites. A1–D1 represent the classical view and Hatschek's interpretation according to which the notochord derives from the roof of the archenteron and is thus of endodermal origin. In A2–D2 the germ layer assignment of the notochord is modified to show the notochord as a derivative of the mesoderm (red). Recent studies on amphioxus show that both notochord and mesoderm derive from endoderm as a secondary germ layer. The notochord is flanked by mesoderm from early in development (A2). According Onai et al. (2015) in amphioxus, mesoderm and endoderm do originate from the same single cell layer (archenteron). The dorsal mesoderm includes somites and the notochord. Mesoderm differentiates into the musculature, somites and notochord, from which muscle fibers extend to the neural tube. Mesodermal origin of the notochord (A2), temporary integration of the notochord into the roof of the archenteron (B2, C2) and subsequent dissociation from endoderm (D2) is also assumed to hold for mammals. Labels: cc, coelomic cavity; ec, ectoderm; en, endoderm; ep, epidermis; gv, gut volume; md, mesoderm; nc, notochord; nf, neural fold; np, neural plate; nt, neural tube; pm, parietal mesoderm; so, somite; vm, visceral mesoderm; yc, yolk cells. (Modified after Hatschek, B., 1893. *The Amphioxus and its Development*. Swan Sonnenschein & Co, London, 181 p. Labeling follows Hertwig, O., 1915. *Lehrbuch der Entwicklungsgeschichte des Menschen und der Wirbeltiere*. 10th Edition, Verlag von Gustav Fischer, Jena, 782 pp; see also Figure 3.1.)

B. Notochord as Mesodermal

Must we assume that the notochord is endodermally derived? The answer is no for several reasons. Importantly, and a potential source of confusion in determining origins, mesoderm is a secondary germ layer-derived from the endoderm, which is a primary germ layer (Hall 1997, 1998b, 2015; Hashimshony et al. 2015). Indeed, the term endomesoderm is often used for the earliest cell layer that has not yet differentiated into mesoderm and endoderm, but that is destined to give rise to both. The notochord may be set-aside at this earliest stage (Figure 3.2A2–D2).

According to the classical *enterocoel hypothesis*, first proposed by Lankester (1877) in the context of determining the significance of germ layers, mesoderm in coelomates can arise as an outpocketing from the embryonic gut or archenteron. Enterocoely is now known to be a basic chordate developmental process (Holley 2007; Lowe et al. 2015), seen in amphioxus and in lampreys (Adachi and Kuratani 2012). Other views, based on sea urchin development, consider that mesoderm initially arises at the ectoderm–endoderm boundary at the equator of the blastula as the result of inductive interactions initiated by the endoderm (reviewed by Hall 1998a, b). In amphioxus, mesoderm that forms the anterior somites does indeed arise together with the notochord from the roof of the archenteron (Onai et al. 2015), while posterior somites pinch off sequentially from the tail bud (Lowe et al. 2015, p. 458). Although it is generally accepted that endoderm can give rise to mesoderm, there is a continuing debate over whether the *endoderm to mesoderm transition* also applies to the notochord precursor cells.

The notochord can be considered as endodermal in origin because the cells derive from the roof of the archenteron (Lankester 1877; Hertwig 1915; Goodrich 1930; Remane 1936) (Figure 3.1). The alternative view is to assume that the cells that will form the notochord also undergo an endoderm to mesoderm transition. Thus, not only the dorso-lateral parts but the medial roof of the archenteron would give rise to mesoderm. The two alternative scenarios, which are depicted in Figure 3.2 are

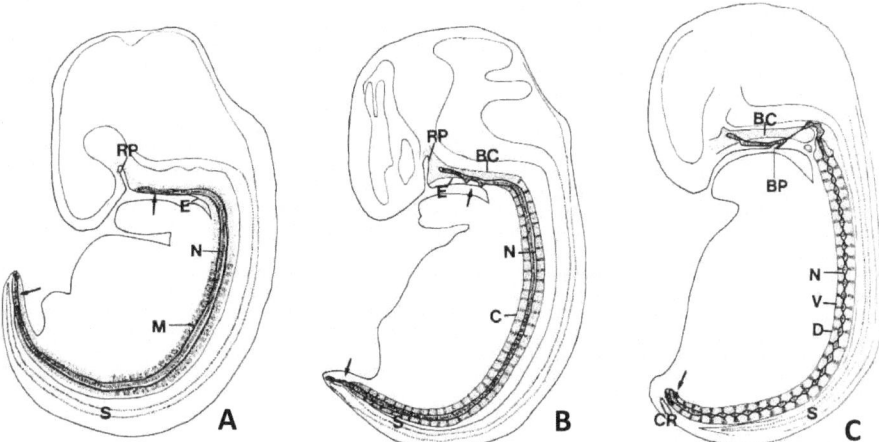

FIGURE 3.3 Relationship of the notochord to the pharyngeal endoderm and adjacent axial structures during human development, shown as diagrammatic mid-sagittal sections. (A) Mid-sagittal section through a five-week-old embryo demonstrates the relationship of the notochord (N) to the pharyngeal endoderm (E), to adjacent axial mesenchyme (M) of the vertebral column, and to the neuroectoderm of the spinal cord (S). Arrows indicate the areas of closest association. RP, Rathke's pouch invagination. (B) Mid-sagittal section through a five-six-week-old embryo. Initial separation of the notochord (N) from the pharyngeal endoderm (E) and partial incorporation of the notochord into the developing base of the chondrocranium (BC) is indicated by an arrow. C, somites. (C) Mid-sagittal section through a 12-week-old fetus. The notochord (N) degenerates inside the vertebral bodies (V) and widens in the intervertebral disks (D). In the head region, partial degeneration of the notochord inside the basis of the chondrocranium (BC) and development of the bursa pharyngea (a remnant of the notochord as a cyst in the posterior wall of the nasopharynx found in some 4% of adult humans) are shown. In the regressed tail region where the sacrococcygeal vertebrae are forming, (CR), the arrow indicates branching of the notochord. (Reproduced with the permission of UPV/EHU Press from Babic, M. S., 1991. Development of the notochord in normal and malformed human embryos and fetuses. *Int J Dev Biol* 35: 345–352.)

both used in the literature. Schubert et al. (2001) and Bertrand et al. (2017) solve the conflict by considering the tissues that will give rise to the paraxial mesoderm and to the notochord as *mesendoderm* (endomesoderm; see above), within which the axial dorsal region is fated to become the notochord, and the dorsal paraxial mesendoderm is fated to give rise to the somites. A discussion of notochord origin in different groups of vertebrates led LaFlamme et al. (1988) to suggest (i) that the notochord is a discrete early differentiating product of migrating cells during gastrulation, and (ii) that germ layer derivation may be less important than the specific cellular mechanisms acting in notochord formation.

In zebrafish, the notochord anlage is centered in the dorsal midline with somitic mesoderm on either side of the notochord primordium (Shih and Fraser 1995). A similar close arrangement of notochord and somites was reported in the African clawed toad *Xenopus laevis* by Keller (1976). Studies of the fate map in zebrafish embryos emphasize a mesodermal origin of the notochord. Cell tracing, the close similarities between the notochord and cartilage, and the active skeletogenic function of noto-chord support this view (Kimmel et al. 1990).

The T-box gene *Brachyury*,[1] which is expressed in the notochord of all chor-dates (Figure 3.4), is required for mesoderm and notochord development in mouse embryos. In zebrafish, the *Brachyury* homologue *no tail* (*ntl*) is equally required for notochord development (Schulte-Merker et al. 1992). The genes *floating head* and *spadetail* are expressed in the midline mesoderm of zebrafish. *Floating head* expres-sion is required for notochord development, while *spadetail* expression is required for muscle development (Amacher and Kimmel 1998). *Floating head*, the zebrafish homologue of *not*, which is a homeobox gene expressed in the amphibian dorsal organizer (see below) through production of the protein XNOT, is, as in zebrafish, required for notochord development (Talbot et al. 1995).

C. Notochord as Ectodermal

The hypothesis that the notochord is a derivative of the ectoderm has been less frequently proposed or discussed. A strong proponent for an ectodermal origin of the notochord in mammals is Huber (1918) who concluded his detailed studies on guinea pig (*Cavis porcellus*) development as follows (note that Huber uses "entoderm" where we would use endoderm):

> Since the entoderm takes no active part in the histogenesis of the head process, chordal canal, and chordal plate and since the chordal plate becomes only partially and tempo-rarily incorporated in the entoderm; there seems no justification for classing the chorda dorsalis as an entodermal derivative. And since the head process, the anlage of the chordal canal and derived structures, has its anlage in the cranial portion of the primi-tive node, a region of active ectodermal cell proliferation; and since the chordal canal and plate retain their continuity with the primitive node, which serves as a growth zone; there seems justification in regarding head process-chordal canal, and derived structures, chordal plate and chorda dorsalis as a derivative of the ectoderm in the sense that the mesoderm is derived from the ectoderm of the primitive streak region of the embryonic shield.
>
> *(Huber 1918, p. 262)*

Huber's view is supported by Jurand (1974) who points out that the notochord prolif-erates from the dorsal lip of the blastopore in amphibians and from the cranial end of the primitive streak (Hensen's node) in amniotes. If this is so, then the notochord should be regarded as mesodermal, but Jurand (1974) argues it is not.

The notochord does not derive from preformed mesoderm but develops inde-pendently, although simultaneously with the mesoderm from the cephalic end of the primitive streak. Jurand (1974) concludes that because of the ectodermal

---

[1] As a gene and transcription factor, *Brachyury* (from the Greek for short tail) creates the usual problems associated with names for the same gene in different vertebrates. In humans, the protein *Brachyury* is encoded by the *T-box transcription factor T* (*TBXT*) gene. Until 2018, TBXT was known as T. *No tail* (*Ntl*) has been proposed as the name for *Brachyury* in zebrafish but *No Tail* is more usually used for a zebrafish mutant of *Brachyury*. We use *Brachyury* for the gene across the vertebrates, in the hope of avoiding confusion.

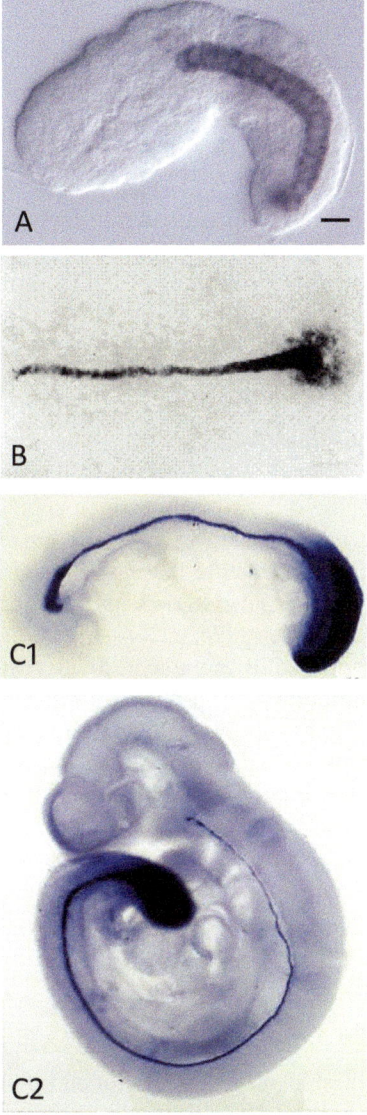

FIGURE 3.4   (A) Expression of *Brachyury* in the notochord of an early hatchling of the lar-vacean urochordate, *Oikopleura dioica*. Scale bar = 10 μm, anterior to the left. (B) *Brachyury* expression visualized by *in situ* hybridization in a 5-somite-stage zebrafish (*Danio rerio*) embryo. Anterior to the left. Expression is along the entire anterior-posterior axis but is high-est in cells of the posterior presumptive notochord (right). (C) Whole mount *in situ* hybridiza-tion for the expression of *Brachyury* in the notochord of mouse embryo at E8.5 (C1, anterior to the left) and E10.5 (C2, left lateral view with the head at the top). Note the enhanced expression of *Brachyury* in the tail bud. ((A) With permission from Bassham and Postlethwait, 2000. *Dev Biol* 220: 322–332; (B) with permission from Schulte-Merker, S., et al., 1992. *Development* 116: 1021–1032. (C1 and C2) With permission from Barrionuevo, F., et al., 2006. *Dev Biol* 295: 128–140.)

position of the cells in the blastopore lip and in Hensen's node, it is valid to view the notochord as of ectodermal origin. If we take the origin of the notochord further back in development, the study of Nieuwkoop and Ubbels (1972) on early development in urodeles (*Ambystoma*) would lend support to an ectodermal origin of the notochord in non-amniotes as well as in mammals. In *Ambystoma*, the animal cap of the blastula (which gives rise to atypical ectoderm if isolated) furnishes all ecto-neurodermal derivatives, the mesodermal structures, and even contributes to endodermal structures. The notochord would then be initially of ectodermal origin, independent from subsequent connections to endo- or mesoderm, a conclusion that depends critically on the accuracy of the identification of the germ layers at their earliest development.

D. Chordamesoderm and the Organizer

What Goodrich and Remane regarded as the location of notochord extension is now considered as the point of origin of notochord development (Scott and Stemple 2004; Stemple 2005a). Fate mapping of embryonic cells in zebrafish (Kimmel et al. 1990), chickens (Resende et al. 2010) and mice (Smits and Lefebvre 2003) identifies the notochord as a mesodermal midline structure that grows out from a region that received signals from the *dorsal organizer* (also known as Spemann's organizer and in avian embryos, as Hensen's node [Hensen 1876, p. 266] and discussed in the next section).

This growth process, reviewed by Stemple (2005a), requires the transition of dorsal organizer cells into chordamesoderm during early gastrula stages. By this process, the chordamesoderm that gives rise to the notochord, becomes morphologically and molecularly distinct from the surrounding mesoderm. Cells rearrange and converge toward the dorsal midline, in cellular movements that transform the chordamesoderm into an elongated stack of cells. Stemple (2005a) emphasizes that notochord specification is a stable property of the dorsal organizer. A strong argument for this model is the observation that transplantation of prospective chordamesoderm into an equivalently staged host embryo initiates formation of a notochord (Shih and Fraser 1996; Saúde et al. 2000).

E. The Dorsal Organizer

The dorsal organizer was first identified as the dorsal lip of the blastopore in experimental studies of embryos of two species of urodeles, the northern crest newt *Triturus cristatus* and the marbled newt *T. taeniatus* by Hilde Mangold and Hans Spemann (Spemann and Mangold 1924). They used the term *Organisatoren* (organizers) in the title of their paper. So important was this discovery and the concept of Organizers that Spemann received the 1935 Nobel Prize in Physiology or Medicine "for his [sic] discovery of the organizer effect in embryonic development." Hilde Mangold, Spemann's student, whose dissertation research formed the basis of the discovery, died tragically in a kitchen stove fire in 1924, age 26. Because Nobel prizes are not given posthumously, Spemann alone received the prize, although the concept of the organizer was developed by both researchers and consequently is often known as the Spemann-Mangold organizer.

Just a few years later, Conrad Waddington, who studied with Spemann in Germany, showed that Hensen's node, which is the tip of the primitive streak in birds and the node (primitive knot) in mammals, are homologous structures (Waddington 1930, 1932, 1937). In teleosts, such as zebrafish, the corresponding (homologous) structure is the center of the embryonic shield (Oppenheimer 1934, 1939). Therefore, organizers are universal in initiating vertebrate embryonic development. As regards the notochord, therefore, *the function of all dorsal organizers is the same; they provide signaling that induces cells that give rise to the notochord* (Stemple 2005a, and see Section 3.5 for the tail bud as an organizer of the most caudal region of the A–P axis). Since cells in the immediate vicinity of the organizer are instructed to differentiate, the literature not only talks about induction but also about an organizer that provides cells (Stemple 2005a, b). Anderson and Stern (2016) review the function of the organizer in different classes of vertebrates.

Transplantation of an organizer not only induces notochord development, it also induces the formation of a second embryonic axis in the ectopic location, as first demonstrated by Spemann and Mangold (1924). As even further demonstration of equivalent and conserved inductive functions, organizers can be transplanted from embryos of one class of vertebrates into embryos from another class where they fulfill the same inductive functions, first demonstrated by Waddington (1937) using rabbit and chicken embryos.

In chicken embryos at Hamburger Hamilton (HH) stage 4 (Hamburger and Hamilton 1951), lateral and medial mesoderm cells from Hensen's node contribute to the notochord (Selleck and Stern 1991; Figure 3.5). Taking this into account, the first step in notochord development is the transition of cells from dorsal organizer tissue into chordamesoderm. Next, during early gastrula stages, the chordamesoderm becomes morphologically and molecularly distinct from other mesoderm as a rod of single cells that expresses a set of genes including *Echidna hedgehog*, *Sonic hedgehog*, *col2a1* and *Brachyury* (Stemple 2005a).

F. THE ANTERIOR NOTOCHORD

In chicken embryos, the main part of the notochord is laid down by Hensen's node as it migrates posteriorly (Bancroft and Bellairs 1976). Still, a part of the chicken notochord is formed anterior to Hensen's node. This part of the notochord, known as the *notochord head process* or anterior extension of the primitive streak (Bancroft and Bellairs 1976), was designated as the chordamesoblast condensation by Jurand (1962); see also Jurand (1974) for germ layer identifications in mammals.

In mammals at least, cell labeling, orthotopic transplants, and transplantation to ectopic sites and all other results indicate that the definitive endoderm is derived from the same cell population as is the notochord, namely, the anterior extension of the primitive streak or head process (Lawson et al. 1991). It is this very first anlage of the notochord that was regarded as endodermal by Goodrich (1930) and Remane (1936) and as a mesodermal condensation and specialized median portion of the mesoderm by Hamburger and Hamilton (1951) and by Nelsen (1953). Whether this part of the notochord is endodermal or mesodermal remains a matter of contention. However, we do know that, subsequent to its specification, the notochord itself

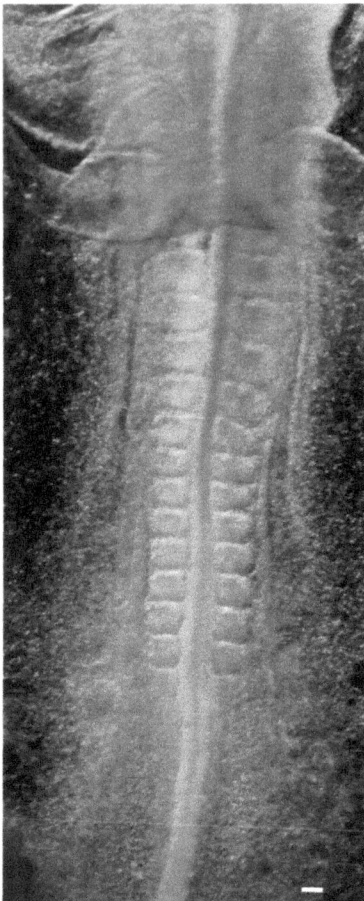

FIGURE 3.5 Injection of DiI into the lateral portion of Hensen's node in a chicken embryo results in labeling of the medial portion of the somites as seen in this embryo viewed from above (anterior at the top). This particular group of labelled cells also contributed progeny to the notochord posteriorly (bottom of photograph) and to endoderm underlying the more posterior somites (out of focus fluorescence underlying the last five somites). Scale bar = 100 μm. (With permission from Selleck, M. A, and Stern, C. D., 1991. *Development* 112: 615–626.)

functions as a primary embryonic organizer in all vertebrate embryos, inducing neural ectoderm and the neural tube, and in many groups initiating vertebral chondrogenesis (Chapter 1, and see Hall 2015, p. 651, and Anderson and Stern 2016).

3.4   THE NOTOCHORD IN HUMAN EMBRYOS

Interest in the mammalian notochord is fueled by the fact that the notochord persists lifelong inside mammalian intervertebral disks. Its function in adult mammals and its contribution to disk maintenance is increasingly being studied (Mwale 2014). Recent studies also focus on early notochord development in mammals.

Cox and Serra (2014) summarize the early formation of the human notochord as follows: (i) The notochord forms during embryonic gastrulation; (ii) cells that migrate through the primitive streak form endoderm and mesoderm; (iii) some of the migrating cells form the notochord process, and (iv) the cells integrate transiently with the endoderm to form the notochord plate.

Balmer et al. (2016) provided the following description that supports the transient integration of notochord precursor cells into the endoderm in mouse embryos and which is very similar to the modified scenario pictured in Figure 3.2A2–D2. Studies on the node and notochord plate precursors establish that these cells originate from the population of cells emerging at the most anterior part of the primitive streak at the early-streak stage (E6.25). The cells become morphologically distinct from the rest of the mesoderm and are referred to as axial mesoderm, chordamesoderm, or mesoderm; Kunz (2004) uses the term "prechordal plate" for these cells in teleost fishes. Axial mesoderm cells produce three distinct cell populations along the anterior-posterior axis: (i) the prechordal plate; (ii) the anterior head process, and (iii) the node-derived notochord precursors. Together, cells from the three populations form the notochord plate at the surface of the mouse embryo that is contiguous with the gut endoderm. The prechordal plate (to be distinguished from notochord plate) represents the anterior-most mesoderm. It gives rise to populations of cells in the forebrain and in the rostral hindbrain. Cells of the anterior head process participate in the formation of the most anterior part of the notochord.

Likewise, Moore et al. (2016) describe for human embryos that the notochord plate buds off from the endoderm to form the notochord that comes to lie between the roof of the primitive gut and the floor of the developing neural tube. If we accept this scenario, then the early observations about the notochord folding off from the endodermal roof of the archenteron would be correct. The results of cell lineage tracing demonstrate, however, that these are *mesenchymal cells* that reintegrate into the endoderm and subsequently bud off again from the endoderm to form the notochord; i.e., a notochord of mesodermal origin buds off from the endoderm (Balmer et al. 2016). Interestingly, these observations are in line with the description of early notochord development by Hertwig (1915) concerning intermediate fusion of chordamesoderm and endoderm and subsequent separation.

However, perhaps we can put an endodermal versus mesodermal debate about the origin of the notochord to rest at least for humans and perhaps for mammals. As the notochord has all the characters of a primary skeleton, including but not limited to, production of skeletal collagens, transformation into musculature in amphioxus, transformation into cartilage in many vertebrates (Chapters 6 and 10) and mineralization of its own sheath in teleosts (Chapter 7), the notochord, certainly functions as, and perhaps is best considered as, a mesoderm-derived structure, as is generally the case for elements of the endoskeleton. Then, endoderm-derived cells that contribute to the notochord undergo a transition from endoderm to mesenchyme, a transition normally regarded as a property of mesoderm (Hall 2014). Such a transition follows the evolutionary trail blazed by mesoderm, which is a secondary germ layer derived from endoderm or neural crest ectoderm. No wonder the debate over whether the notochord is endodermal or mesodermal has persisted for so long.

3.5   THE TAIL BUD, SECONDARY NEURULATION AND SECONDARY NOTOCHORD EXTENSION

Neurulation is usually depicted as the transformation of the neural plate into a closed neural tube by the elevation and fusion of the neural folds (Figure 3.1C), a process that is initiated anteriorly and posteriorly along the A–P body axis and that occurs after the germ layers have segregated into ectoderm (from which the neural tube arises), mesoderm (which gives rise to the somites) and endoderm. This is correct as far as it goes, but it does not go far enough.

The process just described represents *primary neurulation*, which is one of the two forms of neurulation that occur in embryos of all vertebrates, and which takes place *after the germ layers have differentiated*. Primary neurulation does indeed produce the majority of the neural tube along the A–P axis. The most caudal neural tube, however, forms by a different mechanism known as *secondary neurulation*, which does not involve the formation of germ layers. Rather, the neural tube along with the most caudal somites and vascular tissues arises from a blastema of cells, referred to as a *tail bud* (Figure 3.4C). Secondary neurulation has been demonstrated in lampreys, teleosts, amphibians, birds and mammals (including humans), which is consistent with secondary neurulation as a developmental process that arose early in vertebrate evolution (Beck and Slack 1998, 1999; Handrigan 2003; Hall 2005).

The mechanism of secondary neurulation is induction and transformation of epithelial cells into a tail bud that is composed of mesenchymal cells. Early studies on secondary neurulation paid little attention to the formation of the posterior-most part of the notochord. However, within the tail bud reside not only future neural, muscular, vascular and tail gut cells but also future notochord cells. The region within the tail bud that contains future notochord cells is known as the chordoneural hinge (Figure 3.6). The mesenchymal tail bud cells begin to differentiate the caudal neural tube, which develops cranially to meet the major portion of the neural tube that is developing in a posterior (caudal) direction. The point of contact — the chordoneural hinge — as in any situations where two elements have to align and meet, is a site of potential deviations in development, such as an abnormal hole in the neural tube that can result in neural tube defects such as spina bifida in humans.

Although the terms primary and secondary neurulation were not coined until 1928 (Holmdahl 1928), the existence of these two modes of development was known far back as 1884, when, on the basis of his analysis of the development of germ layers, Kölliker (1884) proposed that the most caudal part of the nervous system arises from mesoderm and not from neural ectoderm. Bijtel (1931) extended Kölliker's finding when he used vital dye-staining to demonstrate that tail somites in the Mexican axolotl, *Ambystoma mexicanum*, arise from the caudal neural plate and not from mesoderm. Schoenwolf and Nichols (1984) and Schoenwolf et al. (1985) then described caudal development in mice and avian embryos as based on tail buds rather than differentiated germ layers (see also Beddington and Robertson 1998).

Again using mouse embryos, Griffith et al. (1992) demonstrated that muscle, cartilage, neural cells, and pigment all differentiate in culture from what appears to be homogeneous tail bud mesenchyme. Because these tissue derivatives represent multiple germ layers, Griffith and colleagues concluded that the tail bud comprises

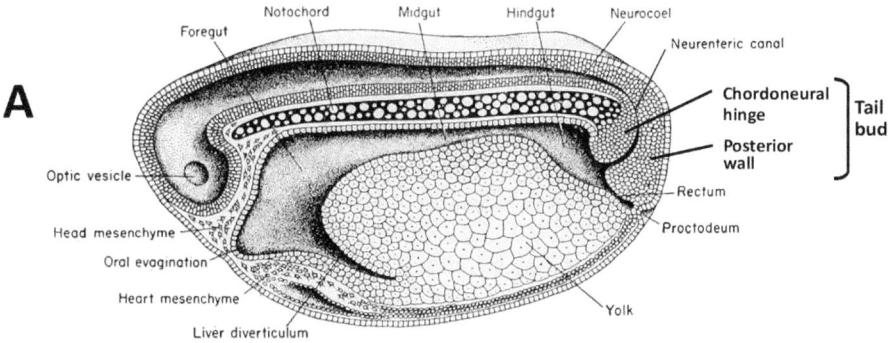

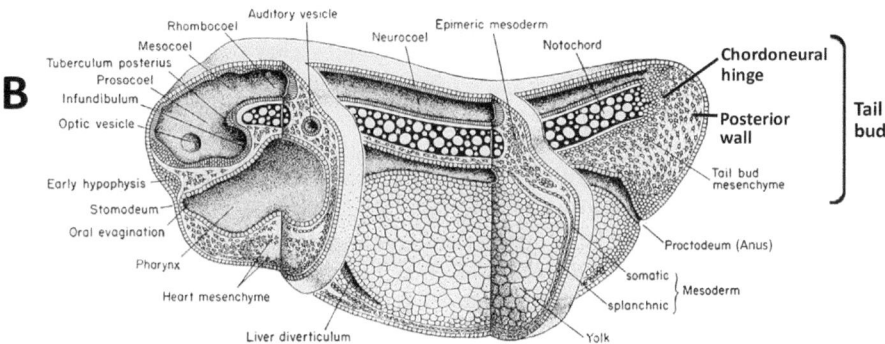

FIGURE 3.6 Late neurula stage (A) and tail bud stage (B) embryos of the northern leopard frog, *Rana pipiens*. (A) Shows the early tail bud that consists of the chordoneural hinge and the posterior wall. In the sagittal plane shown here the neurenteric canal appears to subdivide the two parts of the tail bud but the tail bud continues laterally. The chordoneural hinge provides cells for the floor of the spinal cord, for elongation of the notochord and for the roof of the hindgut. (B) Tail bud stage open anus and fully developed tail bud. Notice that neither the elongating spinal cord nor the elongating notochord are closed tubes at their distal ends. (Modified after Rugh, R., 1951. The Blankiston Company, Philadelphia. Labeling of the tail bud follows De Robertis, et al., 1994. *Development* 117–124.)

cells representing three unseparated germ layers. Consistent with this interpretation, Kanki and Ho (1997) showed that the cells within zebrafish tail buds are pluripotential and contribute to caudal trunk tissue and to the tail bud. Davis and Kirschner (2000) used photoactivation of fluorescence labeling of groups of cells to show that the tail bud in *Xenopus laevis* contributes cells to the neural tube, notochord, somites (muscle cells) and other structures. Their conclusion was that segregation of germ layers is delayed in the tail bud but there is no evidence for a late development of germ layers or indeed for any specification of germ layers in tail buds, although two phases of gene expression in the *Xenopus* tail bud were shown by Beck and Slack (1998, 1999).

The epithelium on the tail bud in mouse embryos develops into a ridge, named the ventral ectodermal ridge (VER) by analogy to the apical ectodermal ridge (AER) of limb buds. Removal of the VER between 9 and 12.75 days of gestation demonstrated that (i) survival and growth of tail mesenchyme *depends* on the presence of the VER, (ii) initiation of chondrogenesis *depends* on the presence of the VER until 10.5 days of gestation (when the VER is at its maximal extent), and (iii) that initiation of myogenesis is *independent* of the presence of the VER (Hall 2000).

Studies on the tail bud of rats, mice, *Xenopus* and zebrafish also revealed the contribution of tail bud cells to extension of the notochord (Tucker and Slack 1995; Yamanaka et al. 2007; Tamplin 2009; Corallo et al. 2013; Row et al. 2016) in a process best described, by analogy with secondary neurulation, as *secondary notochord formation* (Gajovic et al. 1989) (Figure 3.6). Tail bud contribution to the extension of the notochord is foremost observed in species that develop a long tail; in chicken embryos, structures that develop from the tail bud regress in the course of development (Griffith et al. 1992). Still, using Japanese quail /domestic chick chimeras, in which cells from quail embryos are grafted into the equivalent position in chick embryos, Teillet et al. (1998b) showed that the most caudal cells [which they call the cordoneural hinge cells (chordoneural hinge, Pasteels 1937)] give rise to caudal notochord, floor plate cells of the neural tube and endoderm. Two important conclusions drawn by Teillet and colleagues are: (i) "secondary neurulation taking place in the tail bud in which the neural tube forms from a solid cord of cells similar to the neural keel of the teleost embryo" and (ii) that "the group of cells located at the posterior tip of the notochord and floor plate in the developing tail bud is the equivalent of Hensen's node at earlier developmental stages — that is, to the [dorsal] organizer" (p. 11737) — and give rise to the posterior extension of the notochord in a process that equals secondary neurulation. *That these cells have the properties of an embryonic organizer is important.* The dorsal organizer is discussed in Section 3.3D while the importance and role of organizers in early embryonic development is reviewed by Anderson and Stern (2016).

In *Xenopus* embryos, the anterior two thirds of the axial tail structures are derived from the trunk and are displaced into the tail as a result of anterior movement of the proctodaeum during extension of the body. In *Xenopus*, the tail bud, which only gives rise to the distal third part of the tail, has two regions; (i) the chordoneural hinge, that derives from the organizer and adjacent cells, and (ii) the posterior wall, a derivative of the posterior (caudal) neural plate (Tucker and Slack 1995; Stern et al. 2006). Both notochord and tail bud regions express *Brachyury* but only the chordoneural hinge contributes cells to the extension of the notochord (De Robertis et al. 1994; Yamanaka et al. 2007).

Based on the study of the expression patterns of 12 genes in the tail bud region of *Xenopus* embryos, Beck and Slack (1998) concluded that the notochord represents a novel tail bud region. Yamanaka et al. (2007) and Tamplin (2009) distinguished the tail bud-derived notochord from trunk notochord in mouse embryos (Figure 3.7). In mice, development of the notochord in all regions depends on *Foxa2* and *Brachyury* expression. Development of the tail notochord in mice is dependent from the homeobox transcription factor *Noto* that has similar patterns of expression in the posterior notochord of all studied vertebrates. While the function of *Noto* in mice is restricted

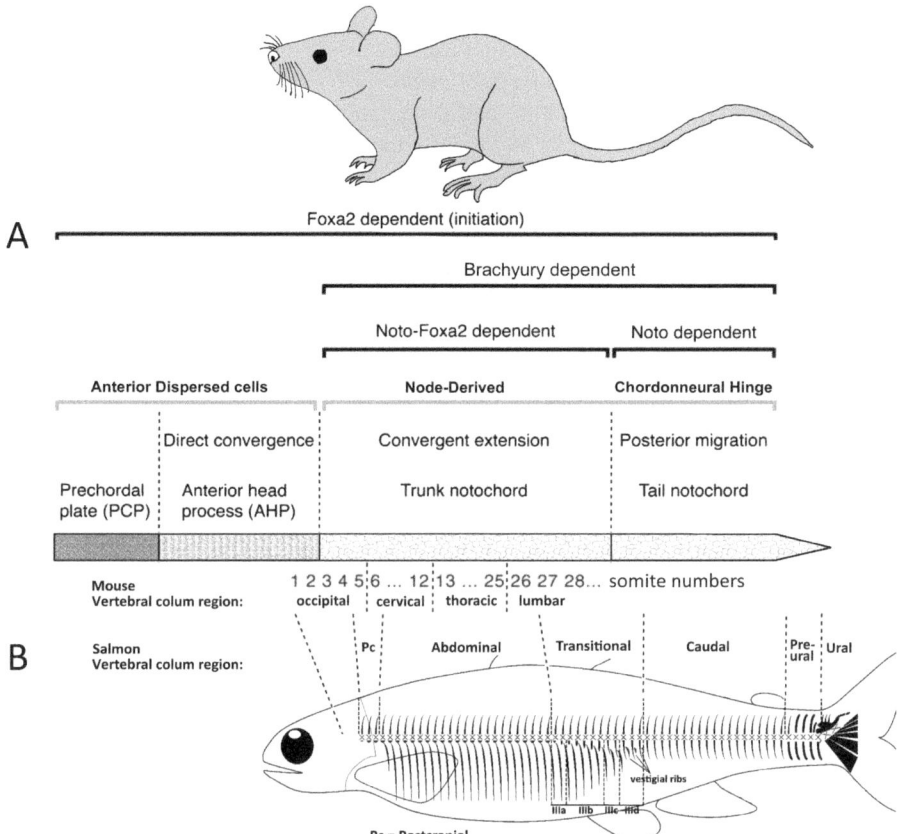

FIGURE 3.7 Regionalization of the notochord and vertebral column. (A) A model of regional notochord morphogenesis based on mouse development. The axial mesoderm is subdivided in four distinct regions that represent the prechordal plate and the notochord proper: Prechordal plate (PCP), Anterior head process (AHP), trunk notochord and tail notochord. The black brackets indicate dependence of the four regions on expression of different transcription factors: *Foxa2*, *Brachyury*, *Noto-Foxa2* and *Noto*. The gray brackets indicate the cellular origin of the four regions and the different morphogenetic mechanisms that generate the notochord. Depending on region, one of three mechanisms (Direct Convergence, Convergent extension or Posterior migration) is responsible for cell condensation. The tail notochord is the product of the tail bud (see text and Figure 3.6). Below the scheme of the notochord, represented by shaded areas, mouse somite numbers in relation to notochord regions and to the regions of the vertebral column are listed. (B) Numbers of vertebral bodies and the regionalization of the vertebral column varies considerably in different vertebrates. Here, as an example, the vertebral column of a Chinook salmon, *Oncorhynchus tshawytscha*, is aligned with the mouse somite numbers. The differences in vertebral column regionalization between the two species are apparent, particular in the caudal region of the vertebral column. *Hoxc-6* expression co-aligns with the occipital-postcranial border in zebrafish and in mice (Section 3.5). Salmon have a large transitional region with vertebral bodies that have intermediate characters between abdominal and caudal vertebrae. In other teleosts, such as zebrafish, this region is short and may only be represented by one vertebral body. ((A) Modified after Yamanaka, et al., 2007. *Dev Cell* 13: 884–896. (B) Modified after De Clercq, et al., 2017. *J Anat* 231: 500–514.)

to the posterior part of the tail, zebrafish mutation floating head (*noto = flh*) results in a complete lack of notochord formation (Talbot et al. 1995; Yamanaka et al. 2007).

A recent analysis by Aires et al. (2019) using *Gdf11⁻/⁻* mutant embryos combined with fluorescent labeling, cell culture and gene analysis has revealed a gene network operating at different stages of tail bud development in mice. *Gdf11* (growth differentiation factor 11, bone morphogenetic protein 11) initiates the transition from trunk to tail development and activates *Lin28* expression (abnormal cell LINeage 28, an RNA-binding protein known to specify stem cells) that promotes proliferation and maintenance of the progenitor cells in the tail bud, while *Hox13* under the control of *Gdf11* overrides *Gdf11* and *Lin28* to prevent further accumulation of progenitor cells. The later role is consistent with the caudal expansion of the spinal cord in *Hoxb13⁻/⁻* embryos which results from incomplete removal of the secondary neural tube (Economides et al. 2003). The comparison of caudal *Hox* gene expression boundaries between mouse and zebrafish is not always evident. Only *Hoxc-6* expression co-aligns with the post-cranial-abdominal border in zebrafish. *Hoxd12a* labels the posterior-most expression boundary, located in the middle of the transitional region between the abdominal and caudal regions of the vertebral column (Figure 3.7). There is currently no strong support for further caudal co-alignment of Hox-gene expression domains in zebrafish or any other actinopterygians (De Clercq et al. 2017).

As is the case for the neural tube, the notochord does not show any structural differences between the regions that arise from different developmental processes. We are unaware of evidence that connects the formation of caudal vertebrae to the existence of secondary neurulation and secondary notochord formation, but such evidence may not have been sought. In basal osteichthyans, diplospondyly is most common in the caudal part of the vertebral column. Moreover, skeletogenesis around the notochord posterior to the caudal vertebrae does often differ, one example being when the notochord extends beyond the most caudal vertebrae as it is the case in basal actinopterygians such as *Amia* and *Lepisosteus* (Schultze and Arratia 1986). In teleosts, the notochord and the spinal cord extend caudally beyond the regions of vertebral centra, the notochord being extended by the opisthural cartilage, recognized by Huxley (1859) as the post caudal cartilage (Arratia et al. 2001). Whether the opisthural cartilage is a cartilaginous transformation of the notochord or a product of the pluripotent tail bud remains an intriguing question (see also Chapter 8).

3.6   EXTERNAL TAILS IN HUMANS

Tails are prominent features of many vertebrates, including mammals, serving many functions including locomotion, balance, grasping. Tails contain notochord and caudal vertebrae, the latter adapted to specific life styles; for example, the prehensile tails of monkeys are reinforced to resist high bending and torsional forces. The red kangaroo, *Macropus rufus* has 21 caudal vertebrae, the long-tailed pangolin, *Phataginus tetradactyla*. Even tailless mammals may have some very reduced caudal vertebrae. Regression of the tail somites during mammalian embryonic development is a normal event associated with development of the tail (Hughes and Freeman 1974; Griffith et al. 1992). What of mammals that do not possess tails?

During the fifth to sixth weeks of gestation of human embryos the tail bud forms a tail, with up to 10–12 vertebrae in the proximal region and mesenchymal tissue distally (Pillai and Nair 2017). In normal development, three to five (usually four) vertebrae fuse to form the coccyx. Obviously humans do not have an external functional tail, so what happens to this embryonic "tail" in human development? Within 14 days — the eighth week of gestation — the tissues of the tail are eliminated by apoptosis and regress leaving only the three to five vertebrae that fuse to form the coccyx or tailbone, the term "tailbone" reflecting that the coccyx is regarded as a vestigial tail, but see below.

As one might expect with processes such as regression and apoptosis, developmental errors can occur. Hughes and Freeman (1974) discussed how regression of the tail somites in animals with reduced tails renders them especially liable to malformations of the lumbosacral spinal cord. One consequence of errors in regression in human embryos can be the development and retention of an *external tail* through to and after birth. This is an extremely rare condition, fewer than 40 cases having been reported with a male:female bias of 2:1 (Pillai and Nair 2017). Such external tails can be removed surgically with no lasting effects (Mukhopadhyay et al. 2012).

External tails have been named *vestigial tails* (vestigial because human ancestors had tails), or in the same vein, they been interpreted as *atavisms*, reflecting the retention of the ancestral ability to form a tail. A second occasional structure, a pseudotail, differs from an external tail in arising by elongation of the coccyx, often in association with spina bifida and in being more painful than an external tail. A pseudotail is neither a vestige nor an atavism but is a developmental abnormality; see Hall (2003) for an analysis of the subtle but real differences between vestiges and atavisms.

Tail development in mice is *Brachyury* and *Noto* dependent (Figure 3.7) and under the regulation of the Wnt family gene, *Wnt3A* and a caudal-type homeobox gene (*Cdx1*). Apoptosis of tail bud cells is initiated in mice when *Wnt3A* is down-regulated (Takada et al. 1994; Greco et al. 1996; Prinos et al. 2001), making mutation of *Wnt3A* a potential candidate in the development of external human tails.

Clearly there is more than absence of resorption and retention of a vestigial tail at play; the tail buds continue to develop beyond the normal stage of regression, producing the range of tissues known to develop from tail buds. External tails contain nerves, blood vessels, muscles, adipose and connective tissues and are covered with skin (Mukhopadhyay et al. 2012). As a consequence, external tails can contract and are moveable. Tails as long as 13 cm have been reported. External tails lack vertebrae, cartilage or bone. As what appears to be an exception, but on closer inspection is not, cartilage and bone were observed in a sacrococcygeal teratoma associated with the external tail in one individual (Mukhopadhyay et al. 2012), raising the likely interpretation that the skeletal cells arose in the teratoma; both cartilage and bone commonly develop in teratomas (Hall 2015).

External human tails do not contain notochord or spinal cord and so do not represent either a caudal extension of secondary neurulation or secondary notochord formation (Figure 3.6); see Hughes and Freeman (1974) for a discussion of this concept in relation to chickens and pig development. Why? An obvious answer is that by the time the distal portions of the tail buds have been removed in normal development — the eighth week of gestation — notochord and spinal cord have completed their development. Indeed, secondary neurulation in human embryos is well underway during the fourth week of gestation (Müller and O'Rahilly 1987; de Bree et al. 2018). That external tails contain neither cartilage nor bone, presumably reflects their origin from the distal portion of the tail bud after regression of the more proximal region that contains vertebral (coccyx) anlagen, although no data exist to support this claim. Whether a feature that lacks notochord, spinal cord or skeletal elements qualifies as a tail also remains for a future tale. And, thereby, hangs our tale and ends this chapter.

3.7 SUMMARY

Whether the notochord arises from endoderm, mesoderm or ectoderm has been debated almost since von Baer identified the notochord (chorda dorsalis) as the precursor of the vertebral column in 1828. One hundred years later Goodrich (1930) concluded that the notochord always arises from the dorsal roof of the endodermal archenteron. Since then, it has been determined that endoderm and mesoderm may both exist in the roof of the archenteron and that the origin of the notochord varies among different groups of vertebrates. Part of the difficulty in assessing the origin of the notochord is that (i) the mesoderm is a secondary germ layer (indeed, often the literature refers to an *endomesoderm*), and (ii) that anterior and posterior regions of the notochord in vertebrates can and do arise from different populations of cells. Similarities between notochord and cartilage encourage us to conclude that the notochord is mesodermal. So too does the formation, during early development, of what is termed the chordamesoderm as notochord precursors segregate from the remaining mesoderm under the influence of the dorsal organizer (Hensen's node).

In the next chapter, we turn to the various functions of the notochord in early development. The number of functions may surprise you. They certainly explain why the notochord has been conserved throughout chordate evolution.

Section II

Function of the Notochord and Notochord Sheath

As discussed in Chapter 4, in all vertebrates on which experimental studies have been undertaken, the notochord — as the primary organizer — (i) specifies and segregates neural from epidermal ectoderm, (ii) establishes the primary antero-posterior (A–P) body axes along the dorsal neural tube, (iii) specifies the dorso-ventral (D-V) polarity of the neural tube, including (iv) specification of the ventral floor plate of the neural tube. The function of the notochord in the early steps of the vertebral centra development differs in teleosts — in which the segmented mineralization of the notochord sheath establishes the vertebral centra anlagen — and in amniotes — in which the notochord induces sclerotomal mesenchymal cells to differentiate into the ventral portion of cartilaginous anlagen of the vertebral centra. Sclerotomal mesenchymal cells also surround the ventral portions of the neural tube, which induces them to form the dorsal vertebral cartilages. Both notochord and floor plate regulate the morphogenesis of the cartilages they induce. Molecular signaling of all these functions is based on pathways based on the genes *Shh*, *Pax*, and Bmps and the Bmp inhibitors noggin and gremlin.

The embryonic development that ensues the inductive function of the notochord becomes increasingly based on the notochord sheath. For that reason, we use Chapter 5 to explore our understanding of the nature, origin and function of the notochord sheath (and associated cells) across the vertebrates. Although there is a long history of debate over whether the sheath is produced by the notochord cells, it is now clear that it is. Production and secretion of the sheath begins when the notochord primordium is organized in the "stack of coins" arrangement, at which stage the cells

DOI: 10.1201/9781315155975-5

express what is now known as a characteristic cluster of notochord genes — *Sonic hedgehog (shh), Echidna hedgehog (Ehh)* and *Brachyury (no tail)*. Notochord cells also express *col2a1* and deposit type II collagen into the developing sheath. Type II collagen is known as "cartilage-type collagen" and so we further explore similarities between the notochord and cartilages and cartilage cells, and the sheath and extra-cellular matrices. Once formed, the notochord sheath comprises vacuolated cells, a notochord epithelium, and a collagen type II-based sheath with inner and outer elastin layers. The vacuolated cells expand the notochord, allowing it to function as a hydrostatic skeleton in chordate embryos and many adult chordates. Although the notochord sheath is acellular, cells can invade the sheath. We discuss the function of these cells in different vertebrates.

4 Function of the Notochord in Early Embryonic Development

CONTENTS

4.1 INTRODUCTION

The notochord, an embryonic midline structure present in all vertebrates, is essential for specification of neural ectoderm and establishment of the primary A–P body axis, roles established by the classic studies of Hilde Mangold and Hans Spemann on the notochord as the primary organizer, as discussed in Chapter 3, and by a series of studies during the 1990s that revealed the molecular basis of this primary role, reviewed by Takahashi et al. (1999) and Stemple (2005a, b) and discussed below.

Early in its development, the notochord produces secreted factors that signal to surrounding tissues. The notochord is the signaling center that establishes *the antero-posterior midline axis* and the metameric organization of the embryos of all vertebrates. The notochord also initiates the signaling that establishes the *dorso-ventral polarity of the midline structure* in vertebrate embryos (Yamada et al. 1991, 1993; Johnson et al. 1994), including polarity of the neural tube, somites (Pourquie et al. 1993; Bumcrot and McMahon 1995) and division of each somite into sclerotome (skeletal-forming) and myotome (muscle-forming; Munsterberg and Lassar 1995; Christ et al. 2004; Figure 4.1). In this phase, the notochord also *determines ventral fates within the spinal cord*, controls aspects of left-right (L-R) asymmetry (Danos and Yost 1995; Lohr et al. 1997), induces pancreatic cell fates (Kim et al. 1997;

DOI: 10.1201/9781315155975-6

57

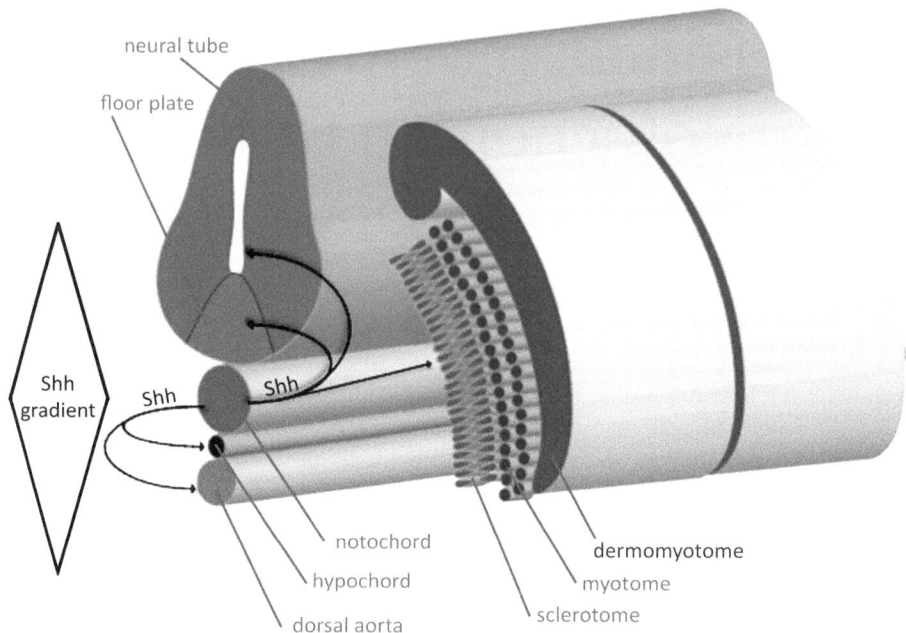

FIGURE 4.1 Schematic drawing of the trunk region of a vertebrate embryo to show Shh signaling (black arrows) from the notochord to target neural and somitic primordia, according to Corallo et al. (2015). Shh signaling patterns the ventral neural tube and induces the specification of the floor plate. The notochord also induces the formation of somites from presomitic mesoderm. The gradient of Shh concentration determines dorsal and ventral somitic fates. Low levels of Shh induce the dorsal dermomyotome while slightly higher levels induce the loss of dermomyotomal markers and activate myogenic differentiation as the myotome. Further increased levels of Shh induce the loss of myotomal markers to specify the sclerotome. Notochord signaling is also critical for the development of the hypochord and the dorsal aorta. (Modified after Corallo et al., 2015. *Cell Mol Life Sci* 72: 2989–3008.)

Cleaver and Krieg 2001), controls the arterial versus venous identity of the major blood vessels (Fouquet et al. 1997; Goldstein and Fishman 1998), and induces varies mesodermal cell types to form cartilage (Stemple 2005a and references therein).

Although initiated by the notochord, as embryonic development ensues, the ventral neural tube and overlying epidermis interact with the notochord to *organize and regionalize the somitic mesoderm.* In summary, and with particular reference to the neural tube and somites, the notochord establishes:

  i. the antero-posterior midline body axis;
 ii. dorso-ventral polarity of the neural tube by specifying the floor plate of the neural tube;
iii. dorso-ventral polarity of the somites by establishing the myotomal and sclerotomal regions of the somites; and,
 iv. along with the floor plate of the neural tube in tetrapods, functions as the inducer of vertebral chondrogenesis from sclerotomal mesoderm.

In tetrapods, chondrogenesis of sclerotomal mesoderm initiates vertebral development. Subsequently, in tetrapods, the notochord and spinal ganglia control the *morphogenesis of the vertebral centra and vertebral arches, respectively* (Section 4.4). The role of the notochord in the development of the vertebrae and vertebral column in teleost fishes and other primarily aquatic osteichthyans is extended compared to its role in tetrapods as mineralization of the notochord sheath initiates vertebral body centrum development (Section 4.4, and Chapters 7 and 8).

4.2 MIDLINE DORSO-VENTRAL ORGANIZATION: THE FLOOR PLATE

Studies undertaken primarily using chick, quail, mouse and zebrafish embryos have shown that the notochord signals to the overlying neural tube to elicit the development of a specialized group of cells in the ventral neural tube (and then in the central nervous system) known as the *floor plate*. This important notochord-neural tube signaling established the dorso-ventral polarity of the neural tube. Signaling from the notochord also is required to pattern the floor plate (Bronner-Fraser and Fraser 1997).

Expression of *Foxa2* is essential for the development of Hensen's node, the notochord (and therefore the floor plate), as first demonstrated in mouse embryos almost 25 years ago in back-to-back papers published in *Cell* by Ang and Rossant (1994) and Weinstein et al. (1994); Hensen's node, the notochord and the floor plate fail to develop when *Foxa2* is knocked out. This need not mean that *Foxa2* controls directly the initiation of all three structures. Given that there is a hierarchy of interactions — the notochord depends on Hensen's node, while the ventral floor plate depends on the notochord — *Foxa2* could be essential only for node development.

4.3 NOTOCHORD-MESODERM SIGNALING SEGREGATES MYOTOME FROM SCLEROTOME

The notochord and the newly established floor plate signal to the somitic mesoderm to initiate segmentation along the embryonic axis and to establish dorsal (myotomal) and ventral (sclerotomal) regions that in turn differentiate as muscle or cartilage (Fan and Tessier-Lavigne 1994, and see the excellent review by Takahashi et al. 1999).

In collaboration with the floor plate of the ventral neural tube, the tetrapod notochord *induces vertebral chondrogenesis* from sclerotomal mesoderm (Section 4.4). Briefly, we summarize the evidence for the signaling molecules Shh, Gremlin1 and Noggin produced by the notochord and their activation of *Pax1*, *Pax9* and *Sox9* in sclerotomal mesenchyme (Figure 4.2). As discussed below and in Chapter 1, in tetrapods, subsequent morphogenetic signals from both floor plate and notochord are required for the organization of chondrogenesis in each vertebra into vertebral arches (floor-plate-mediated) and centra (notochord-mediated; see below).

Direct evidence of notochord *signaling to the myotomal portion* of the somite in quail embryos was provided by Pownall et al. (1996) who demonstrated that both initiation and maintenance of expression of two muscle-specific (and therefore

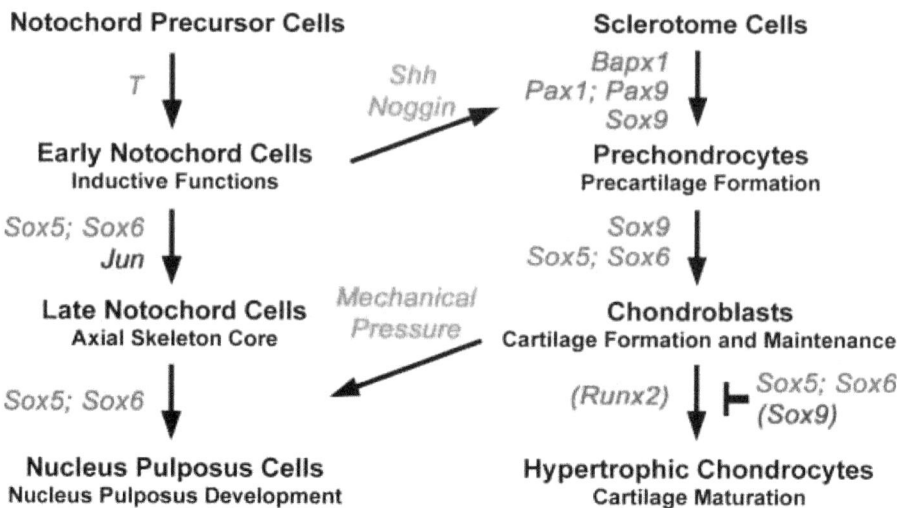

FIGURE 4.2 Cross talk between the notochord and adjacent structures after the initial Shh signaling from the notochord shown in Figure 4.1. *Sox5* and *Sox6* control essential steps in the notochord and chondrocyte differentiation pathways. Chondrocytes (see also Figure 4.5) and notochord cells progress through four major differentiation stages during development. Major transcription factors known to control one or several of these stages are indicated; arrows indicate activation, the turnstile symbol indicates inhibition. Biochemical (Shh and noggin) and physical (mechanical pressure) interactions occur between the two lineages at several stages. (From Smits and Lefebvre, 2003. *Development* 130: 1135–1148, with permission.)

myotomal) genes — *MyoD* and *Myf5* — are controlled by notochord signals *before* the myotome separates from the somites and therefore *before* the myotome segregates into regionalized blocks of muscles acting on each vertebra. Collagen type XV, a component of the notochord sheath in zebrafish, is required to maintain the sheath (perhaps through attachment of the basement membrane to the notochord cells). Moreover, through interaction with Shh, collagen XV is involved in signaling for the onset of muscle development in the myotome (Pagnon-Minot et al. 2008). *Jagged-1–Notch* signaling from the notochord also acts via Shh to integrate notochord and muscle development (Yamamoto et al. 2010).

The notochord in chick embryos is a primary secretory center for Shh, which plays a crucial role in axial patterning. Shh, which is a potent mitogen for somitic cells, regulates expression of *Pax1* (which is a sclerotomal marker) but not *Pax3*, which is a myotomal marker. Shh also maintains the expression of *MyoD* in the myotome. *Shh*-null mice have normal molecular markers in the three portions of the somites — sclerotome, dermatome and myotome. Marcello et al. (1999) separated the paraxial mesoderm from the axial structures and examined the role of *Shh* in the expression of *Pax1*, *MyoD* and *Pax3*, finding that (i) *Pax1* is rescued by *Shh*; (ii) *MyoD* is maintained but not induced by *Shh*, and that (iii) *Pax3* is expressed independently of *Shh*. Along the dorsoventral axis, sclerotome-promoting ventral Shh signals from the notochord and floor plate compete with sclerotome-inhibiting dorsal Wnt signals from the roof plate and surface ectoderm. Thus, high levels of Shh but low levels of Wnts are required to determine sclerotomal fates as opposed to dermomyotomal fates and *vice versa* (Scaal 2016).

Expression of *shha* (the zebrafish homologue of tetrapod *Shh*) in the floor plate and early notochord in zebrafish was first shown by Krauss et al. (1993) and subsequently by Amacher and Kimmel (1998). Haga et al. (2009) demonstrated continuous expression of *shha* in both structures in zebrafish at 20 days post fertilization. Zebrafish mutants that lack the notochord and thus *shha* expression can at least still develop a floor plate. In addition to *shha* zebrafish embryos express two other Hh-related genes, *echidna hedgehog* (*ehh*) *and tiggy-winkle hedgehog* (*twhh*) and it may be that multiple hedgehog genes co-operate in floor plate induction in zebrafish (Corallo et al. 2015). In zebrafish, sonic hedgehog (Shh) secreted from the notochord also induces slow muscles (used for constant movements) that differentiate early and derive from medial adaxial myoblasts, whereas fast muscles (the vast majority of teleost muscles) differentiate later from a separate myoblast pool (Blagden et al. 1997). Like zebrafish, mammals also continue to express *Shh* in later notochord stages and *Shh* is expressed in both prenatal and postnatal nuclei pulposi (Choi and Harfe 2011; Smith et al. 2011).

In upregulating *Pax1*, *Shh* from the entire length of the notochord maintains proliferation and limits apoptosis in sclerotomal mesoderm (Teillet et al. 1998a), roles that are not associated with regional expression of Shh along the length of the notochord. Shh expressed in chicken notochord persists as long as the notochord persists, a difference from mammalian embryos in which the notochord forms the nucleus pulposus. Notochord cells in the nucleus pulposus of adult mice continue to express *Shh* (Choi et al. 2008; Bruggeman et al. 2012). *Sox5* and *Sox6* encode transcription factors that are expressed in the developing notochord and in sclerotome cells from which tetrapod vertebrae develop. In mammals, *Sox5* and *Sox6* are required for notochord sheath formation, for survival of chordoblasts, and for the transformation of the notochord into the nucleus pulposus (Lefebvre 2002; Smits and Lefebvre 2003).

The notochord provides a primary source of Shh for axial development in mouse embryos. Both the notochord and the floor plate produce Shh as a diffusible signal that elicits expression of *Pax1* and maintains expression of *Twist* in sclerotomal mesoderm. Sonic hedgehog can replace the notochord and induce sclerotome from embryonic mesoderm (Fan and Tessier-Lavigne 1994). No evidence for a gradient or regional localization of Shh along the notochord was reported. Through interactions such as these, the notochord enhances sclerotomal differentiation and represses myotomal differentiation.

Studies about the function of *Shh* have focused on mouse, chicken and zebrafish but we know that in amphibians, somites fail to differentiate without the notochord (Malacinski and Youn 1982). In summary, (i) patterning of the ventral neural tube, (ii) induction of the floor plate, (iii) subsequent *Shh* expression by the notochord and the floor plate as ventralizing signals within the neutral tube, (iv) segmentation of the presomitic mesoderm into somites and later myotome, and (v) sclerotome differentiation from somites are all notochord-dependent processes linked to *Shh* signaling in all vertebrates (Corallo et al. 2015; Scaal 2016).

4.4   THE NOTOCHORD AND CHONDROGENESIS OF SCLEROTOMAL MESODERM IN TETRAPODS

Many tissues and organs develop in proximity to somitic (sclerotomal) mesoderm and to the primary body axis and could therefore influence vertebral chondrogenesis. Notochord, neural tube, epithelia, spinal ganglia and migrating neural crest cells are prime candidates that could act individually or in one or more combinations to initiate and/or maintain chondrogenesis within sclerotomal mesoderm or to influence subsequent vertebral morphogenesis and/or growth.

As summarized below and in detail in Hall (1977, 2015), tetrapods initiate vertebral development in sclerotomal mesenchyme in response to signals from the notochord and from the floor plate of the neural tube, the latter having been induced by the notochord (Section 4.2). The notochord signals to the sclerotome to initiate chondrogenesis but does not contribute any cartilage during vertebral chondrogenesis. In contrast, in the largest group of vertebrates, the teleosts, but also in other osteichthyans, vertebral centrum development is initiated by mineralization of the notochord sheath *before* sclerotomal cells surround the notochord. Moreover, mineralization of the notochord sheath is not restricted to teleosts. It is also a pre-teleostean feature (Patterson 1968) and may also occur in other osteichthyans such as tetrapods (discussed in Witten and Hall 2021). Furthermore, sclerotomal cells can be involved in vertebral body chondrogenesis in teleosts. Their role is, in the first place, to form the anlagen of hemal and neural arches and, secondary to the primary role of notochord mineralization, to contribute to the anlagen of vertebral centra (Chapter 7). Arratia et al. (2001) provide an overview of the different modes of vertebra centrum development in actinopterygians. Although the different functions of the notochord in tetrapods and in teleosts have been known since the 19th century, they were forgotten in large parts of the scientific community until quite recently, as will become evident as the chapter proceeds and from the discussions in Chapter 7.

Chondrichthyans have yet another way of creating vertebral centra. Scleroblasts that derive from the basis of the neural and the hemal arches or from the autocentrum (see discussion in Chapter 9), penetrate the outer elastic membrane of the notochord sheath and invade the sheath matrix where they differentiate inside the collagenous notochord sheath into chondrocytes.

A. Removal or Transplantation of the Notochord and/or Neural Tube

Despite this neat and tidy summary of vertebral development in tetrapods — "tetrapods initiate vertebral development in sclerotomal mesenchyme in response to signals from the notochord and from the floor plate of the neural tube" (Hall 1977) — studies conducted in the middle decades of the last century gave contradictory results concerning whether chondrogenesis of chick somitic mesoderm required any influence from ("was induced by") adjacent tissues such as the notochord and/or neural tube. Evidence for self-differentiation *and* for induction of sclerotomal mesoderm was obtained (see Hall 1997, 2015 for discussions).

A classic study published in the mid-1950s appeared to resolve the situation in favor of induction. The evidence came from an extensive series of experiments removing either the notochord and/or the neural tube from embryos *in ovo* or grafting combinations of mesoderm, notochord and/or neural tubes to the chorioallantoic membranes (CAMs) of host chick embryos (Watterson et al. 1954). Removing notochord and neural tube *in ovo* prevented chondrogenesis; somites taken from embryos younger than three days of incubation and grafted to the CAM failed to form cartilage. These experiments were interpreted as providing evidence for an inductive interaction(s) that was completed by the third day of incubation. Pugin (1973) showed that the neural tube from mouse embryos could replace the neural tube from chick embryos and elicit chondrogenesis, results that were taken as evidence for a conserved signaling system between birds and mammals (or at least between chickens and mice).

Each tetrapod vertebra consists of a dorsal component, the vertebral arch, which develops adjacent to the neural tube and spinal ganglia, and a separate ventral component, the vertebral centrum that develops from mesenchyme surrounding the notochord (Figure 4.3 and see Gegenbaur 1862; Lauder 1980; Laerm 1982).

Removing notochord alone from chick embryos prevented vertebral centra from forming but the vertebral arches developed normally. *Removing spinal ganglia alone* prevented the arches from forming but the centra formed normally. Watterson et al. (1954) concluded that notochord and neural tube induce the cartilage that surrounds them in a local and site-specific way. The study provided strong experimental evidence for the biphasic nature of each vertebra (vertebral body), with the centra and arches developing in relative independence from one another. These results, much expanded by Strudel, are discussed in Section 4.5 in the context of morphogenesis of the vertebral column and summarized in Figure 4.4.

A saline extract of neural tubes or notochords from chick embryos was found to be sufficient to induce cartilage formation from somites maintained *in vitro* (Strudel 1953, 1962, 1963). Lash et al. (1957) showed that extracellular matrix products deposited onto Millipore filters could induce cartilage differentiation when somites were cultured on the treated filters. Both these approaches were interpreted as evidence for diffusible chemical signaling between neural tube/notochord and somitic mesoderm and notochord induction of chondrogenesis in sclerotomal mesoderm. Because metachromatic, GAG-rich matrix accumulates around the notochord before chondrogenesis of the adjacent mesenchyme, it was suggested that these notochord matrix products were incorporated to become the matrix of the vertebral cartilage laid down by sclerotomal mesenchymal cells (Hall 1977). What is the nature of the inductive signals?

B. MOLECULAR SIGNALING CASCADES AND VERTEBRAL CHONDROGENESIS

a. Shh, Noggin and Gremlin1

As discussed in Section 4.3, the notochord is a primary source of Sonic hedgehog and of the two Bmp antagonists — Noggin and Gremlin1 — all three of which promote the expression of the sclerotomal genes *Bapx1*, *Pax9* and *Sox9* (Figure 4.1 and see Smits and Lefebvre 2003, and Senthinathan et al. 2012 for overviews). Noggin expression

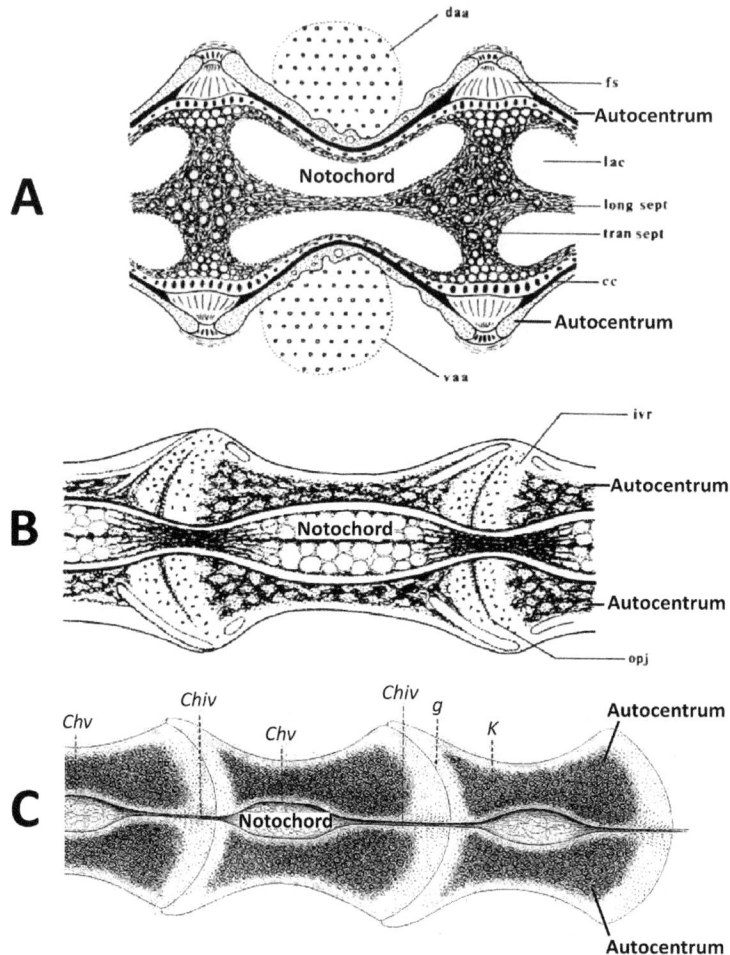

FIGURE 4.3 The notochord and autocentra in three vertebrates, depicted as longitudinal sections: (A) Typical teleost (advanced actinopterygian), (B) Early vertebral bodies of the gar, *Lepisosteus* (basal actinopterygian), (C) Vertebral bodies of a grass snake, *Natrix natrix* (Squamata). Autocentra, which are formed around the notochord, may consist of intramembranous bone, cartilage, mineralized cartilage or endochondral bone, and are present in all three species as indicated (Autocentrum). Notice that the vertebral body design of *Lepisosteus* (B) is similar to that of the reptile (C) but differs from the teleost design (A). Joint surfaces in *Lepisosteus* are opisthocoelous, those of *Natrix* are procoelous. Procoelous and opisthocoelous joints can occur in the same species. Sauropod dinosaurs had opisthocoelous joints in the anterior vertebral column and procoelous joints in the tail region. Abbreviations: ac, autocentrum; cc, chordacentrum; Chiv, intervertebral notochord; Chv, intravertebral notochord; daa, dorsal arch anlagen; fs, fibrous sheath; g, joint surface; ivr, intervertebral ring; k, calcified cartilage (*Knorpelknochen*); lac, lacuna; long sept, longitudinal septum; opj, opisthocoelous joint; tran sept, transverse septum. (A and B Modified from Laerm, 1982. *J Paleontol* 56(1): 191–202. C after Gegenbaur, 1862. Wilhelm Engelmann, Leipzig.)

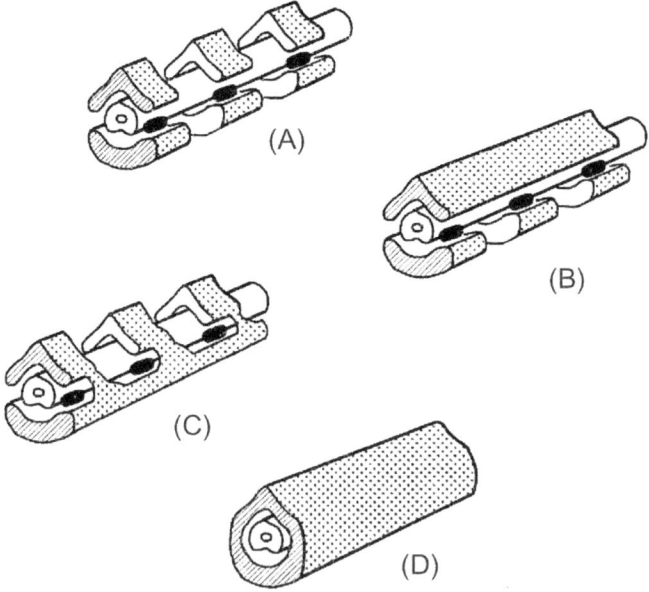

FIGURE 4.4 Spinal ganglia and notochord influence vertebral morphogenesis. (A) The axial structures of a chick embryo show the segmental arrangement of the spinal ganglia (black), neural arches and centra of the developing vertebrae (stippled), the neural arches lying above and the centra below the spinal cord (white) and notochord. (B) Excising the spinal ganglia results in development of a continuous rod of neural arches but does not influence morphogenesis of the centra. (C) Excising the notochord results in the development of a continuous rod of centra but does not influence morphogenesis of the neural arches. (D) Excising both spinal ganglia and the notochord results in the development of a continuous tube of cartilage around the spinal cord. (From Hall, B. K., 2015. *Bones and Cartilage: Developmental and Evolutionary Skeletal Biology*. Second Edition. Academic Press, London.)

in somites is under the control of a neural-tube-derived factor, whose effect can be mimicked experimentally by Wnt1. *Noggin* is downstream of both *Wnt* and *Shh* and antagonizes *Bmp4* in regulating somitic development (Hirsinger et al. 1997).

Stafford et al. (2011) examined mouse embryos in which one or both of the Bmp antagonists — Noggin or Gremlin1 — had been knocked out. Sclerotomal development failed to occur in the double knockouts, whereas myotomal development (visualized using expression of *Pax3*, *Myf5* and *Lbx1*) was not blocked. Additional studies were consistent with Noggin and Gremlin1 maintaining a zone of mesoderm in which Bmp was inhibited, an inhibition that was required for Shh signaling to initiate sclerotomal development. Therefore, complete differentiation of vertebral cartilage requires Bmp signaling, with Shh enhancing the response of somitic cells to Bmp (Murtaugh et al. 1999; McMahon et al. 1998).

b. *Pax1, Pax9* and *Sox9*

As discussed in Section 4.3, *Pax1*, a sclerotomal marker that maintains proliferation and limits apoptosis in sclerotomal mesoderm, is upregulated by notochord *Shh*. *Pax1* is not expressed in the notochord and so is an effective marker distinguishing sclerotome from notochord (and from myotome). Presomitic mesodermal cells respond to *Shh* by forming cartilage. *Shh* and Bmp are produced by the notochord and by the ventral neural tube (the floor plate) in both control embryos and in embryos from which the notochord has been removed, notochord signaling being subordinate to ventral neural tube signaling.

Mice in which both *Pax1* and *Pax9* have been knocked out lack vertebral centra, intervertebral disks and the proximal portion of the ribs. Undifferentiated sclerotomal mesenchyme is present but exhibits no evidence of patterning or segmentation and is removed by apoptosis as development proceeds (Balling et al. 1996). *Pax1* and *Pax9* act synergistically in vertebral development. *Pax1* is expressed in sclerotomal mesoderm before chondrogenesis is initiated, in the perichordal zone of the vertebral column of embryonic but not adult mice, and in association with formation of the intervertebral disks (Deutsch et al. 1988). *Pax9* is expressed in the vertebral column, visceral pouches, tail, head and limbs; the mutant *Danforth's short tail* shows loss of caudal expression of *Pax9* and therefore lacks notochord inducer activity.

The paired type homeodomain transcription factor *Uncx4.1*, which acts upstream of *Pax9*, is required before vertebral arches can form from lateral sclerotomal cells. Mice homozygous for a targeted mutation in *Uncx4.1* die perinatally with severe malformations of the axial skeleton — all lateral sclerotomal derivatives (vertebral arches) fail to form because of inhibition of chondrogenesis. The defect is early — skeletal anlagen form but fail to condense — further reinforcing the roles of these genes at the condensation stage of arch formation during vertebral chondrogenesis (Leitges et al. 2000; Mansouri et al. 2000).

Pax1 is involved in ventralizing the sclerotome; loss-of-function of *Pax1* is associated with sclerotomal and vertebral anomalies. Levels of both *Pax1* and *Pax9* (and of *Sox9*) are reduced in sclerotomal mesenchyme associated with vertebral defects in chicken embryos exposed to the folic acid agonist homocysteine (Kobus et al. 2013).

4.5   NOTOCHORD SIGNALING AND VERTEBRAL MORPHOGENESIS IN TETRAPODS

A. NOTOCHORD, NEURAL TUBE AND *PAX1*

The first studies on a role for the notochord in formation of the vertebral centra and on the modularity of vertebral centra on the one hand and vertebral arches on the other, involved removing the notochord and/or neural tube from embryos of the spotted salamander *Ambystoma punctatum* and the Mexican axolotl *A. mexicanum*. Morphogenesis of the vertebral arches was disrupted after notochord removal, often resulting in formation of a fused rod of arches (Hörstadius 1944; Kitchin 1949). A similar situation was shown in chick embryos by Watterson et al. (1954), as discussed in Section 4.4A.

A classic series of studies undertaken by Georges Strudel between the 1950s and 1960s and summarized in 1973 established the concept of a patterning relationship between notochord and the spinal ganglia on the one hand and the vertebral centra and arches on the other (Strudel 1953, 1955, 1967, and 1973). See Hall (1977) for a detailed analysis of these studies and see Section 4.4B below for the morphogenetic roles of notochord and spinal ganglia in tetrapod vertebral development. The results and conclusions from these extensive studies are twofold (Figure 4.4):

i. Excision of the notochord does not inhibit chondrogenesis of sclerotomal mesoderm but does affect morphogenesis of the vertebral centra, which fail to segment. The vertebral arches form normally. The conclusion is that notochord is required for segmentation of vertebral centra.
ii. Excision of the spinal ganglia does not inhibit chondrogenesis of sclerotomal mesoderm but does affect morphogenesis of the vertebral arches, which fail to segment to form individual arches. The centra form normally. The conclusion is that the spinal ganglia are required for segmentation and perhaps for formation of the arches (Strudel 1953, 1955, 1967, 1973).

Using a similar approach, Teillet and Le Douarin (1983) confirmed that segmentation of somitic mesoderm was either disrupted or failed to occur after the notochord and/ or neural tube were removed from chick embryos. Using a different technique with chick embryos — transplantation of neural tube and notochord into unsegmented lateral plate mesoderm — Fraser (1960) demonstrated that somite-like structures formed in the lateral plate, a result taken as further evidence for a role for the neural tube and or the notochord in early somite segmentation.

These roles of notochord, neural tube and spinal ganglia in patterning the somites have been challenged by recent studies in chick embryos in which notochord, neural tube and/or spinal ganglia were removed and/or transplanted, and *Pax1* (a member of the paired box of homeobox genes) was used as a marker for sclerotome segmentation. These results were interpreted as supporting autonomous segmentation of the somitic precursors of the chick vertebral column through midline registration of the paired sclerotomes, independent of influences from notochord or neural tube (Senthinathan et al. 2012), a topic taken up again at the end of Chapter 9. The structural marker used for segmentation was the formation of segmented vertebral centra or intervertebral disks. One can imagine that earlier aspects of sclerotome segmentation might not be revealed by *Pax1* expression. If that is not the case, then this study stands in contrast to the earlier studies by Strudel.

Expression of *Pax1* and the winged helix transcription factor *Mfh1* in murine sclerotomal mesoderm both depend on *Shh* from the notochord for their expression. *Pax1* and *Mfh1* influence vertebral column development synergistically, double knockout (*Pax1⁻/Mfh1⁻*) early embryos lack the dorso-medial elements of the vertebrae. Older embryos fail to form vertebral centra or intervertebral disks (Furumoto et al. 1999).

B. Notochord and Spinal Ganglia

Trunk neural crest cells migrate segmentally from each somite in vertebrate embryos to produce the spinal ganglia, which also are organized segmentally along the dorso-lateral margin of the developing spinal cord. Although the early notochord itself does not show a periodic pattern of gene expression, nevertheless, as reviewed by Fleming et al. (2004), the notochord is a segmented structure.

If neural crest cells are removed from early amphibian or avian embryos, spinal ganglia either fail to form or form but are very small. The arches of each vertebra develop but not as individual arches with one pair of arches per vertebra but as an unsegmented, continuous cartilaginous "roof" over the spinal cord. Vertebral centra, however, develop with the normal pattern of segmentation, which is one centrum per vertebra (Figure 4.4; Detwiler 1937; Detwiler and Van Dyke 1934; Strudel 1967). In a second experimental approach, somites were taken from chick embryos and grafted to the CAMs of host embryos (Williams 1942). Somites chondrified but only formed segmented arches when spinal ganglia were grafted along with them. Somites grafted with the notochord alone fail to form segmented centra (Figure 4.5).

Both sets of experiments indicate that in chick embryos, segmented spinal ganglia and unsegmented notochord both impose segmentation onto developing vertebrae, spinal ganglia controlling morphogenesis of the vertebral arches and the notochord controlling morphogenesis of vertebral centra. After reviewing this and other evidence, Stern (1990) concluded that "the notochord is the archetypal segmented structure in the trunk of vertebrates" and, using the type of evidence outlined above, concluded that "even in those animals in which the notochord is not overtly segmented, this structure may imprint segmented information onto the adjacent paraxial mesoderm." (p. 111). More recent evidence supporting this conclusion is discussed in Chapter 9.

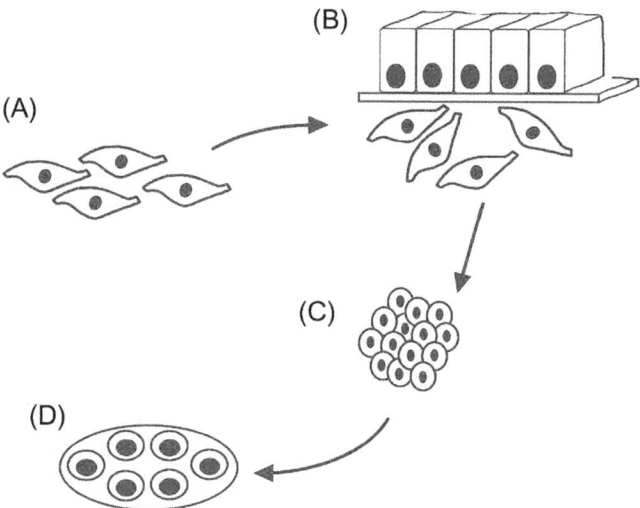

FIGURE 4.5 The four phases of cartilage cell differentiation (also see Figure 4.1). (A) Origination of chondrogenic mesenchyme. (B) Epithelial mesenchymal interaction; epithelium shown as a layer of laterally connected cells on a basement membrane; mesenchyme shown as a meshwork of isolated cells (also see (A)). (C) Condensation in which cells aggregate together. (D) Differentiation and deposition of the extracellular matrix. Condensation and differentiation (C, D) are both multistep processes under different phases of genetic control. (See Chapter 8 for notochord–cartilage relationships.) (From Hall, B. K., 2015. *Bones and Cartilage: Developmental and Evolutionary Skeletal Biology.* Second Edition. Academic Press, London.)

There is, however evidence for segmentation of the notochord in bony fish (teleosts). One type of evidence is the transient but regionalized expression of Hox genes in the zebrafish notochord discussed in Section 4.6. The second, as discussed in detail in Chapter 7, is the origin of vertebral centra from segmental mineralization of the notochord as a chordacentrum. Related to the process of defining vertebral centra and intervertebral joints, the earliest segmentally expressed gene detected is expression of *entp5* in the dorsal part of the notochord epithelium, in zebrafish three days post fertilization (Lleras-Forero et al. 2018). *entp5* codes for the enzyme ectonucleoside triphosphate diphosphohydrolase 5. At eight days post fertilization, *cyp26b1* (which codes for a cytochrome enzyme) is segmentally expressed (Pogoda et al. 2018). Collagen type II can be detected at about 15 days post fertilization (Bensimon-Brito et al. 2012b) and the gene *tiggy-winkle hedgehog* (*twhh*) at 20 days post fertilization (Haga et al. 2009). In Atlantic salmon, segmented expression of alkaline phosphatase and a gene coding for collagen type XI (*col11a2*) and segmented differentiation of the notochord epithelium are linked to vertebral centra and intervertebral joint formation (Grotmol et al. 2005, Wang et al. 2014, and see Chapter 7 for further details).

4.6   SEGMENTAL HOX GENE EXPRESSION IN THE TELEOST NOTOCHORD

Zebrafish (*Danio rerio*) have 32 Hox genes, several of which are not found in tetrapods. Expression is regional but in zebrafish, anterior *Hox* genes are compacted over a shorter antero-posterior region than in tetrapods; there is co-linear expression in the central nervous system and paraxial mesoderm of zebrafish embryos (Prince et al. 1998a). Expression of Hox genes in paraxial (somatic) mesoderm is confined to the most anterior (rostral) portion of each somite, indicating a metameric (segmented) pattern. No expression was reported in the notochord, which is consistent with lack of Hox gene expression in tetrapod notochords. A significant difference between zebrafish and tetrapods is that *Hoxd*11, 12 and 13, which are crucial for digit formation in tetrapods and expressed in the anterior half of limb buds, are not expressed in zebrafish fin buds (Sordino et al. 1995). Regulatory elements associated with *Hoxd*12 are conserved in tetrapods but not found in fish, findings used to help understand the evolution of tetrapod limbs (Hérault et al. 1999).

In a second study, however, Prince et al. (1998b) reported an absence of morphological regionalization of the zebrafish embryo notochord but provided evidence from *in situ* expression of *regional but transient expression* of four of the 32 Hox genes along the notochord. *Hoxb*1, *Hoxb*5, *Hox*6 and *Hoxc*8 are expressed in an antero-posterior sequence with the most 3-prime genes expressed most anteriorly. Expression of each gene is transient but with a consistent anterior expression boundary throughout the time of expression. For each gene, the anterior boundary of expression in the notochord lies posterior to the expression boundary in the central nervous system and somitic mesoderm, i.e., registration of each Hox gene is organ-specific. Expression in the notochord is lost with further development but continues in the CNS and somitic mesoderm. Expression of other genes in the zebrafish notochord — *shha, no tail, cola1* — are along the full length of the noto-chord, without evidence of regionalization.

Given the absence of Hox gene expression in the notochords of tetrapods, noto-chord expression in zebrafish embryos, even transiently, is unexpected. Within the zebrafish notochord, *Hoxb1*[1] is expressed in the notochord at the boundary between somites one and two, *Hoxb5* at the boundary between somites four and five, *Hoxc6* at the boundary between somites five and six, and *Hoxc8* at the boundary between somites six and seven. Prince and colleagues concluded that these patterns of expres-sion are evidence of cryptic "segmentation," concluding that "The expression patterns of these Hox cluster genes in the zebrafish are the most direct molecular evidence for a system of antero-posterior regionalization of the notochord in any vertebrate studied to date" (Prince et al. 1998b, p. 517).

However, Hox genes are only partly responsible for the regionalization of the vertebral column in zebrafish; anterior expression boundaries of Hox genes are concentrated in the anterior-most part of the vertebral column, as discussed by De Clercq et al. (2017), the posterior end of the postcranial region in Chinook salmon (*Oncorhynchus tshawytscha*) co-aligns with the anterior expression bound-ary of *hoxc6* in zebrafish. Only *Hoxc6* expression co-aligns with the postcranial-abdominal border. In contrast, the *hox12a* anterior expression boundary is located in the middle of the transitional region between the abdominal and the caudal vertebral column and not co-aligned with any sharp anatomical boundary. The transitional region in Chinook salmon and other teleosts shows a gradual ante-rior to posterior change of the parapophyses — the transverse processes from the vertebral centra — to vertebral arches and regression of ribs to vestigial ribs (De Clercq et al. 2017). The gradual change in structures in the transitional region chal-lenges the correspondence of Hox-gene expression patterns to sharp morphological boundaries. Except for the *hoxc6* anterior expression boundary, there is currently no strong support for co-alignment of Hox-gene expression domains and verte-bral column regions in actinopterygians. Moreover, the differentiation of the tele-ost caudal fin endoskeleton represents pronounced axial skeleton regionalization (Bensimon-Brito et al. 2012a) but no Hox gene expression boundaries are assigned to this region of the vertebral column. It has indeed been shown that caudal fin loss in sunfish (*Mola mola*) is not associated with the loss of a specific Hox gene (Pan et al. 2016).

4.7 SUMMARY

This chapter treats the various functions of the notochord in early embryonic devel-opment, after the notochord as the primary organizer specifies neural tissue and the A–P body axis, which was the topic of the previous chapter. Five functions of the notochord are discussed:

---

[1] Notochord expression of Hox genes is not restricted to zebrafish. Zebrafish (*Danio rerio*), Japanese medaka (*Oryzias latipes*) and striped bass (*Morone saxatilis*) share patterns of expression and function of *Hoxb1* but changes in *cis*-regulatory regions vary from species to species (Hurley *et al.* 2007).

1. Specification of the dorso-ventral organization of the neural tube, including specification of the ventral floor plate;
2. Specification of each somite into separate muscle-forming (myotome) and skeletal-forming (sclerotome) components;
3. Initiation of the differentiation of cartilage (chondrogenesis) in sclerotomal mesoderm in tetrapods but not in teleosts, including discussions of the major genes involved (sonic hedgehog, pax genes and the Bmp inhibitors Noggin and Gremlin1);
4. Specification, along with the neural tube and spinal ganglia, of the morphogenesis of vertebral centra and vertebral arches, again under the control of sonic hedgehog, pax genes, Noggin and Gremlin1; and
5. A discussion of whether the teleost notochord is segmented.

5 Notochord Cells and Notochord Sheath Formation

CONTENTS

5.1 INTRODUCTION

This chapter explores both the classical and recent literature concerning the state of knowledge of the notochord sheath and notochord cells in different classes of vertebrates with a special emphasis on the very small notochord of amniotes during early development.

Once the "stack of coins" stage is established (as shown in Figures 1.5 and 2.1), notochord cells subdivide and differentiate into a core of highly vacuolated notochord cells and an epithelial layer of peripheral cells, as illustrated in Figure 1.2. High pressure inside the vacuoles of the central notochord cells is counteracted by the notochord sheath establishing the mechanical properties of the notochord as a flexible, rigid, non-compressible rod (Adams et al. 1990). Three basic components of the notochord have been identified in all vertebrate groups:

    i. vacuolated cells;
   ii. a notochord epithelium; and
  iii. a collagen type II-based notochord sheath with inner and outer elastin layers (Schmitz 1998a, b; von Ebner 1896; see Figure 1.4).

As discussed in Chapter 6, the structure and function of the notochord in adult vertebrates of different species is quite diverse. Goodrich (1930) pointed out that the fate of the notochord and its sheath varies in different groups of vertebrates and that the

DOI: 10.1201/9781315155975-7

structural importance of the notochord is inversely proportional to the extent of development of the vertebral bodies outside the notochord. While this is true, we will see that, despite a variable fate of the notochord in adult vertebrates, key characters are present in almost all groups. Biomechanical properties — flexible but incompressible — remain as an important character of the notochord as part of the intervertebral disks in mammals.

Similar to research on basal chordates (Holland and Holland 2017), research on the notochord has a long and rather disjunct history. Decades of intense research alternate with decades of rather little interest in the subject. Detailed studies about the notochord in urochordates, cephalochordates and osteichthyans were published by Kölliker (1859), Klaatsch (1895), von Ebner (1896), Schauinsland (1903) and others. Many studies are summarized in Goodrich's textbook (Goodrich 1930).

5.2 A STACK OF COINS AND THE FIRST LAYER OF THE NOTOCHORD SHEATH

In the mouse, according to Jurand (1974), prenotochord cells arise by proliferation from the pluripotent ectoderm in the area of the gastrula known as Hensen's node. With cell image tracing, Charles B. Kimmel and co-workers determined that from the onset of gastrulation in zebrafish within the dorsal mesoderm, the domain of prospective notochord cells is already a lineage compartment separate from the somitic domains (Kimmel and Warga 1986; Melby et al. 1996). Toward the end of gastrulation (late mid-gastrula stage), notochord cells in the African clawed frog *Xenopus laevis* are already morphologically distinct; notochord cells are larger and have fewer and shorter cell protrusions than the cells of the surrounding somitic mesoderm (Keller et al. 1989). At this stage, although a notochord sheath is not yet present, axis formation and elongation of the notochord are underway.

The first extension of the notochord takes place before the development of a sheath. Two general processes that elongate the embryo were described by Glickman et al. (2003). One is mesodermal convergence, i.e., the narrowing of the tissues with respect to the embryonic axis. The second is mesodermal extension. Convergence toward the midline extends the anterior to posterior size of the embryonic axis. Especially observed in the elongating notochord anlage is a process that has been designated as medio-lateral intercalation behavior. In the midline of the embryo, notochord cells move toward each other (mesoderm convergence) but the cells do not just condense, they intercalate with each other. This of course also elongates the notochord. Medio-lateral intercalation behavior of notochord cells has been studied in detail in zebrafish and in the African clawed frog by Keller et al. (1989) and by Glickman et al. (2003).

Further movements of the cells toward the midline eventually complete intercalation and generate the stack of coins that typifies early notochord cells prior to differentiation into inner vacuolated cells and the outer notochord epithelium (Figure 5.1). The "stack of coins", which is also referred to as a "pile of coins", is a hallmark of early notochord development (Figure 5.1) and one of the most highly conserved notochord characters. During the "stack of coins" arrangement, each of the flattened cells represents the entire diameter of the notochord and the cells are already surrounded by the outer elastin layer; elastin secretion must have started when the cells underwent medio-lateral intercalation. The stack of coins characterizes notochord development in urochordates, cephalochordates and vertebrates. Only in amniotes

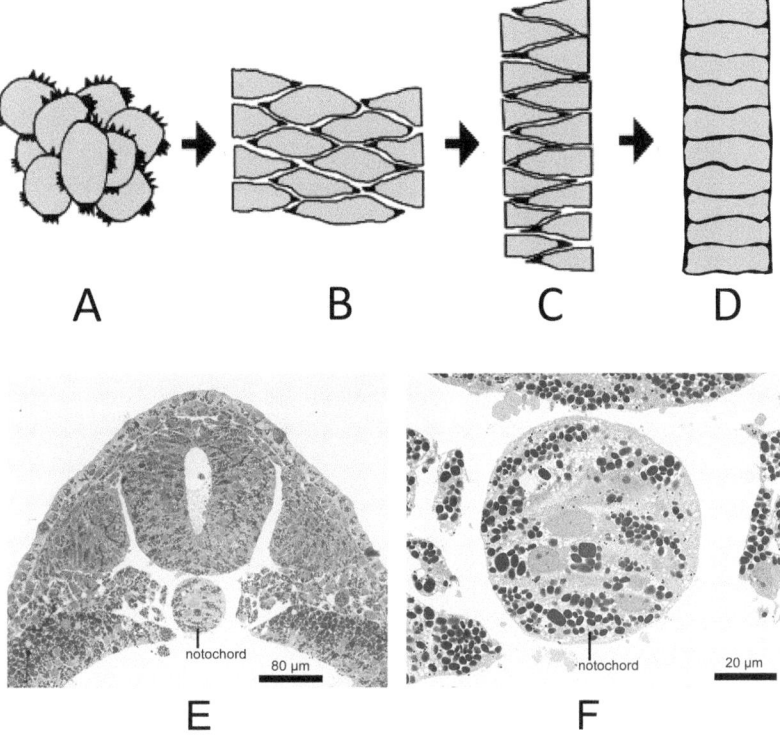

A   B   C   D

E   F

FIGURE 5.1 Movement and intercalation of notochord cells is shown diagrammatically in A–D showing medio-lateral intercalation of the early mesenchymal cells (A) that develops into the notochord, cell movement toward the midline and intercalation (B, C) to generate the stack of coin arrangement (D) that typifies the early notochord prior to differentiation into inner vacuolated cells and the outer notochord epithelium. As discussed in the text, intercalation of mesenchymal cells into notochord cells has been studied in detail in zebrafish *Danio rerio* and in the African clawed frog *Xenopus laevis*. The anterior notochord of a Bichir (*Polypterus senegalus*) at stage 23 is shown in transverse transmission electron micrographs in (E) and (F). The cells of the notochord have converged in the midline but not yet developed vacuoles. Similar to other somatic cells, notochord cells in this early stage are loaded with dark yolk granules (black in F). The stack of coins (D) appears at stage 25 in the Bichir according to Diedhiou and Bartsch (2009). (A-D) Modified and extended after Keller, R., et al. 1989. *J Exp Zool* 251: 134–154 and Glickman, N. S., et al. 2003. *Development* 130: 873–887. (E and F) Provided by PEW.

can the "stack of coins" no longer be recognized (Jiang and Smith 2007; Stach 1999). In the European lancelet, *Branchiostoma lanceolatum*, 20–40 flattened cells occupy the length of one myotome (Welsch 1968). In the tunicate *Ciona* and other ascidians, the current literature reports a fixed number of 40 notochord cells in the "stack of coins" (Takahashi et al. 1999; Jiang and Smith 2007; Reeves et al. 2017). In contrast, Michael J. Katz reports flexible numbers, "about 40 cells" (Katz 1983, p. 7). According to Kovalevsky's studies on the vase tunicate, *Ciona intestinalis*, cells propagate further; he illustrated a late larval stage prior to disintegration of the notochord. At this stage, *Ciona* has produced 45 notochord cells (Figure 2.1; Kovalevsky 1866a, Taf. II, fig. 24).

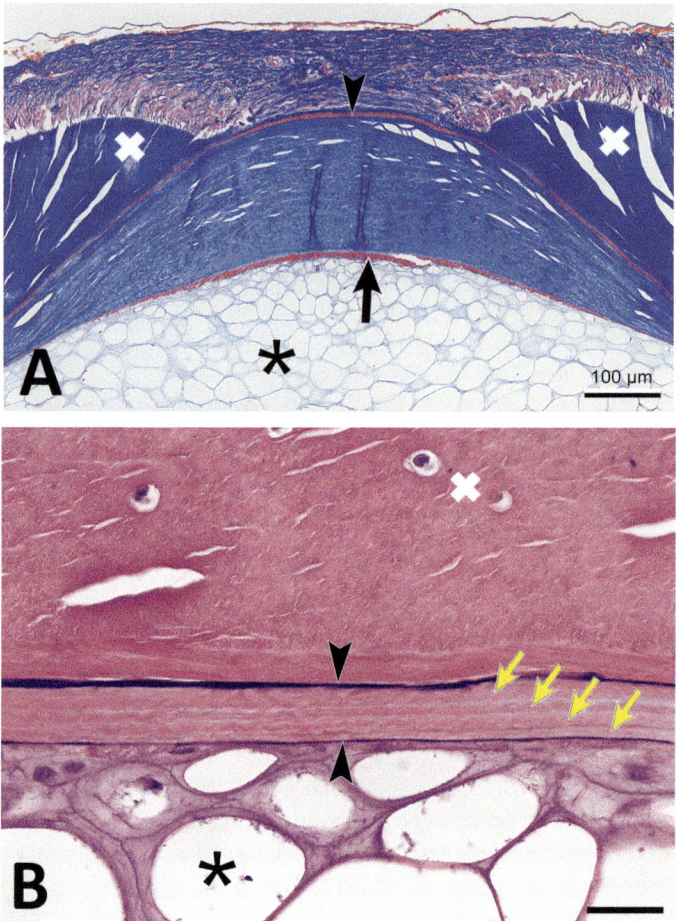

FIGURE 5.2 The notochord with its elastin layers in the early seawater stage of a juvenile Atlantic Salmon, *Salmo salar*. (A) Shows the extended notochord sheath in the dorsal intervertebral space, where the notochord sheath is part of the intervertebral ligament. Azan staining visualizes the outer elastin layer of the notochord sheath as a red line (black arrowhead). Notice the distance of the outer elastin layer from the cells of the notochord epithelium (black arrow). This suggests that connective tissue cells outside the notochord also contribute to the extension of the elastin layer. (B) Shows the outer and the very thin inner elastin layer of the notochord sheath (arrowheads). Demonstration of the inner layer requires special elastin staining, here visualized with Verhoeff's elastin stain. Outer and inner elastic layers are indicated by black arrowheads. Notice the stratification of the notochord sheath proper (yellow arrows) which relates to changes in collagen fiber bundle direction (see also Grotmol et al. (2006)). The notochord sheath does not contain cells. Black asterisks, vacuolated chordocytes; white crosses, bone. ([(A) From Witten, et al. 2019. *J Exp Biol* 222: jeb188763, scale bar = 100 µm. (B) Provided by PEW, scale bar = 20 µm.])

Strikingly, a "stack of coins" arrangement also typifies early cartilage development (Figure 1.5). The arrangement has been documented for chondrocytes of cephalochordates, jawless and jawed vertebrates using the Florida lancelet *Branchiostoma floridae*, the sea lamprey *Petromyzon marinus*, the cichlid *Hemichromis bimaculatus*, and the zebrafish *Danio rerio*. This pattern of early chondrocyte arrangement is essentially not distinguishable from the arrangement of the early notochord cells (Figure 1.5; Mookerjee et al. 1953; Huysseune 1990; Kimmel et al. 2001; Jandzik et al. 2015). The "stack of coins" arrangement is another character that fuels discussion about a close developmental and evolutionary relationship between notochord and cartilage, a topic discussed in detail in Chapter 8.

5.3 THE NOTOCHORD SHEATH

A. NOTOCHORD SHEATH FORMATION

After differentiation of mesenchymal cells into notochord cells (see Chapter 3 for a discussion of the ectodermal, mesodermal or endodermal origin of the notochord), the first layer of the notochord sheath is secreted by the notochord, while its cells are flat and arranged as a stack of coins. At this stage, the cells express characteristic notochord genes such as *sonic hedgehog (Shh)*, *Echidna hedgehog (Ehh)*, *Brachyury (Tbxt)* and *type II collagen (col2a1)* (Corallo et al. 2015) (see also Table 5.1). The question as to which cells produce the notochord sheath has been discussed for more than 100 years (Welsch et al. 1998).

This first secreted layer of the notochord sheath is the elastin-rich membrane that later becomes the outer elastic layer of the notochord (Klaatsch 1895). Lack of structures around the notochord at this stage is taken as evidence that the notochord sheath and its outer elastic membrane are indeed products of the notochord cells (Klaatsch 1895; Schauinsland 1903; Goodrich 1930; Sagstad et al. 2011). In chicken embryos, Bancroft and Bellairs (1976) detected fibrillar notochord sheath material in the intracellular spaces between the notochord cells, which they took as evidence that the notochord secretes the sheath. Later, however, cells that surround the notochord also may synthesize elastin and laminins. Chordamesoderm expresses mRNA for four laminin chains — α-1, α-4, β-1 and γ-1 that also are constituents of the notochord sheath. As these laminins can also be produced for the notochord sheath by mesenchymal cells that surround the notochord (Stemple 2005a), it appears reasonable to assume that this is also the case for later elastin production. In intervertebral spaces/joints, for example, the outer elastin layer is considerably thickened on top of a massive notochord sheath, as shown in the Atlantic salmon notochord in Figure 5.2. For reasons of spatial arrangement alone, and although the notochord epithelium continues to produce elastin, it would be difficult to imagine that the elastin on the outside of the massive notochord sheath is still only produced by the notochord epithelium.

The next step of differentiation separates the cells of the notochord into inner vacuolated cells and the outer cell layer, the notochord epithelium. As has been shown for zebrafish, in the stack of coins phase, notochord cells ubiquitously express collagen type II (*col2a1*) under the control of *sox9*, as it is also the case for cartilage. Notch signaling subsequently initiates differentiation of the notochord epithelium. Prospective cells of the notochord epithelium continue to express *col2a1* and move to the periphery of the notochord. Cells that become vacuolated remain in the center and cease *col2a1* expression. The differential expression of *col2a1* in stack of coins cells is evident at 17hpf. At 26 hpf, col2a1 positive cells have moved to the periphery and form the notochord epithelium (Figure 5.3; Dale and Topczewski 2011).

TABLE 5.1
Classes of Molecules and Cell Structures Found in the Notochord, Notochord Sheath, Nucleus Pulposus and Annulus Fibrosus, Summarized from the Literature

Notochord	Species	Reference
Collagens (Col.), fibrillar		
ColA (col2a1 precursor)	*Branchiostoma floridae* (lancelet)	Zhang and Cohn (2006)
ColA precursor (notochord or somites?)	*Ciona intestinalis* (tunicate)	Reeves et al. (2017), Zhang and Cohn (2006)
Col. I	*Homo sapiens*	Roughley (2004)
Col. I	*H. sapiens*	Richardson et al. (2014)
Col. I later	*Gallus domesticus*	Ghanem (1996)
Col. I (antibody, strong)	Amphioxus	Bŏcina and Saraga-Babić (2006)
Col. II	*Perca fluviatilis* (European perch)	Schmitz (1995)
Col. II	*G. domesticus* (chicken)	Smith and Watt (1985), Text Chapter 6
Col. II	*G. domesticus*	Ghanem (1996)
Col. II	*Xenopus laevis* (African clawed toad)	Smith and Watt (1985)
Col. II	*Danio rerio* (zebrafish)	Scott and Stemple (2004), Corallo et al. (2013)
Col. II	*D. rerio*	Cerdà et al. (2002)
Col. II	*D. rerio*	Bensimon-Brito et al. (2012b)
Col. II	*H. sapiens*	Roughley (2004), Richardson et al. (2014)
Col. II	*X. laevis*	Smith and Watt (1985)
Col. II	*G. domesticus*	Smith and Watt (1985)
Col. II	*Acipenser* (sturgeon)	Schmitz (1998a, b)
Col. II	Vertebrates	Cole (2011)
Col. II	*Mus musculus* (mouse)	Barrionuevo et al. (2006)
Col. II protein early	Hagfish	Zhang and Cohn (2006)

(Continued)

TABLE 5.1 (*Continued*)

Classes of Molecules and Cell Structures Found in the Notochord, Notochord Sheath, Nucleus Pulposus and Annulus Fibrosus, Summarized from the Literature

Notochord	Species	Reference
Col. II	Lamprey	Kimura and Kamimura (1982)
Col. III	*H. sapiens*	Richardson et al. (2014)
Col. III	*H. sapiens*	Roughley (2004)
Col. I–III	*H. sapiens*	Mwale(2014)
Collagens (Col.), non-fibrillar		
Col. V	*H. sapiens*	Roughley (2004)
Col. VI	*H. sapiens*	Roughley (2004)
Col. XII	*H. sapiens*	Roughley (2004)
Col. IX	*G. domesticus*	Text Chapter 5
Col. IX	*D. rerio*	López-Cuevas et al. 2021
Col. IX	*H. sapiens*	Roughley (2004)
Col. X aging nucleus pulposus Also Col. IV–V, IX–XII	*H. sapiens*	Mwale(2014)
Col. X	*G. domesticus*	Linsenmeyer (1986), Hayashi (1992)
Col. X	*H. sapiens*	Roughley (2004)
Col. XI	*Salmo salar*	Wang et al. (2014)
Col. XI	*H. sapiens*	Roughley (2004)
Col. XIV	*H. sapiens*	Roughley (2004)
Col. XXVII	*D. rerio*	Christiansen et al. (2009)
Elastin		
Elastin	*Geotria australia* (lamprey)	Schinko et al. (1992)
Elastin	*Salmo salar* (Atlantic salmon), *D. rerio., G. domesticus*	Sagstad et al. (2011)
Elastin	*H. sapiens*	Roughley (2004 p. 2695), Mwale (2014)
Elastin	*Oikopleura* (tunicate)	Welsch (1969)
Elastin	*Perca flavescens* (yellow perch)	Schmitz (1995)
Proteoglycans, aggregating		
Keratan sulfate	*X. laevis, G. domesticus*	Smith and Watt (1985)

(Continued)

TABLE 5.1 (*Continued*)
Classes of Molecules and Cell Structures Found in the Notochord, Notochord Sheath, Nucleus Pulposus and Annulus Fibrosus, Summarized from the Literature

Notochord	Species	Reference
Aggrecan Chondroitin sulfate proteoglycan 1 Cartilage-specific proteoglycan core protein (CSPCP) Molecule: aggrecan + chondroitin sulfate + keratan sulfate	*Homo sapiens* (humans)	Roughley (2004)
Aggrecan	*S. salar*	Ytteberg et al. (2010)
Aggrecan	*G. domesticus*	Domowicz et al. (1995)
Versican	*H. sapiens*	Roughley (2004)
Hyaluronan	*H. sapiens*	Roughley (2004)
Hyaluronic acid (GAG)	*G. domesticus*	Tool (1972)
Chondroitin-4-sulfate	*S. salar*	Hanneson et al. (2015)
Chondroitin-6-sulfate (intervertebral spaces only)	*S. salar*	Hanneson et al. (2015)
Chondroitin sulfate	*Myxine* (Hagfish)	Ueoka et al. 1999
unsulfated GAG's ivs only	*S. salar*	Hanneson et al. (2015)
Heparan sulfate (GAG)	*G. domesticus*	Mathews (1971; 1975), Hay and Meier (1974)
Link protein	*H. sapiens*	Roughley (2004)
Keratan sulfate (all cells)	*S. salar*	Hanneson et al. (2015)
Keratan sulfate (from mesenchymal cells that migrate into the nucleus pulposus; ageing, maturation)	*H. sapiens*	Roughley (2004)
Proteoglycans, fibril (Col 2) associated		
Decorin	*H. sapiens*	Roughley (2004)
Biglycan	*H. sapiens*	Roughley (2004)
Fibromodulin	*H. sapiens*	Roughley (2004)
Lumican	*H. sapiens*	Roughley (2004)
Proteoglycans, pericellular, Perlecan	*H. sapiens*	Roughley (2004)
Glycoproteins		
Fibronectin	*D. rerio*	Cerdà et al. (2002)
Fibronectin	*H. sapiens*	Roughley (2004)
Laminin	*D. rerio*	Scott and Stemple (2004), Cerdà et al. (2002)
Laminin, inner sheath as basal membrane	*D. rerio*	Parsons et al. (2002)
Tenascin	*D. rerio*	Cerdà et al. (2002)

(Continued)

TABLE 5.1 (*Continued*)
Classes of Molecules and Cell Structures Found in the Notochord, Notochord Sheath, Nucleus Pulposus and Annulus Fibrosus, Summarized from the Literature

Notochord	Species	Reference
Emelin3	*D. rerio*	Corallo et al. (2013)
Oxytalan fibers	lamprey	Schinko et al. (1992)
Sialoglycoprotein	*Latimeria*	Griffith et al. (1975)
Chondromodulin-1 (antiangiogenic factor)	*D. rerio*	Sachdeva et al. (2000)
Energy Storage (Glycogen and Lipids)		
Glycogen	*Ichthyophis* (caecilian)	Welsch and Storch (1971)
Glycogen	*Dendrodoa grossularia* (tunicate)	Welsch and Storch (1971)
Glycogen	*Petromyzon* (sea lamprey)	Schwarz (1961)
Enzymes		
4-chloro-5-bromoindoxyl a. e. (lysosomal) Acid phosphatase (lysosomes) a-naphthyl-acetate esterase (lysosomes)	*Ichthyophis* (caecilian)	Welsch and Storch (1971)
H+ATPase	*D. rerio*	Ellis et al. (2013)
Collagenase MMP1	*H. sapiens*	Roughley (2004)
Collagenase MMP8	*H. sapiens*	Roughley (2004)
Collagenase MMP13	*H. sapiens*	Roughley (2004)
Gelatinase MMP2	*H. sapiens*	Roughley (2004)
Gelatinase MMP9	*H. sapiens*	Roughley (2004)
Stromelysin MMP3	*H. sapiens*	Roughley (2004)
Disintegrin and Metalloprotease (ADAM) family		
ADAMTS4	*H. sapiens*	Roughley (2004)
ADAMTS5	*H. sapiens*	Roughley (2004)
Cytoskeleton		
Actin	*Canis lupus* (wolf)	Hunter et al. (2003)
Vimentin	*C. lupus*	Hunter et al. (2003)
Connexin	*C. lupus*	Hunter et al. (2003)
Intermediate Filaments (Cytokeratin)		
Intermediate filaments	*Myxine* (Hagfish)	Koob and Long (2000)
Intermediate filaments	*Petromyzon* (sea lamprey)	Schwarz (1961)
Intermediate filaments	*Acipenser* (sturgeon)	Schmitz (1998a, b)

(*Continued*)

TABLE 5.1 (*Continued*)
Classes of Molecules and Cell Structures Found in the Notochord, Notochord Sheath, Nucleus Pulposus and Annulus Fibrosus, Summarized from the Literature

Notochord	Species	Reference
Intermediate filaments	*Perca fluviatilis* (European perch)	Schmitz (1995)
Intermediate filaments	*Protopterus* (lungfish)	Schmitz (1998a, b)
Intermediate filaments (tonofilaments)	*Ichthyophis* (caecilian)	Welsch and Storch (1971)

Keratins

Keratin (endo B) embryonic	*X. laevis*	LaFlamme et al. (1988)
Keratin (skin)	*Acipenser* (sturgeon)	Schmitz (1998a, b)
Keratin (skin)	*Perca flavescens*	Schmitz (1995)
Keratin	*D. rerio*	Kague et al. 2021
Keratin (skin)	*Protopterus* (lungfish)	Schmitz (1998a, b)

Minerals

Hydroxyapatite (pathology)	*Ovis* (sheep)	Melrose et al. (2009)
Hydroxyapatite (late)	*Acipenser* (sturgeon)	Leprévost et al. (2017)
Hydroxyapatite	*S. salar*	Wang et al. (2013)
Apatite or calcium carbonate	Chimeras	Francois (1966)
possibly apatite (calcium staining)	Urodeles	Danto et al. (2019)
Apatite or calcium carbonate	Mammals	Williams (1908)
Calcium and phosphate detected	*D. rerio*	Bensimon Brito et al. 2012a
mineralized notochord sheath	*D. rerio*	Fleming et al. 2004,
mineralized notochord sheath	*Oryzias latipes* (medaka)	Yu et al. (2017)
calcified notochord sheath	*Polypterus*	Bartsch and Gambela (1992)
mineralized notochord sheath	basal osteichthyans	Witten and Hall (2021)

Cell Structures (Desmosomes = Internal Filaments)

Massive endoplasmic reticulum (sheath)	*Petromyzon*	Schwarz (1961)
Rough endoplasmic reticulum (sheath) (NE)	*Acipenser*	Schmitz (1998a, b)
Rough endoplasmic reticulum	*Oikopleura* (tunicate)	Welsch and Storch (1969)
endoplasmic reticulum	Gymnophiona (legless amphibians)	Welsch and Storch (1971)

(*Continued*)

TABLE 5.1 (*Continued*)
Classes of Molecules and Cell Structures Found in the Notochord, Notochord Sheath, Nucleus Pulposus and Annulus Fibrosus, Summarized from the Literature

Notochord	Species	Reference
Desmosomes	*Scyliorhinus* (catshark)	Restovic et al. (2016)
Desmosomes (NS)	*Petromyzon*	Schwarz (1962)
Desmosomes (vacuolated cells)	*Petromyzon*	Schwarz (1962)
Desmosomes (all cells)	*Perca flavescens*	Schmitz (1995)
Desmosomes (all cells)	Teleosts	Schmitz (1995)
Caveolae	*D. rerio*	Lim et al. (2017)
Caveolae	*Perca fluviatilis*	Schmitz (1995)
Caveolae	*Acipenser breviostratus* (shortnose sturgeon)	Schmitz (1998a)
Caveolae	*Acipenser breviostrat* (African lungfish)	Schmitz (1998a)
Caveolae	*D. rerio*	Garcia et al. (2017); Lim et al. (2017)
Caveolae inferred from caveolin-1	*H. sapiens*	Lotz and Hsieh (2014)
Pinocatic Vesicles	*Latimeria*	Lock (1980)
Golgi	*Oikopleura* (tunicate)	Welsch and Storch (1969)
Vacuoles and Lysosomes		
Vacuoles/lysosomes	*D. rerio*	Ellis et al. (2013)
Vacuoles/lysosomes (NE)	Gymnophiona	Welsch and Storch (1971)
Vacuoles	*Oikopleura* (Tunicate)	Welsch and Storch (1971)
Vacuoles (in muscle cells)	*Branchiostoma* (lancelet)	Stach (1999)
Vacuoles	*Perca flavescens*	Schmitz (1995, 1998b)
Vacuoles	*Acipenser*	Schmitz (1998a, b)
Vacuoles	*Protopterus* (lungfish)	Schmitz (1998a, b)
Vacuoles	*Myxine* (Hagfish)	Koob and Long (2000)
Vacuoles	*Triturus* (newt), *Ambystoma* (salamander)	Mookerjee et al. (1953)
Tissues		
Cartilage (intravertebral)	*Eurycea bislineata* (salamander)	Wake and Lawson (1973)

(*Continued*)

TABLE 5.1 (*Continued*)
Classes of Molecules and Cell Structures Found in the Notochord, Notochord Sheath, Nucleus Pulposus and Annulus Fibrosus, Summarized from the Literature

Notochord	Species	Reference
Cartilage (intervertebral)	Chicken	Linsenmayer et al. (1986)
Cartilage	*Gekko* spp.	Jonasson et al. (2012)
Cartilage	Amphibians	Lawson (1966)
Muscle cells	*Branchiostoma* (lancelet)	Welsch (1968), Stach (1999)
Stack of coins	*Branchiostoma*	Stach (1999)
Stack of coins	all non-amniotes	Stach (1999)
Transcription factors		
Brachyury TBXT	all chordates	Witten and Hall (2021)
Sox5	Mouse	Smits and Lefebvre (2003)
Sox6	Mouse	Smits and Lefebvre (2003)
Sox9	*D. rerio*	Tamplin (2009)
Sox9	*Mus musculus* (house mouse)	Barinuevo et al. (2006)
Sox9	Vertebrates	Stemple (2005a)
Sox9 (co-expressed with col 2)	Lamprey	Zhang (2006)
Signaling Factors / Morphogens /Enzymes		
Shh	*D. rerio*, chicken	Scott and Stemple (2004)
Twhh	*D. rerio*	Du and Dienhard (2001), Haga et al. (2009)
Calymin	*D. rerio*	Cerdà et al. (2002)
Sclerostin	*O. latipes*	Ofer et al. (2019)
Retinoic acid	*D. rerio*	Pogoda et al. (2018)
Cytochrome enzyme, Cyp26b1	*D. rerio*	Pogoda et al. (2018)
Ectonucleoside triphosphate diphosphohydrolase, Entp5	*D. rerio*	Lleras-Foreo (2018)

This table is comprehensive but inevitably some molecules or structures may have been missed or may be mentioned/discussed in the body of the text but not listed in the table. We welcome input from readers to remedy any such lapses (peckhardwitten@aol.com; bkh@dal.ca).

In zebrafish, cells of the notochord epithelium show a perfect hexagonal arrangement (Bevilacqua et al. 2019).

The notochord epithelium secretes the basal lamina that later becomes the notochord sheath proper (Fleming et al. 2015; Ward et al. 2018; Wopat et al. 2018; see Section 6.3A for different names used for the notochord epithelium). Some authors describe the basal lamina, and not the earlier produced elastin layer, as the first step of notochord sheath development (Mansfield et al. 2015; Corallo et al. 2015). Based on the composition of the early notochord sheath, Stemple (2005a, b) views the sheath as a thick basement membrane.

The type II collagen layer is the principal component of the notochord sheath. This layer contains cross helical arranged fibers that encircle the notochord. In the African clawed frog *Xenopus laevis*, these fibers lie at an angle of 54° to the long axis of the notochord (Koehl et al. 2000). In Atlantic salmon (Figure 5.2B), cross-arranged multiple layers of fiber bundles alternate at angles between 70° and 110° to the long axis of the notochord (Grotmol et al. 2006). Studies on humans have shown that, in addition to type II collagen, the sheath — represented by the annulus fibrosus — also can contain type I collagen (Roughley 2004; Richardson et al. 2014). Type I and type III collagen have both been detected in the notochord sheath of birds (Ghanem 1996). Other typical constituents are chondroitin sulfates and glycosaminoglycans, components that are shared with cartilage (Table 5.1 and see Chapter 8).

The main components of the notochord sheath in *X. laevis* are type II collagen, keratan sulfate and an 86-kD glycoprotein (Smith and Watt 1985). For a stable notochord sheath, type II collagen and elastin fibers must be enzymatically cross-linked. Lysyl oxidases are Cu^{2+}- dependent enzymes that catalyze the self-assembly of tropocollagen into collagen microfibrils (Mwale 2014; Forlino and Marini 2016). Lysyl oxidases also catalyze the important cross-linking between elastin and collagen (Gansner et al. 2007; Trapani et al. 2017). The importance of lysyl oxidases for a stable notochord is demonstrated by the fact that inhibition of the enzyme or mutations that effect the enzyme in zebrafish result in undulating notochords and a failure of vacuolated cell differentiation,

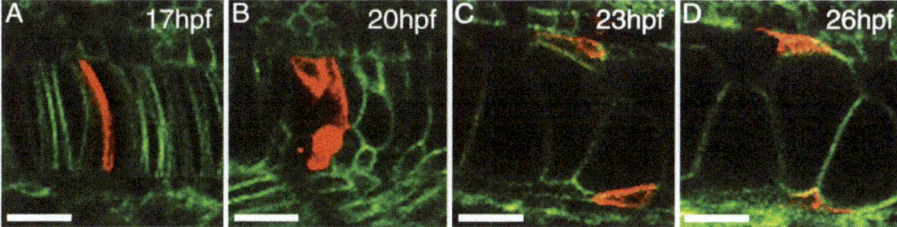

FIGURE 5.3 Differentiation of the notochord epithelium in a zebrafish embryo between 17- and 26-hours post fertilization (hpf), initiated by Notch signaling. (A) The differential expression of col2a1 at the stack of coins stage (green fluorescent marker) is evident at 17hpf. Prospective cells of the notochord epithelium continue to express *col2a1* (red fluorescent marker) and move to the periphery of the notochord (B–D). Cells that become vacuolated chordocytes remain central and stop the expression of *col2a1*. (D) The differential expression of *col2a1* in stack of coins cells is evident at 17hpf (A). At 26 hpf, *col2a1*-positive cells have moved to the periphery and form the notochord epithelium (D). Scale bar = 20 μm. (Used with permission from Dale, R. M., and Topczewski, J., 2011. *Dev Biol* 357: 518–531.)

suggesting an interdependency of notochord sheath production and vacuolated cell differentiation (Anderson et al. 2007; Gansner et al. 2007; Gray et al. 2014).

A notochord sheath has been found in all species that have been examined histologically (Klaatsch 1893a, b, 1895; Ellis et al. 2013). Whether the notochord of all basal chordates is based on type II collagen, as it is in vertebrates, remains to be elucidated. The notochord sheath of amphioxus does show strong binding of antibodies directed against type I collagen (Bŏcina and Saraga-Babić 2006). Reeves et al. (2017) reported the presence of Col2A1 in the notochord of *Ciona intestinalis* but these data derive from whole mount *in situ* hybridization and so do not precisely answer the question of whether the Col2A1 is expressed in the notochord or in the surrounding mesenchyme.

The differentiated, healthy, notochord sheath, as seen in Figure 1.4, has four layers, from inside to outside:

  i. a laminin-rich basal lamina of the notochord epithelium;
 ii. a thin (sometimes transient) elastin-rich layer;
iii. a middle layer with a composition very similar to cartilage matrix (Chapter 8), and
 iv. an outer layer rich in elastin fibers.

The classical names for layers ii, iii and iv are *elastica interna*, *elastica media* and *elastica externa* (Welsch et al. 1998; Arratia et al. 2001). A list of synonyms used in the literature around 1900 to describe the different layers of the notochord sheath can be found in Gadow and Abbott (1895, p. 165). The inner elastin layer is sometimes not separated in the literature from the basal lamina but Bruns and Gross (1970) clearly show these two layers in their studies of tadpoles of the American bullfrog, *Rana catesbeiana*. Kryvi et al. (2017) provide a detailed description of the development of the notochord in Atlantic salmon, from early stages to the notochord in adult animals.

The four-layered notochord sheath provides a casing that withstands pressure generated by the vacuolated notochord cells and can serve as an endoskeleton, especially in the absence of vertebral bodies. The notochord sheath can be further reinforced by a collagen layer that forms around the notochord sheath proper [sometimes designated as the outer (external) notochord sheath, also designated as autocentrum] and that is produced by fibroblasts (Welsch et al. 1998). This external layer may remain non-mineralized (if vertebral centra are lacking), may mineralize after mineralization of the notochord sheath (for example, in teleosts) or may mineralize independently of notochord sheath mineralization (for example, in urodeles).

The thickness of the notochord sheath varies considerably, depending on whether it is an embryonic structure or whether it functions as the main axial skeletal support in adults. The location of the sheath along the notochord has a great influence on its thickness, which increases between vertebrae in species where it becomes part of the intervertebral ligament (see Box 5.1). Intravertebrally, the sheath often does not further increase in thickness and may remain as it was in the early embryo. Figure 5.4 shows the rapid increase in thickness of the notochord sheath in sturgeon from 4.7 to 173 µm within 50 days post-hatching. In the same species, in individuals of 108 cm total length, the thickness of the notochord sheath increases to 8 mm, an increase by a factor of 46 (Leprévost et al. 2017). Notice that the thickness of the notochord

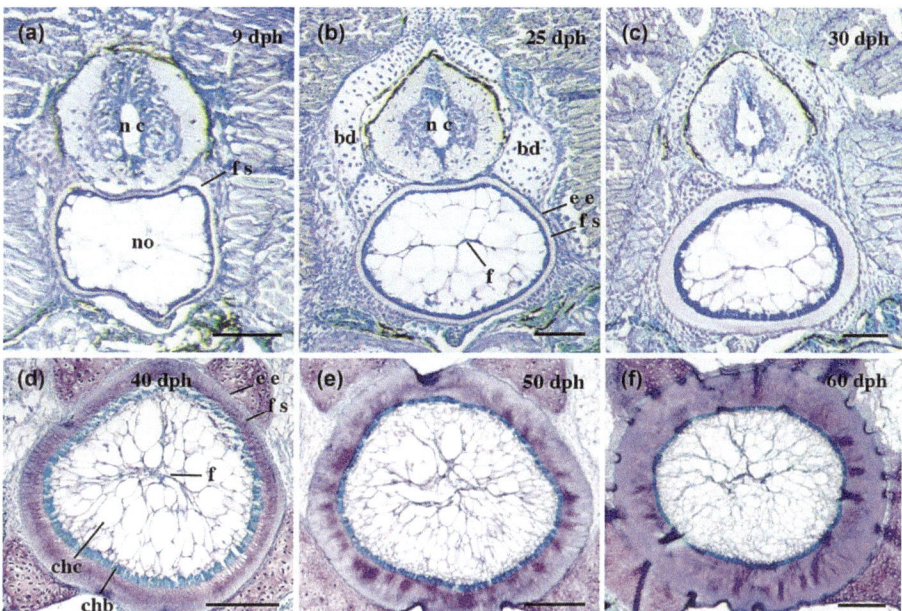

FIGURE 5.4 Transverse, sections through the abdominal region in a growth series of the Siberian sturgeon, *Acipenser baerii*, shows the rapid increase in the thickness of the notochord sheath. Notice that, different from lungfish, no cells reside inside the massive notochord sheath. (a) Nine days post-hatching (dph), 1.5 cm total length (TL) TL; (b) 25 dph, 2.7 cm TL; (c) 30 dph, 5.3 cm TL; (d) 40 dph, 7.4 cm TL; (e) 50 dph, 9.7 cm TL; (f) 60 dph, 12 cm TL. Abbreviations: bd, basidorsal; chb, chordoblast; chc, chordocyte; e e, elastica externa; f, funiculus; f s, fibrous sheath; n c, neural canal; no, notochord. Scale bars: a–c = 100 μm; d–f = 200 μm. Paraffin embedding, Toluidine blue staining. (Used with permission from Leprévost et al, 2017. *J Morph* 278: 1586–1597.)

sheath in juvenile Atlantic salmon (37 cm total length), as shown in Figure 5.2B, is about 14 μm, a value already reached by the sturgeon at 25 days-post-fertilization. Evidently, if neither bone nor cartilage develops around the notochord, the thickness of the sheath increases.

BOX 5.1   THICKNESS OF THE NOTOCHORD SHEATH IN A UROCHORDATE AND IN A SAMPLE OF VERTEBRATES

The urochordate *Oikopleura dioica* has a notochord sheath of 0.25 μm thickness.

In a pig embryo of 7.8 mm, the notochord is fully formed and only 1–1.5 μm thick (Williams 1908).

Adult Bibron's thick-toed geckos (*Chondrodactylus bibronii*) have a notochord sheath in the tail of about 8 μm thickness (Jonasson et al. 2012)

In an adult wild-type zebrafish, the thickness of the notochord sheath in vertebral locations is about 1 μm in contrast to 45 μm in the location of the intervertebral space. Measurements are taken from Figure 7 in the study by Gistelink et al. (2016).

In Atlantic salmon (*Salmo salar*) prior to hatching, the notochord sheath in prospective intervertebral joints is about 40 μm thick, and in prospective vertebral locations, it is about 15 μm thick (Nordvik et al. 2005). The thickness of the notochord sheath does not increase further, the sheath remaining at about 14 μm in vertebral locations in an animal of 37 cm total length. In contrast, the sheath increases strongly in the intervertebral spaces to a thickness of 185 μm, a pattern similar to that seen in Atlantic salmon (above; measurements taken from Figure 9 in a study by Witten et al. 2019).

A 6-cm-long specimen of the shortnose sturgeon *Acipenser brevirostrum* has a notochord sheath of about 50 μm thickness (Schmitz 1998a). 50 days post-hatching the notochord sheath in a Siberian sturgeon (*Acipenser baerii*) is already about 115 μm thick (Figure 5.4). Measurements are taken from Figure 3 in a study by Leprévost et al. (2017).

The thickness of the notochord sheath in a 7-year-old (108 cm total length) shortnose sturgeon matches the thickness of the notochord sheath in adult coelacanths. Smith (1953) and Schultze and Cloutier (1991) document for *Latimeria*, a notochord sheath thickness of about 8 mm in individuals of 139 and 156 cm total length, respectively.

B. Invasion of Cells into the Notochord Sheath

The notochord sheath proper (elastica media) usually lacks cells. It is acellular and has been compared to acellular cartilage (Cole and Hall 2004b).

The acellular character of the notochord sheath is assumed to be the basic chordate condition since invertebrate chordates and extant jawless vertebrates have an acellular notochord sheath (Klaatsch 1893a; Welsch 1968; Schinko et al. 1992). The sheath can, however, contain cells in some groups of vertebrates. To put things straight, this is about cells that reside *inside* the matrix of the notochord sheath, not about *notochord sheath cells*, which are synonymous with notochord epithelial cells, which are the cells that produce the notochord sheath (see Section 6.3A).

We note that this can be confusing. As far as we know, cells that reside inside the matrix of the notochord sheath have been described in several groups of vertebrates: chondrichthyans, dipnoans, urodele amphibians (for which see Figure 5.5 and 5.6), legless amphibians (caecilians or gymnophiona) and specimens of old sturgeons (*Acipenser* spp.) (Goodrich 1930; Schmitz 1998a; Bartsch 1989; Arratia et al. 2001; Danto et al. 2019). Bartsch (1989) reports that cells inside the notochord sheath have also been observed at the tip of the notochord in amniote embryos, which is the portion of the notochord enclosed by the cartilage of the skull base. Within these groups, all available information suggests that the cells inside the matrix do not derive from the notochord epithelium but invade the notochord sheath from the surrounding mesenchyme, i.e., their origin is external to the notochord. (Invasion of

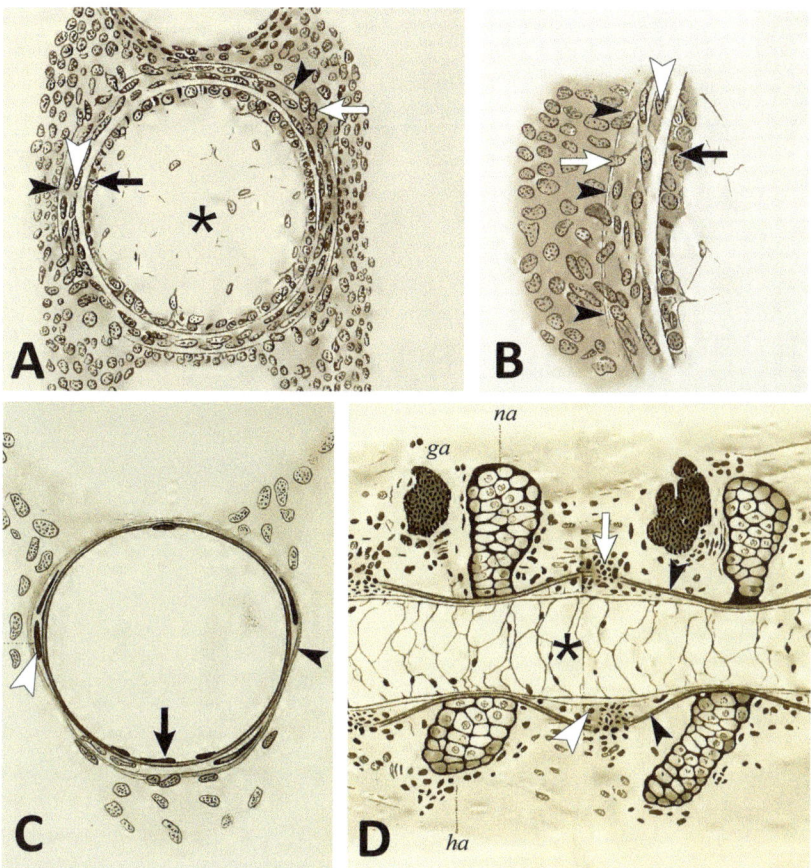

FIGURE 5.5 Invasion of skeletogenic cells into the notochord sheath in the smooth-hound shark *Mustelus mustelus* (A, B) and in the common newt *Lissotriton vulgaris (Triton vulgaris)* (C, D). (A, B), cross section of the developing vertebral column in a 35-mm-long embryo of the smooth-hound shark *Mustelus mustelus*, according to Hasse (1893). Skeletogenic cells (white arrowheads) move into the notochord sheath by fenestration of the outer elastic membrane (black arrowheads). The vertebral body is established between the notochord (black asterisk) and the outer elastic membrane of the notochord sheath. No cells from the notochord epithelium (black arrows) move into the notochord sheath. (A) shows the complete notochord, (B) shows a greater magnification with details of cells that move through the outer elastic membrane and cells that have established themselves below the outer elastic membrane. The extent to which this process contributes cells to the vertebral body can variety between different elasmobranch species (Ridewood 1921). (C, D) transverse and sagittal sections, respectively, through the developing vertebral column of a 6 mm total length embryo of the common newt *Lissotriton vulgaris (Triton vulgaris)* according to Hasse (1892). Similar to elasmobranchs, cartilage precursor cells (white arrow in D) invade the notochord sheath (white arrowheads in C and D) by fenestration of the outer elastic membrane of the notochord (black arrowheads). These cells will differentiate into the cartilage of the intervertebral joint that later constricts the notochord (black asterisk). Unlike elasmobranchs, the invading cells do not contribute to the formation of vertebral bodies. Notice in D that no cartilage cells from the bases of the hemal (ha) or neural (na) arches invade the notochord sheath. ga, spinal ganglion. (Figures modified after Hasse, C., 1892. *Z Wissen Zool* 53 (Suppl): 1–20 and Hasse, C., 1893. *Z Wissen Zool* 55: 533–542.)

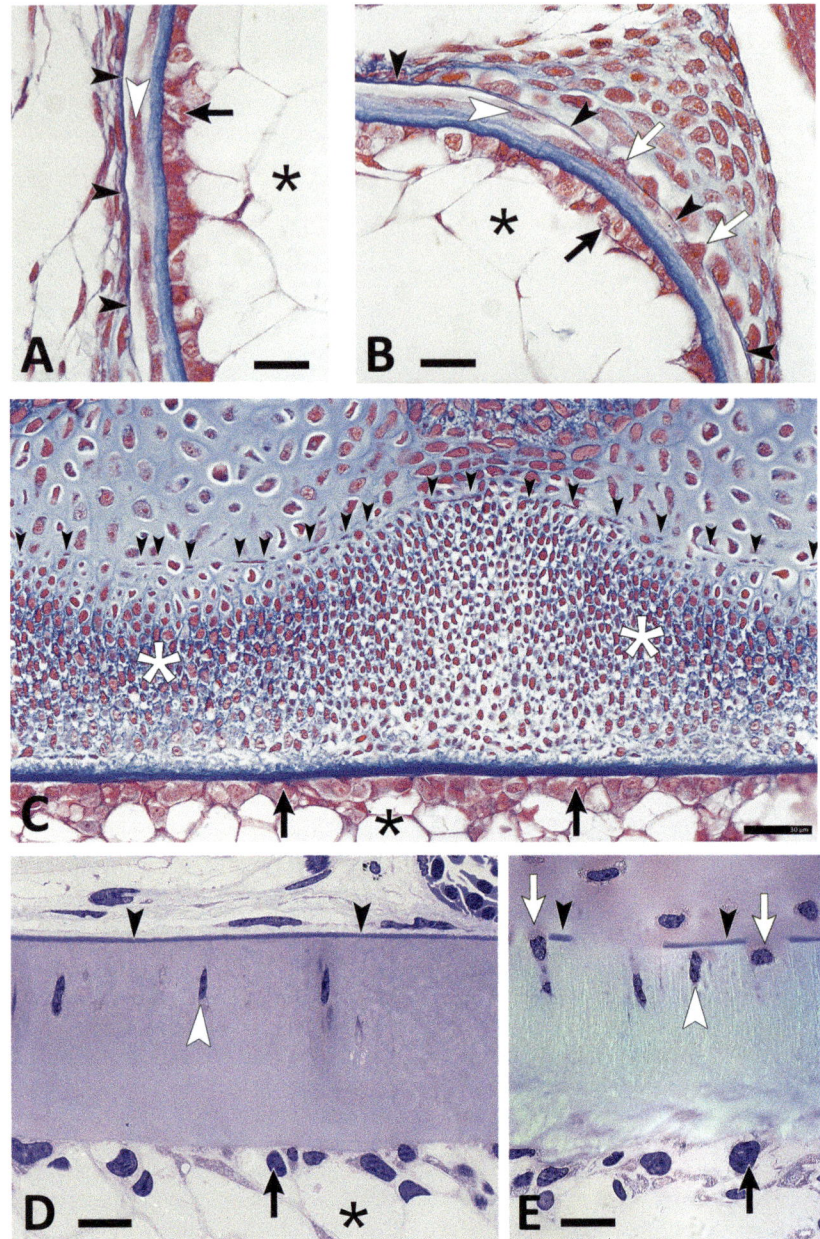

FIGURE 5.6 Invasion of skeletogenic cells into the notochord sheath in embryos of the lesser spotted dogfish, *Scyliorhinus canicula*. (A and B) Cross section through the abdominal region of a two-month-old embryo of 24 mm total length. (A) Skeletogenic cells (white arrowheads) reside below the outer elastin layer (black arrowheads) of the notochord sheath which is produced by the cells of the notochord epithelium (black arrow). Black asterisk, vacuolated notochord cells. (B) Skeletogenic cells (white arrows) invade the notochord sheath, the outer elastin layer (black arrowheads) of the notochord sheath having become fenestrated.

skeletogenic cells into the notochord sheath in embryos of the lesser spotted dogfish, *Scyliorhinus canicula*, is illustrated in Figure 5.6A–C and discussed in some detail in the legend for that figure.) The comments of Klaatsch (1893b) that according to his observations, there is absolutely no evidence for a chordal origin of the cells inside the notochord sheath of the East African lungfish *Protopterus amphibius,* remain undisputed. (Invasion of skeletogenic cells into the notochord sheath in embryos of the Australian lungfish *Neoceratodus fosteri* is illustrated in Figure 5.6D and E and discussed in some detail in the legend for that figure.) Schmitz (1998a) studied the ultrastructure of cells in the sheath of *Protopterus* and confirmed that the cells have characters of fibroblasts and not of chordoblasts.

Cells can invade the notochord sheath from what has been described in chondrichthyan embryos as the arcualia or the bases of the hemal and neural arches, depending on the developmental stage (see also Arratia et al. 2001). The classical literature refers to arcualia-derived cartilage precursor cells that fenestrate the outer elastic membrane and move into the notochord sheath deposited by the notochord epithelium. The literature that rejects the arcualia hypothesis (Williams 1959; Danto et al. 2017) views the cells as derived from the mesenchymal autocentrum that forms around the notochord sheath and as unconnected to the bases of the hemal and neural arches; see Chapter 9 for a more detailed discussion about arcocentra and autocentra. In chondrichthyans, cellular invasion into the notochord sheath is massive. An entire cellular connective tissue forms below the outer elastic membrane of the notochord as shown for sharks in Figures 5.5 and 5.6 and for a holocephalan by Pears et al. (2020). In contrast, in basal osteichthyans, only few cells reside inside the notochord sheath, while in tetrapods invading cells form the cartilage of intervertebral joints (Figure 5.6).

FIGURE 5.6 (*CONTINUED*)   The invading cells appear to be motile as cells are located below sections of the outer elastin layer (white arrowhead) that are not fenestrated. See also Figure 1.4 for a complete cross section of the *Scyliorhinus* notochord, Figure 5.5 for Hasse's observations on embryos of the smooth-hound shark and see Goodrich (1909, p. 99) for similar observations on *Scyliorhinus canicula*. (C) Sagittal section through the caudal region of a three-month-old, 52 mm total length lesser spotted dogfish embryo. Notice the massive accumulation of skeletogenic cells (white asterisks) below the remnants of the outer elastic membrane of the notochord sheath (black arrowheads). Two cartilaginous elements are present per segment in the animals' caudal region (diplospondyly). Skeletogenic cells close to the cartilages undergo chondrogenic differentiation. Black asterisk, vacuolated notochord cells; black arrows, notochord epithelium. (D and E) Cells inside the notochord sheath in an embryo of the Australian lungfish *Neoceratodus fosteri* shown as cross sections through the abdominal region at embryonic stage 50 (about 24 mm total length) as staged by Kemp (1981). (D) Elongated cells (white arrowhead) reside inside the massive notochord sheath. Black arrow, notochord epithelium; black asterisk, vacuolated notochord cells; black arrowhead, outer elastin layer of the notochord sheath. Schmitz (1998a) describes similar cells in the notochord sheath of the West African lungfish *Protopterus annectens* as fibroblasts. (E) Cartilaginous cells move into the notochord sheath (white arrows). Subsequently, hyaline cartilage extends into the notochord sheath. Similar to elasmobranchs and urodeles, the elastic membrane (black arrowheads) of the notochord sheath becomes fenestrated. Polarized light shows the orientation of the cells (white arrowhead) parallel to the orientation of the fibers of the notochord sheath. Scale bars in A, B, D, E = 20 μm, Scale bar in *C* = 100 μm. (A–C) Paraffin embedding, Azan staining. (D, E) Epon embedding, Toluidine Blue staining, (E), polarized light microscopy. (Images provided by PEW.)

Interestingly, and despite a large phylogenetic distance, the mode of invasion appears to be similar in chondrichthyans, dipnoans and amphibians. For the South American lungfish *Lepidosiren paradoxa*, Kerr (1909) described the colonization of the notochord sheath by "immigrant amoeboid cartilage cells from the arcualia" (p. 11). von Ebner (1896) described the same process for chondrichthyans and the West African lungfish *Protopterus annectens*. According to Hasse (1893), rounded cells inside the notochord sheath of *Protopterus sp.* are only present in the dorsal part. Cells inside the lateral portion of the notochord sheath are flat and the ventral sheath is acellular. This is different from the South American lungfish *Lepidosiren paradoxa* in which cells move into the notochord sheath from all four entry points, the bases of the paired future hemal and neural arches (Arratia et al. 2001).

For amphibians, invasion of cells was observed in the Mexican burrowing caecilian *Dermophis mexicanus* by Wake and Wake (2000). In contrast, Lawson (1966) did not find evidence of cell invasion into the notochord sheath in the Frigate Island caecilian *Hypogeophis rostratus*. Hasse (1892) shows clear evidence for cells below the outer elastic layer in the notochord sheath of the smooth newt *Triton taeniatus* (now *Lissotriton vulgaris*).

Despite the fact that cartilage cells — respectively cartilage precursor cells — migrate into the fibrous sheath in elasmobranhs and in dipnoans, Arratia et al. (2001) cautioned not to forget that subsequent vertebral centrum formation in both groups is different. Most elasmobranchs develop chordacentra based on the cells that invaded the sheath. In contrast, extant dipnoans do not form chordacentra. In dipnoans, the notochord may become thicker due to the presence of cartilaginous cells in its matrix, but unlike most elasmobranchs, the cellular fibrous sheath does not calcify. In contrast to extant sarcopterygians, some urodeles and some caecilians, invasion of cells into the notochord sheath was not observed by Klaatsch (1893a) in basal actinopterygians, such as sturgeons (*Acipenser* spp.), gars (*Lepisosteus* spp.) and in teleosts (but see above for other reports). Moreover, the notochord attracts chondrocytes in amniotes, as discussed in Section 8.6C in the context of a discussion of intervertebral and intravertebral cartilages.

During the development of chicken intervertebral joint cartilage, cells move into a matrix generated by the notochord.

5.4 NOTOCHORD CELLS

A. The Notochord Epithelium

After the "stack of coins" phase, notochord cells differentiate into central vacuolated cells and a peripheral cell layer (Figures 1.2, 5.2, 5.6). Many terms are used in the literature for the peripheral cell layer:

- inner epithelial sheath (Kölliker 1861)
- epitheliomoph layer (Grassi 1883)
- chorda-epithel (Gegenbaur 1862)
- notochord epithelium (Jurand 1962)
- peripheral notochord cells (Gardiner 1983)
- basal cell layer (Schmitz 1998a, b)

- outer sheath cells (Ellis et al. 2013)
- outer-epithelial-like cells (Corallo et al. 2015)
- chordoblasts (Fleming et al. 2015)
- notochord sheath cells (Wopat et al. 2018)

To find so many different names for one skeletal cell type is curious and unfortunate. For example, bone forming cells are osteoblasts; they do not have ten different names.

Many names perhaps result from uncertainty about the nature and the function of the cells in the peripheral layer of the notochord. Similar to research on basal chordates (Holland and Holland 2017), research on the notochord has a rather disjunct history. Decades of intense research alternate with decades of rather little interest in the subject. According to current knowledge these cells have several functions. They (i) secrete the notochord sheath and (ii) give rise to the centrally located vacuolated cells. Being highly active cells, they contain abundant rough endoplasmic reticulum (rER). For mammals, the rER has been described as encircling the mitochondria (Trout et al. 1982b).

Recent studies show that the notochord epithelium (iii) also controls the segmented mineralization of the notochord sheath in those osteichthyans that establish vertebral body anlagen by sheath mineralization prior to cartilage or bone formation around the sheath (see Chapter 7). There is general agreement that in cephalochordates, tunicates and vertebrates (including amniotes), the outer cells of the notochord form a cell layer that has characters of a typical epithelium (Carlson 1973; Jurand 1974; Annona et al. 2015). At the same time, many studies that characterize these cells look at anamniote vertebrates, many of which maintain a prominent uninterrupted notochord in later life stages. In zebrafish, the differentiation of the notochord epithelium depends on notch signaling. In the absence of notch signaling, only vacuolated notochord cells develop (Trapani et al. 2017).

The notochord epithelium produces laminin as the main component of the basal lamina (see Table 5.1). Laminins are heterotrimeric glycoproteins that provide connection between cell membranes and the extracellular matrix (Trapani et al. 2017 and references therein). Four laminin chains — α-1, α-4, β-1 and γ-1 — participate in the formation of the notochord sheath with the chordamesoderm expressing mRNA for each chain (Parsons et al. 2002; Stemple 2005a). In zebrafish, 72 hours post fertilization, the notochord epithelium also expresses *col92a* while the vacuolated cells express *col8a1a* (Garcia et al. 2017). Eventually, the notochord epithelium produces a type II collagen-based thick notochord sheath and a thin inner elastic membrane (Miller and Mathews 1974; Kimura and Kamimura 1982; Smith and Watt 1985; Schmitz 1995; Roughley 2004; Scott and Stemple 2004; Zhang et al. 2006; Cole 2011).

Based on studies using transmission electron microscopy (TEM), Schmitz (1995) showed that the cell layer in the yellow perch *Perca flavescens* is indeed a stratified squamous epithelium with cells that are interconnected by desmosomes and gap junctions. Likewise, also based on TEM, Bruns and Gross (1970) clearly distinguish the basal lamina of the notochord epithelium from the inner elastin layer of the notochord sheath in the American bullfrog (*Rana catesbeiana*). With the same technique, Bancroft and Bellairs (1976) distinguish the basal lamina laid down by notochord epithelium from the notochord sheath in chicken. Additional studies confirm that the innermost part of the notochord sheath has indeed the character of a basement membrane which qualifies the peripheral cell layer of the notochord as epithelium (Bruns and Gross 1970; Hall 2015, p. 76; Schmitz 1998b; Bensimon Brito et al. 2012a; Wang et al. 2014; Pogoda et al. 2018).

The structural components of the notochord sheath are products of the notochord epithelium (see the previous section). At the same time, these products also characterize epithelial cells. Laminins, heterotrimeric extracellular glycoproteins, the characteristic products of epithelial cells are principal components of this basal lamina (Corallo et al. 2015). Other typical components are keratins. In the notochord cells of amphioxus proteins such as cytokeratins, desmin and vimentin along with microtubule components (ß-tubulin) have been identified with antibodies directed against the human protein variants (Bočina et al. 2006). Schmitz (1998b) successfully labeled the peripheral cytoplasm of notochord cells in the West African lungfish (*Protopterus annectens*) and in the shortnose sturgeon (*Acipenser brevirostratus*) with three antibodies directed against human skin keratins. Likewise, Godsave et al. (1986) labeled cytokeratins in notochord cells of *Xenopus laevis* with antibodies directed against mammalian cytokeratins. As we will see in Chapter 6, keratin expression is not restricted to peripheral notochord cells. In the maturing and aging notochod, cells that are located in the center of the organ start to express keratin.

The cells of the notochord epithelium usually do not contain vacuoles (Lawson 1966; Ellis et al. 2013) but can contain small vacuolated cells, for example, in the bullfrog (Bruns and Gross 1970). The presence of prevacuolated cells next to the notochord epithelium indicates that the notochord epithelium not only gives rise to the notochord sheath matrix but also to the vacuolated cells in the notochord lumen (Schmitz 1995).

Cartilage-cell-like signaling pathways and the production of cartilage proteins such as type II collagen are general characters of the vertebrate notochord (Kaneko et al. 2016; see also Chapter 8). Reports about the presence of type II collagen exist for *Entoshenus japonicus*, lamprey (Kimura and Kamimura 1982); *Myxine glutinosa* hagfish (Zhang and Cohn 2006); *Acipenser brevirostratus*, sturgeon (Schmitz 1998a, b); *Xenopus laevis*, African clawed frog (Smith and Watt 1985); *Danio rerio*, zebrafish (Cerdà et al. 2002; Bensimon Brito et al. 2012a); *Gallus domesticus*, chicken (Smith and Watt 1985; Ghanem 1996); *Homo sapiens*, human (Roughley 2004; Cox and Sera 2014) and *Mus musculus*, house mouse (Barrionuevo et al. 2006; Peck et al. 2017). Fibrillar collagen type II precursor proteins have been identified in basal chordates such as *Branchiostoma floridae* and *Ciona intestinalis* (Zhang and Cohn 2006; Reeves et al. 2017). It is safe to conclude that type II collagen expression is one of the highly conserved notochord characters.

B. Vacuolated Cells and the Expansion of the Notochord

Notochord cells do not proliferate during the stack of coins phase (Boeke 1902), but proliferation resumes when the cells differentiate into peripheral notochord epithelium cells and centrally located vacuolated cells. The development of cells with large vacuoles expands the notochord and provides the basis for its function as a hydrostatic skeleton in chordate embryos and in many adult chordates (Nordvik et al. 2005; Annona et al. 2015). The adult hydrostatic notochord skeleton may be restricted to intervertebral spaces, for example in mammals, or it may be a large and uninterrupted structure, for example in lungfish and sturgeons (see Chapter 6).

The notochord contains two types of vacuoles. The early and permanent notochord cells contain intracellular vacuoles. Later, the notochord also develops extracellular vacuoles (Trout et al. 1982a, Kryvi et al. 2017, and see Chapter 7). Vacuolation of the centrally located cells starts with small intracellular vacuoles that fuse into large central vacuoles (Malacinski and Youn 1982). The differentiation proceeds from cranial to caudal in actinopterygian fishes and in amphibians (reviewed by Kocher 1957) which is indicative of a primitive osteichthyan differentiation pattern (actinopterygians and sarcopterygians). In mice, vacuolation becomes visible in 13-day-old embryos. The first vacuoles appear in intervertebral areas, in the prospective locations of mammalian intervertebral disks, and in places where the notochord will be retained. On day 14, the cells in intervertebral regions are prominently vacuolated (Paavola et al. 1980). In later stages, the notochord epithelium produces additional vacuolated cells. In the yellow perch (*Perca flavescens*), the transition from the prevacuolar cells to a fully formed (although small) vacuole appears rapidly since very few cells with intermediate stages of vacuole formation are observed (Schmitz 1995).

Studies on the zebrafish notochord provide evidence that vacuoles are best considered as lysosome-related vesicles (Ellis et al. 2013). The vesicular transport system in notochord cells is understood in considerable detail. Figures 5.7 and 5.8 show the subcellular details of the components of the system, which are the endoplasmic reticulum, Golgi complex, and clathrin-coated invaginations/vesicles. Vesicle coat protein complex I and II- (COP I and COP II)- dependent transport of vesicles from the endoplasmic reticulum to the Golgi complex is shown in both figures, major genes that affect the vesicular transport system in Figure 5.7, major mutations affecting the complex in Figure 5.8. Table 5.2 elaborates in greater detail the nature and functions of the components of the vesicular transport system as understood from studies on developing notochord cells in the Japanese medaka *Oryzias latipes* and the zebrafish *Danio rerio*. Mutations affecting genes that regulate the vesicular transport system, effects of these mutations on notochord cells and on notochord morphology, vertebral centra, neural arches, tail morphology and skull development all are summarized in Table 5.2.

This long list of genes and their effects illustrates the critical role played by the vesicular transport system in notochord cell differentiation and notochord morphogenesis during embryogenesis. Notochord defects in zebrafish and medaka are often observed in the context of biomedical research that uses these small teleosts as models to study genetic factors that cause human diseases. We usually have no knowledge of whether notochord defects similar to those seen in fish embryos also occur in human embryos. After hatching zebrafish and medaka are essentially embryos outside the egg (Witten et al. 2017). It is thus possible to observe notochord defects that cannot be observed in humans or other mammals. It may also be possible that notochord defects disturb development in zebrafish and medaka to a larger extent than in mammals. Free-living teleost embryos must use their notochord as a functional axial skeleton until mineralized vertebral centra develop (Bensimon-Brito 2012a). Moreover, the notochord has a central function in the patterning of the vertebral column (Lleras-Foreo et al. 2018), whereas notochord competencies have been transferred to the somites in amniotes (Stern 1990; Ward et al. 2018), and see Section 9.3. Interestingly, mutations in collagen type I can severely disturb notochord development in zebrafish with the consequence of distorted bone formation around a deformed notochord (Gistelinck et al. 2016).

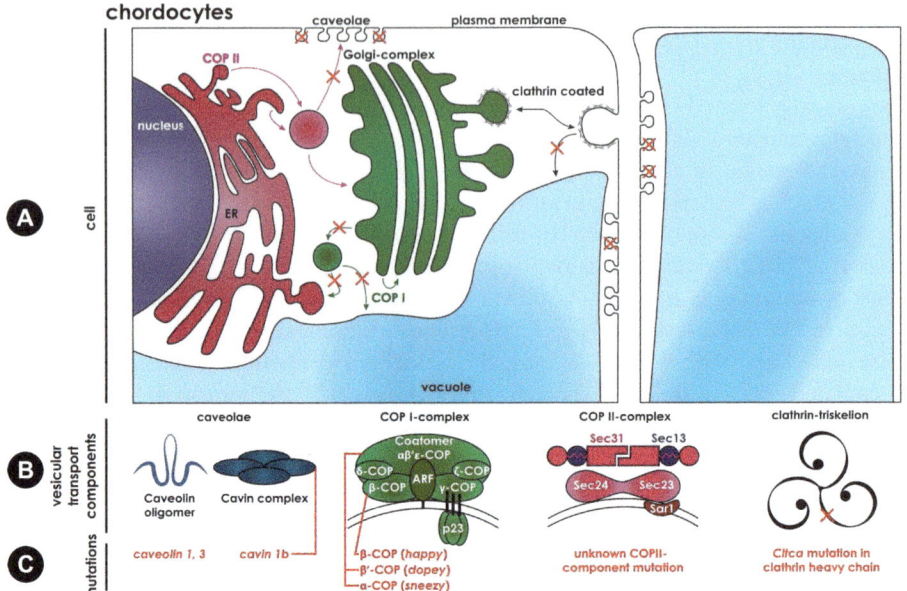

FIGURE 5.7 The cellular vesicular transport system in chordocytes shown in portions of two adjacent chordocytes at two different magnifications (A), as the components of the vesicular transport system (B), and as genes/mutations that affect the transport system (C, red letters). (A) The cellular level to show elements of the system — nucleus, endoplasmic reticulum (ER), Golgi-complex and a clathrin-coated invagination (pit) of the cell membrane. The chordocytes' vacuoles is shown in blue, the cell to the right (at lower magnification) showing that the vacuole fills most of the cell. Caveolae pits are shown in the plasma membrane. Cytosolic coat protein complex II (COP II)-dependent transport of vesicles from the endoplasmic reticulum to the Golgi complex is indicated by purple arrows, COP I-dependent transport (COP I) by green arrows, clathrin-coated vesicular transport by black arrows, with red crosses to show which pathways or components (caveolae) fail when a mutation occurs. (B) The components of the vesicular transport system are shown. Images of caveolae, (dark blue), COP I (green), COP II (purple) and clathrin (black) are adapted from Bastiani and Parton (2010), Nickel et al. (2002), Gomez-Navarro and Miller (2016), Edeling et al. (2006), Robinson (2015), respectively. The parallel curved black lines underneath the COP I-complex components represents the plasma membrane phospholipid bilayer of the transport vesicle. The red cross on the clathrin-triskelion shows where the *Cltca* mutation forms a truncated protein (Edeling et al., 2006). (C) Genes that affect the vesicular transport system; also see Figure 5.8. Details of the genes and their actions may be found in Table 5.2. (Figure from De Clercq (2018) PhD thesis, Massey University, Manawatū, Palmerston North, New Zealand. 271 pages. Used with permission from Adelbert De Clercq.)

Generation of the vacuoles from vesicles also requires endosomal trafficking, which is regulated by the vacuole specific, Ras-related protein RAB32A and the membrane bound proton pump H^+-ATPase (Ellis et al. 2013). Despite the presence of H^+-ATPase, which acidifies lysosomes, notochord vacuoles are not acidic. Ellis et al. (2013) suggest that a sodium/proton exchanger (Na^+/H^+) removes the protons from the lumen of vacuoles. In this model, Na^+ functions as an osmolyte that draws

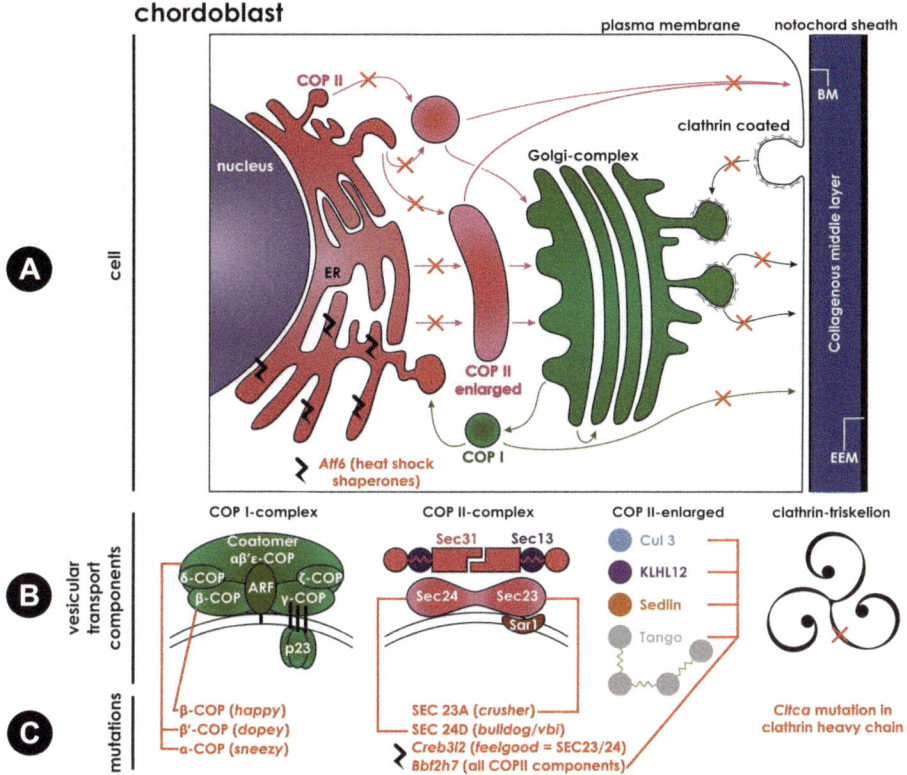

FIGURE 5.8 An overview of mutations in cellular vesicular transport mechanisms in chordoblasts. (A) A schematic representation of the cellular elements of the system in a portion of a chordocyte (left, see also Figure 5.7) alongside the notochord sheath on the right. The basement membrane (BM) of the notochord sheath is directly opposed to the plasma membrane, the collagenous middle layer and external elastic membrane (EEM) forming the matrix of the notochord sheath. Black lightning bolts in the ER indicate where the ER proteins unfold in response to stress in the ER. Cytosolic coat protein complex II (COP II)-dependent transport of vesicles from the endoplasmic reticulum to the Golgi complex (purple arrows), COP I-dependent transport (COP I, green arrows), clathrin-coated vesicular transport (black arrows), and pathways that fail when a given mutation occurs (red crosses) are shown. (B) The components of the vesicular transport system adapted from sources cited in Figure 5.7B. COP I (green), COP II (purple), COP II-enlarged (multi colors) and clathrin (black). (C) Some mutations that effect the components of the vesicular transport shown in B (see Table 5.2 for details).The basement membrane (BM) of the notochord sheath is directly opposed to the plasma membrane, the collagenous middle layer and external elastic membrane (EEM) forming the matrix of the notochord shea The black thunderbolt next to the protein-coding gene *Creb3l2* (*Cyclic AMP-responsive element-binding protein 3-like protein 2*) and the transmembrane bZIP transcription factor *Bbf2h7* (I) in the COP II-complex indicates that these mutations are involved in the unfolded protein response (Table 5.2). (Figure from De Clercq (2018) PhD thesis, Massey University, Manawatū, Palmerston North, New Zealand. 271 pages. Used with permission from Adelbert De Clercq.)

TABLE 5.2

The Vesicular Transport System in Zebrafish and Medaka Notochord Cells: Components, Genes and Mutations

Components	Function of the Components	Mutations	Species	Effects of Mutation(s)	References
Unfolded Protein Response (UPR)	Folding proteins correctly in the ER lumen and evacuating proteins correctly when ER-stress occurs, as when ER bulges due to the presence of too many proteins. Transcriptional regulation of COPII coat components (SEC23/SEC24/SAR1, SEC13/SEC 31, TANGO, SEDLIN, KLHL12, CUL3).	$Atf6\alpha/\beta$ mutants	Medaka	ER-stress occurs Reduced induction of ER chaperones HSP90, GRP94, Lectins, HSP70 and HSP40 family members	Ishikawa et al. (2013) Sitia and Braakman (2003)
	$Bbf2h7$ is a transducer of *all* COP II components (SEC23/SEC24/SAR1, SEC13/SEC31, TANGO, SEDLIN, KLHL12, CUL3) and is similar to the *feelgood* mutant in zebrafish.	$Bbf2h7$ mutants	Medaka	Cartilage in skull underdeveloped Short tail Bent notochord Incomplete bone formation Detachment of chordoblasts from basal lamina ER-stress due to build-up of proteins in the ER No secretion of collagen type xx and IV Down-regulation of *all* COPII coat components	Ishikawa et al. (2017)
	Is similar to $Bbf2h7$ mutant in medaka. However, $Creb3l2$ only regulates a two COP II components;. SEC23A/B and SEC24D.	*Feelgood* ($Creb3l2$) mutant	Zebrafish	Cartilage in skull under develops Short tail Bent notochord ER-stress due to build-up of proteins in the ER No secretion of collagen type-II Down-regulation of SEC23A/B and SEC24D COPII components	Melville et al. (2011)

(Continued)

TABLE 5.2 (Continued)
The Vesicular Transport System in Zebrafish and Medaka Notochord Cells: Components, Genes and Mutations

Components	Function of the Components	Mutations	Species	Effects of Mutation(s)	References
Vesicle coat protein complex I (COP I)	COP I-coated vesicles recycle membrane bound components form the Golgi complex to the ER (retrograde transport). these vesicles have been associated with limited anterograde transport and transport from the Golgi-complex to intracellular vacuoles. COP I components are P23, ARF, COP β, COP γ, COP δ, COP ζ, COP αβ'ε and are highly and consistently upregulated in chordocytes to develop and maintain the vacuoles.	Sneezy (Cop α) Happy (Cop β) Dopey (Cop β') mutants	Zebrafish	Aberrations in chordamesoderm differentiation Vacuoles of chordocytes fail to form ER and Golgi-complex disrupted ER-stress due to build-up of proteins in the ER Collagen (middle) layer in notochord sheath is thin and malformed	Coutinho et al. (2004) Kondylis et al. (2009) Nickel et al. (2002) Odenthal et al. (1996) Stemple (2005) Vacaru et al. (2014)
Vesicle coat protein complex II (COP II)	COP II-coated vesicles (50-90 nm) transport proteins from the ER to the Golgi-complex (anterograde transport). COP II components are SEC23/SEC24/SAR1, SEC13/SEC31 and are highly upregulated in chordoblasts and chondroblasts (cells with high demand for excretion to the ECM). Components such as TANGO, SEDLIN, KLHL12, CUL3 interact with SEC proteins to build larger vesicles to cope with large protocollagen type II molecules (300 nm).	Sec23a (crusher) mutant	Zebrafish	Short tail Skull underdeveloped and looks collapsed ('crushed') Collagen and proteoglycan secretion impaired ER-stress due to build-up of collagen type II in the ER Higher proteasome activity HSP70 (UPR-chaperone) upregulated	Kondylis et al. (2009) Lang et al. (2006)
	= vbi mutant in medaka	Sec24d (bulldog) mutant	Zebrafish	Short tail Skull underdeveloped Bent notochord Impaired chondrocyte maturation ER-stress due to build-up of collagen type II in the ER of chondrocytes UPR components are upregulated	Melville et al. (2011) Sarmah et al. (2010)

(Continued)

TABLE 5.2 (Continued)

The Vesicular Transport System in Zebrafish and Medaka Notochord Cells: Components, Genes and Mutations

Components	Function of the Components	Mutations	Species	Effects of Mutation(s)	References
	= *bulldog* mutant in zebrafish	*Vbi* (*Vertebra imperfecta*) mutant	Medaka	Centrum mineralization delayed Centrum mineralization is irregular Chordacentra fuse ventrally Neural and haemal arches malformed Irregular migration and positioning of osteoblasts lining the notochord ER-stress due to build-up of collagen type II in the ER of chondrocytes HSP47 (UPR-chaperone) is upregulated	Ohisa et al. (2010)
Clathrin coat complex	Clathrin-coated vesicles transport membrane bound proteins and enzymes from the Golgi-complex to the plasma membrane, and transport extracellular signals to the Golgi complex or to other intracellular vesicles.	*Cltca* mutant	Zebrafish	Bent notochord Vertebral centra form to conform to the shape of the bent notochord Vacuoles in the chordocytes collapse	Ellis et al. (2013)
Caveolae	Plasma membrane pits of 60-80 nm which are 'cup'-shaped and important in cellular signaling pathways (extra- to intracellular signaling), endocytosis, cellular attachment and detachment, fatty acid regulation of the plasma membrane, mechanosensing and mechano-protection of cells. Highest number of caveolae are recorded in zebrafish chordocytes. Components are Caveolin 1, 2, 3 and Cavin-1.	*Caveolin 1* (*Cav1*) *Caveolin 3* (*Cav3*) *Cavin-1b* mutants	Zebrafish	Shorter fish Compromised swimming upon mechanical stress Notochord lesions where mechanical stress highest Reduced numbers of caveolae Aberrant morphology of caveolae Delamination of chordocytes Collapse and fragmentation of vacuoles Cell rupture Release of dense material from the ECM	Bastiani and Parton (2010) Garcia et al. (2017) Lim et al. (2017)

Source: Used with permission from Adelbert De Clercq.

Table based on Appendix 7 from De Clercq (2018) with some modifications. De Clercq A (2018) *The Effect of Incubation Temperature on Early Malformation, Regionalization and Meristic Characters of the Vertebral Column in Farmed Chinook salmon (Oncorhynchus tshawytscha)*. PhD thesis, Massey University, Manawatū, Palmerston North, New Zealand. 271 pages.

water into the vacuoles and thus inflates their lumen. Inflation of the vacuoles is a morphogenetic force that elongates the embryo of vertebrate and non-vertebrate chordates (Adams et al. 1990; Jiang and Smith 2007). For *Xenopus laevis* embryos in stages 18–20, values of hydrostatic pressure calculated by Adams et al. (1990) range between 0.5×10^6 and 1.2×10^6 Nm^{-2} but increase in stages 23–26 to 2.4×10^6 Nm^{-2}. This is an increase in the range of 2–4 times and a pressure in the magnitude range of plant cells (Nobel 1970). *In vivo*, the osmolarity in the developing notochord of *Xenopus* was estimated at a magnitude higher than epidermal cells. Waddington and Perry (1962) suggested that glycosaminoglycans (GAG) in the vacuoles function as osmolyticum, a hypothesis endorsed by Grodzinsky (1983). Ellis et al. (2013) tested for the presence of GAG in the lumens of zebrafish vacuoles and did not find evidence for their presence. Undisputed is, of course, the abundance of GAG in the notochord sheath and in the intercellular matrix.

Typical membrane structures of vacuolated cells are caveolae, which are submicroscopic plasma membrane invaginations that are highly enriched with cholesterol, their main constituents being caveolin-1 and caveolin-2 (Sinha et al. 2011); see Figures 1.3, 5.7 and 5.8. Caveolae are not only found in notochord cells. Adipocytes, muscle cells (skeletal, cardiac, and smooth), fibroblasts, capillary endothelium and type I pneumocytes possess caveolae. Notably, caveolae are also a feature of cartilage cells, described for birds and mammals (Schwab et al. 1999; Hollins et al. 2002). Lotz and Hsieh (2014) discuss the function of caveolae for cells in the human nucleus pulposus. Acute increases in cell volume or stretch leads to a rapid loss of caveolae, and so they are implicated in a membrane-mediated mechanical response triggered by tyrosine phosphorylation (Alenghat and Ingber 2002). Caveolae are moved from cell membrane to the endoplasmic reticulum by the vesicular transport system, the structure, components and functions of which in notochord cells are understood in some detail, as discussed above.

Caveolae are required to buffer fluctuations in cell membrane stress induced by acute membrane tension and osmotic shock, such that loss of caveolae compromises buffering of cell membrane tension (Parton and Simons 2007; Table 5.2). These observations suggest that caveolae may participate in the response of nucleus pulposus cells to hydrostatic pressure. Cavin-1B is a coat protein required for caveolae formation. Zebrafish CRISPR/Cas9-generated mutations that lack Cavin1b have a notochord with reduced mechanical properties (Lim et al. 2017). Likewise, the conserved vacuolated plasma protein Caveolin-1 is required for notochord development and its mechanical stability (Garcia et al. 2017; Table 5.2). Caveolin-1 has been identified in mammalian nucleus pulposus cells, its expression decreasing with age (Heathfield et al. 2008). A high density of caveolae is found in the notochord of posthatching zebrafish (Lim et al. 2017; Witten and Hall 2021). Similar high numbers of caveolae have been found in the membrane of prevacuolated cells in the yellow perch (*Perca flavescens*) (Schmitz 1995). Caveolae have been found as membrane structures of vacuolated notochord cells also in juvenile sturgeons (*Acipenser brevirostratus*) and in the West African lungfish (*Protopterus annectens*) (Schmitz 1998a). Waddington and Perry (1962) clearly show caveolae in vacuolated notochord cell membranes of young salamander tadpoles (*Pleurodeles sp.*) but label them as discharging membrane vesicles. Despite the presence of abundant desmosomes, no caveolae are reported from the notochord of *Petromyzon fluviatilis* (Schwarz 1961).

Desmosomes (maculae adherentes), as shown in Figure 1.3, are also typical noto-chord cell membrane structures. Desmosomes are strong cell to cell connections. Like caveolae, desmosomes characterize tissues that are subjected to mechanical stress, for example the epidermis. Cytokeratine (intermediate filaments) are impor-tant desmosome substructures which is in agreement with the presence of cytokera-tins in notochord cells (Schmitz 1998a; LaFlamme et al. 1988). Desmosomes connect the notochord cells of *Branchiostoma*. Thus, Stach (1999) considers the possibility that these structure are relicts of epithelial cells of the archenteron from which the notochord arises during development. The presence of hemidesmosomes that anchor the notochord cells to the sheath, as in other chordates, suggests a function of these epithelial cell structures (Mansfield et al. 2015).

Desmosomes and hemidesmosomes are hallmarks of notochord cells, particularly of vacuolated notochord cells. Notochord-derived tumors (chordomas) also contain vacuoles that connect with desmosomes (Peña et al. 1970). The presence of desmo-somes can be inferred based on gene expression data or based on immunodetection of desmosome-related proteins such as desmoplakin and placophilins (Tan et al. 2015). Primarily desmosome detection depends, however, on the use of TEM. Table 5.3,

---

TABLE 5.3
Identification of Desmosomes in Notochord Cells based on Transmission Electron Microscope Studies

Agnathans

Hagfish — Brown hagfish *Paramyxine atami* (Welsch et al. 1998)
  — *Atlantic hagfish, Myxine glutinosa* (Koob and Long 2000)
Lamprey — European river lamprey, *Petromyzon fluviatilis* (Schwarz 1961)

Chondrichthyans

Dogfish — small-spotted catshark, *Scyliorhinus canicula* (Restovic et al. 2016)

Sarcopterygians

Lungfish — West African lungfish, *Protopterus annectens* (Schmitz 1998a).
Caecilian — Koh Tao island caecilian, *Ichthyophis kohtaoensis* (Welch and Storch 1971)
Common frog — *Rana temporaria* (Fox 1973)
Northern leopard frog — *Rana pipiens* (Overton and Mapp 1974)
African clawed frog — *Xenopus laevis* (Godsave et al. 1986; Honer and Komnick 1990)
American bullfrog — *Lithobates catesbeianus* (Bruns and Gross 1970)
Domestic chicken — *Gallus domesticus* (Bancroft and Bellairs 1976)
Human — *Homo sapiens* (Trout et al. 1982a; Krstic 1985; Lehtonen et al. 1995)

Actinopterygians

Sturgeon — shortnose sturgeon, *Acipenser brevirostratus* (Schmitz 1998a).
Zebrafish — *Danio rerio* (Burger et al. 2014)
Japanese medaka — *Oryzias latipes* (Ekanayake and Hall 1991)
Striped bass — *Morone saxatilis* (Nowroozi et al. 2012)
Yellow perch — *Perca flavescens* (Schmitz 1995)

with no claim of being complete, shows the studies that demonstrate desmosomes in vacuolated cells in different classes of vertebrates. It appears reasonable to assume that vacuolated notochord cells connected by desmosomes is a common vertebrate feature.

5.5 THE AMNIOTE NOTOCHORD IS SPECIAL

The amniote (reptiles, birds and mammals) notochord is special because it is small and is described as a transitory structure in countless textbooks and scientific articles. While a transitory existence is true for chickens, ducks and gulls (neognathous birds) we do not know if this is true for all birds, particularly for all palaeognathous birds. The mammalian notochord remains and expands into the intervertebral spaces, as does the notochord in birds prior to its replacement. This is important because studies on the notochord in chicken and mouse embryos initiated some 50 years ago provided the first knowledge that the notochord secretes collagen and glycosaminoglycans and because this knowledge was applied to the notochord of all vertebrates. However, as discussed in earlier chapters, the nature and function of the notochord in amniotes, fish, urodele and anuran amphibians have evolved and diverged. In this section, we look back at the knowledge accumulated for amniotes, the paradigm shift it produced in our understanding of notochord structure and function and in the relationship between notochord, mesodermal cells and vertebral chondrogenesis.

Despite being tiny and not fully functioning as hydrostatic skeletons, amniote notochords are "good" notochords. The use of chicken embryos as models in embryology beginning in the 1960s resulted in substantial knowledge about the cellular and biochemical composition of the notochord. One of the most significant discoveries was that notochord anlagen express *col2* very early in development and that notochord cells synthezise and deposit type II collagen (the product of the *Col2a1* gene) into the notochord sheath (Table 5.1). Acquisition of this fundamental knowledge was facilitated by the development of TEM, radioisotope labeling, and the ability to localize isotope label in thin histological sections.

A. The Notochord and what we Consider as Cartilage Matrix Components

The demonstration of the role played by the notochord and spinal cord in evoking chondrogenesis *in vivo* discussed in Chapter 4, and the accumulating evidence that normal constituents of cartilage ECM such as chondroitin sulfate, collagen and hyaluronan can influence the rate of synthesis of cartilage ECM through feedback inhibition and stimulation (Hall 1973) led to a search for ECM around the notochord and spinal cord, a search that demonstrated that the notochord and the ventral portion of the spinal cord in chick embryos as young as HH 10 (embryonic stages according to Hamburger and Hamilton 1951) synthesize and accumulate glycosaminoglycans (GAGs) into an ECM. Table 5.4 presents a summary of the major events in the formation of notochord and spinal cord ECMs in chick embryos and the changes occurring in sclerotomal mesenchyme at the same stages.

The first studies utilized autoradiographic techniques to visualize glycosamino-glycans around the notochord when chondrogenesis is commencing within adjacent sclerotomes. Then, ^{35}S injected into the albumen of eggs was detected in the noto-chord and primitive streak in embryos as young as HH 31, *long before somite chon-drogenesis* (Franco-Browder et al. 1963). Notochord and spinal cord from embryos of HH 11 and older are surrounded by sulfated GAGs. ^{35}S was localized over the spinal cord, notochord and immediately adjacent ECM when these regions were organ cul-tured, but Lash and his colleagues could not judge where this material was synthe-sized (Lash 1963; Lash et al. 1964).

The ECM of the notochord in embryonic chicks and mice was then described as containing perinotochordal fibrils, interpreted as arising from surrounding sclero-tomal cells and from the notochord sheath; microfilaments were described on the notochord and between the notochord and spinal cord in chick embryos at HH stage 11 (42 hours) of incubation (Table 5.4). Amorphous material attached to these micro-fibrils and interpreted as *glycosaminoglycans* was described as either:

i. 15–20 nm diameter, unbanded and beaded fibrils located close to the noto-chord, sensitive to removal by collagenase and interpreted as *collagen fibrils* or
ii. 10 nm tubular banded fibrils, located some distance from the notochord, sensitive to removal by hyaluronidase and amylase but not by collage-nase and interpreted as fibrils of *hyaluronic acid* (O'Connell and Low 1970; Frederickson and Low 1971; Minor 1973; Jurand 1974; Bancroft and Bellairs 1976).

Using an experimental approach, notochords dissected from two-day-old chick embryos were exposed to the enzyme trypsin to remove ECM and basement mem-brane. When maintained in organ culture for three to four days, microfibrils accumu-lated adjacent to a reconstituted basal lamina. Addition of ascorbic acid (a co-factor for collagen synthesis)[1] at 100 mg/mL coupled with longer culture times, resulted in the deposition of sheets of cross-striated fibrils, some as wide as 150 nm. The 51 nm axial periodicity of these fibrils (also known as D-period) is within the known size range of collagen fibrils (Linsenmayer et al. 1973; Trelstad et al. 1973). Both these fibrils and those surrounding the ventral surface of embryonic spinal cords were shown to be type II collagen on the basis of binding to antibodies against type II collagen.

Application of electron microscopy and histochemistry (O'Connell and Low 1970; Kvist and Finnegan 1970) demonstrated *sulfated GAGs* (hyaluronan and chondroitin sulfate) adjacent to the notochord at HH 16 and 17, respectively (Table 5.4). By HH 17, there was 2.5 times more hyaluronate than chondroitin sulfate. By HH 28 the ratio had reversed. Therefore, it was suggested by Toole (1972) that hyaluronan plays a role in sclerotome aggregation. The ratio of hyaluronan to chondroitin sulfate is high at HH 23 but low thereafter, the decline correlating

---

[1] Hydroxylation of proline and lysine is inhibited in ascorbic acid (vitamin C) deficiency. Consequently, biosynthesis of collagen and elastin is impaired (Barnes et al. 1969; Wu et al. 1989).

TABLE 5.4

Major Events in the Formation of Notochord and Spinal Cord Extracellular Matrices in Chick Embryos over the First 13 Days of Incubation (HH Stages 3+ −39)

HH Stage	Age	Notochord, Spinal Cord	Sclerotome
3+	12 hours	^{35}S in presumptive notochord	
10	36 hours	Collagen, GAGs as ECM	Intact mass of ovoid cells
11	42 hours	Basement membrane present	Nest of cells breaking up
12	47 hours		Mesenchymal cells beginning to migrate medially
13	50 hours	Youngest age shown to produce collagen when isolated *in vitro*	Sclerotomites starting to form
17	60 hours	Considerable hyaluronan around notochord. Treatment with collagenase or hyaluronidase prevents induction	
18	3 days	Notochord vacuolated, now induces 100% of cultured somites to chondrify	Sclerotomal cells finished migration
23	4 days		ECM around cells closes to notochord — now prechondroblasts
27–30	5–7 days	More CS than hyaluronan	ECM spreading into peripheral cells
31–36	7–10 days	Ventral spinal cord loses inductive ability	Chondroblasts and chondrocytes
36–39	10–13 days	Notochord loses inductive ability (HH 37.5)	Cell death in area nearest notochord

Source: Adapted from Hall (2015).

Abbreviations: CS, chondroitin sulfate; ECM, Extracellular matrix; GAGs, Glycosaminoglycans; HH, Hamburger and Hamilton Stages of chick development; S^{35}, sulfur^{35} isotope.

with increasing levels of hyaluronidase and with formation of metachromatic ECM in perinotochordal mesenchyme; removal of hyaluronan accompanies formation of ECM; see Goldberg and Toole (1984) for hyaluronan as part of the pericellular coat.

Further application of electron microscopy laid the basis for our understanding of the organization of the notochord ECM, demonstrating:

i. amorphous material (GAGs) on microfibrils around the notochords of two- to four-day-old embryonic chicks;

ii. the importance of the metachromatic ECM (GAGs) between sclerotome and spinal cord and around the notochord (Frederickson and Low 1971; Strudel 1971; Corsin 1974), and

iii. the accumulation of 20–40 nm GAG granules along with 15 nm unbanded
collagen fibrils around the notochord and the ventral portion of the spinal
cord at H.H. 10 (Minor 1973; Bancroft and Bellairs 1976).

The interpretation was that all these *matrix products progressively become the
ECM of the vertebral cartilage.* By HH 17, this ECM is prominent and a basement membrane surrounds notochord and spinal cord, a matrix maturation that
precedes the arrival of migrating sclerotomal mesenchyme around the notochord
at HH 18.

Both ECM and basement membrane are lost following the trypsinization used to
isolate somites from adjacent tissues. The finding that products of the ECM *reform*
when the notochord is maintained *in vitro* provided evidence *for the production of
the ECM by the notochord.* Analysis of enzymatic digests of products from cultures
of isolated notochords and neural tubes reveals mostly chondroitin and heparan sulfates, including shared chondroitin sulfate–protein complexes (Mathews 1971, 1975;
Hay and Meier 1974).

In a series of insightful papers, Nagaswamisri Vasan showed that perinotochordal ECM from chick embryos contains small proteoglycans and cartilage-type
proteoglycans, and that only the large aggregated forms are required for notochord
to induce somitic cartilage (Vasan 1987). Unstimulated somites — i.e., somites not
exposed to notochord — do not contain any link protein for proteoglycan. Adding
notochord to somite cultures activates the synthesis of link protein and stabilizes
the ECM. On the other hand, neither hyaluronan nor hyaluronidase are affected
by adding notochord to somite cultures (Vasan et al. 1986). Proteoglycan mono-
mers from sternal cartilage also evoke chondrogenesis from somitic mesoderm
(Vasan and Miller 1985), further implicating ECM components in the inductive
interaction.

B. The Notochord Makes Cartilage Collagens, or the Other Way Round?

Collagen was also being examined in studies that were among the first to demonstrate a family of collagen molecules and tissue-specific collagen types.

Until the late 1950s, and in large part as a consequence of the methodolo-
gies available, collagen had been demonstrated only in mesenchymal tissues.
Consequently, *collagen was regarded as a mesenchymal protein*, the notochord
was not regarded as mesoderm-derived and epithelia were not thought able to
produce fibrillar collagen. From the early 1960s onward, however, there was a
suggestion that epithelial structures such as notochord and spinal cord might
have the ability to synthesize and export collagen. The initial findings came
from ultrastructural studies as TEM became an important tool revealing cellular
organization.

The ECM of the notochord in embryonic chicks and mice was described as
containing fibrils, some thought to be derivatives of sclerotomal cells, some from
the notochord sheath. Then, came a series of studies by Low and colleagues
describing microfilaments on the notochord and between notochord and spinal
cord in chick embryos at HH 11 (O'Connell and Low 1970; Frederickson and

Low 1971). Amorphous material — interpreted as GAG — is attached to these microfibrils, which, as introduced above, are of two types: 15–20 nm diameter, unbanded and beaded fibrils close to the notochord and sensitive to removal by collagenase, and 10 nm tubular banded fibrils some distance from the notochord, sensitive to removal by hyaluronidase and amylase but not by collagenase.

With advances in knowledge of collagen types, perinotochordal and perispinal chordal fibrils were shown to be type II (cartilage-type) collagen as is the collagen synthesized *in vitro* by spinal cords from two-day-old chick embryos (Linsenmayer et al. 1973; Trelstad et al. 1973). Subsequently, antibodies against type II collagen were shown to bind to embryonic notochord (H. von der Mark et al. 1976; K. von der Mark et al. 1976; K. von der Mark 1980). Notochords from chick embryos were shown to secrete type IX collagen just prior to vertebral chondrogenesis (Carlson et al. 1974; Hayashi et al. 1992). The developmental significance of the discovery that notochord and spinal cord, which promote chondrogenesis in sclerotomal cells, produce the same collagen type (type II) as the tissue they induce is the possibility that type II collagen may play an inductive role in chondrogenesis.

The classical studies that combined chicken embryology with biochemistry revealed key characters of the notochord, characters that we now assign to the notochord of other chordates. Amazingly, these data were not obtained from the large notochords found in chondrichthyans or basal osteichthyans. It was the tiny, transient, chicken notochord and a community of dedicated embryologists who provided this wealth of knowledge. As the notochord is phylogenetically the oldest part of the endoskeleton and older than cartilage, the notochord may have acquired the ability to synthesize and deposit type II collagen before cartilage did. Rather than saying that notochord possesses cartilage-type collagen, we should be saying that cartilage possesses notochord-type collagen.

5.6 SUMMARY

This chapter explores the nature of the notochord sheath across the vertebrates and the evidence that notochord cells (chordoblasts) produce the sheath — although it took 100 years to discover this — while its cells are flat and arranged as a stack of coins. These cells express *col2a1* early in development and then synthesize and deposit type II collagen into the sheath. As the notochord is phylogenetically older than cartilage, the notochord may have acquired the ability to synthesize and deposit type II collagen before cartilage did. In addition to *col2a1*, notochord cells express what are now known to be characteristic notochord genes — *Sonic hedgehog* (*Shh*), *Echidna hedgehog* (*Ehh*) and *Brachyury* (*Tbxt*).

In all vertebrates the notochord sheath comprises vacuolated cells, a notochord epithelium, and a collagen type II-based notochord sheath with inner and outer elastin layers. Movement of prenotochord cells toward the midline and intercalation of those non-dividing cells generates the "stack of coins" arrangement that typifies early notochord (and cartilage) cells prior to resumption of proliferation and differentiation into inner vacuolated cells and the outer notochord epithelium. The vacuoles differ from group to group but all are related to lysosomes. Desmosomes and hemidesmosomes are hallmarks of notochord cells, particularly of vacuolated notochord cells. These cells with large vacuoles expand the notochord and provide the basis for its function as hydrostatic skeleton in chordate embryos and in many adult chordates. A fully formed notochord sheath has four layers, from inside to outside: (i) a laminin-rich basal lamina of the notochord epithelial cells; (ii) a thin (sometimes transient) elastin-rich layer, (iii) a middle layer with a composition very similar to cartilage matrix; and (iv) an outer layer that is rich in elastin fibers. The notochord sheath is acellular, which is assumed to be the basic chordate condition since invertebrate chordates and extant jawless vertebrates have an acellular notochord sheath. However, cells can invade the notochord sheath and these are discussed in different vertebrates. Relationships between notochord and cartilage matrices also are discussed in some detail as are the questions of whether notochord cells can differentiate into cartilage cells — they can — and whether cartilage cells can differentiate into notochord cell — they cannot.

Section III

Nature and Fate of the Notochord and Vertebrae across the Vertebrates

Several widely known, but incorrect, conclusions regarding the notochord, prevalent in the literature, are addressed in Chapter 6.

First, the notochord is not a transitory embryonic structure whose functions are over once the vertebral column develops. Indeed, in the majority of vertebrate species, the notochord remains as part of the vertebral column throughout the life of the individual.

Second, the literature presents a one-to-one relationship between somites and vertebral bodies as a primary 'segmentation' of the vertebrate body axis. On the contrary, somite-induced vertebral arches and notochord-induced vertebral centra are separate developmental modules, which means that vertebrates can and do deviate from a one-to-one somite vertebral body relationship.

The third widespread misunderstanding is that vertebral centra are entirely made from somite-derived sclerotomal cells. Indeed, for the bulk of their volume, they are but in teleosts and other primary aquatic osteichthyans the anlagen are established by notochord sheath mineralization and not by chondrogenesis of sclerotomal mesenchyme.

We conclude that there are four basic types of anlagen for vertebral body centra: (i) mineralization of the notochord sheath, (ii) bone formation around the notochord, (iii) cartilage formation around the notochord and (iv) from cartilaginous cells that invade the notochord sheath. The notochord, as a basic vertebrate character, shows a

DOI: 10.1201/9781315155975-8

considerable degree of variation in adult vertebrates in terms of retention, subdivision or loss, but also concerning size and cellular composition. Only a minority of species maintain, throughout life, a notochord composed of vacuolated cells as described in Chapter 5. Still, all notochords continue to express *Brachyury*, be it as functional axial skeletons or as intervertebral disks. In the majority of vertebrates that develop vertebral bodies, the notochord transforms into a functional component of the intervertebral joints together with sclerotome-derived cartilage. The sclerotome-derived cartilage can also constrict the notochord in the intervertebral spaces. This is most extreme in birds in which the notochord is rapidly replaced by this cartilage. Early in development, there is active expansion of the notochord in the position of the prospective intervertebral spaces. At the same time, the notochord becomes constricted through the formation of vertebral centra. In most, but not all, vertebrate lineages — not only in primary aquatic osteichthyans but also in tetrapods, including ancestral mammals — this process continues and creates typical hourglass-shaped vertebral bodies. Chapter 7 provides the evidence for these conclusions.

6 The Role of the Notochord in Vertebral Body Development

CONTENTS

6.1 DO SOMITES RULE OVER THE NOTOCHORD?

The notochord is often wrongly described as a transitory structure, a placeholder for the future vertebral column (Scott and Stemple 2004; Choi and Harfe 2011; Criswell et al. 2017). After having fulfilled its role as a signaling center and primary axial skeleton in early development, in most vertebrates, vertebral bodies, either cartilaginous or osseous, form around the notochord. In the majority of vertebrate species, the notochord remains as a continuous structure throughout the length of the vertebral column. If the notochord becomes interrupted, as it does, for example, in extant mammals, portions of the notochord remain in the intervertebral region where they contribute to the tissues that articulate the vertebral bodies (Risbud and Shapiro 2011; also see Chapter 7).

A common view is that in all vertebrates, the cells that build the vertebral bodies descend from somite-derived sclerotomal mesenchyme. Somites, and not the notochord, are also supposed to contain the information for the patterning of the vertebral column (Krol et al. 2011). The early German term *urwirbel* (primeval vertebrae) for somites reflects this idea (von Ebner 1888, 1896). The basic condition assumes a one-to-one relationship between somites and vertebral bodies. These views are challenged by data now available and discussed in this chapter.

DOI: 10.1201/9781315155975-9

A. Resegmentation and Leaky-Resegmentation

Vertebral bodies consist of two components, *centra* and *arches*, as illustrated in Figure 6.1. For a review of the resegmentation model, it is important to distinguish between arches and centra. In mammals and birds, centra and arches have a continuous cartilaginous anlage and are thus often addressed together as vertebral bodies. Following the arcualia hypothesis, arches gave rise to centra, providing another reason to unite centra and arches as vertebral bodies.

As noted above, the assumption in vertebral development is a one-to-one relationship between somites and vertebral bodies. In many vertebrates, however, development and counts of centra and arches are different. Vertebral centra and arches may develop at somite boundaries in a process designated as *resegmentation* (Remak 1855) (Figure 6.2). According to this tetrapod-based model, two somites contribute cells to one vertebral body. The anterior half of a vertebral body is derived from cells that come from the caudal part of the anterior somite. The posterior half of the vertebral body is made from cells that derive from the anterior half of the following (caudal) somite. The notochord — which is retained in the intervertebral joints of most vertebrates — is located in a segmental position, while the vertebral bodies occupy an intersegmental position. The resegmentation model is supported by:

i. the observation that posterior and anterior halves of each somite are separated by von Ebner's fissure (von Ebner 1888; Stern and Keynes 1987; Andrade et al. 2007);
ii. from grafting studies with half somites (Huang et al. 2000), and
iii. from studies that show somite subdivision at the molecular level (Christ et al. 2004; Lewis et al. 2009).

Over 130 years ago, von Ebner (1888) considered the separation of somites as an early sign of resegmentation.

Differing from tetrapods, in teleost fish more than two somites contribute cells to one vertebral centrum, a process referred to as *leaky resegmentation* (Morin-Kensicki et al. 2002). According to Dawes (1930), leaky resegmentation also is responsible for the development of the vertebral bodies in the cervical region of mice, while resegmentation is more evident in the thoracic region. This view is confirmed by recent molecular studies, reviewed by Scaal (2016). Uncx4.1-lacZ labeling of caudal somite cells in mice showed that the classical resegmentation model holds true in the thoracic and lumbar regions of the vertebral column. In the cervical region, however, cells from the caudal part of the somite (sclerotomal cells) contribute to both anterior and posterior cervical vertebrae on either side of the intervertebral disk.

Despite acceptance of the resegmentation model for the development of mouse and chicken vertebral columns (Huang et al. 2000; Bruggeman et al. 2012), resegmentation has been disputed. Williams (1908), who studied vertebral column development in pigs, did not find any evidence for resegmentation. According to Verbout (1976), a broader comparative analysis of vertebral body development in different amniote species does not support resegmentation. Verbout points out that a functionally meaningful relationship between muscles and skeletal elements (for which resegmentation has

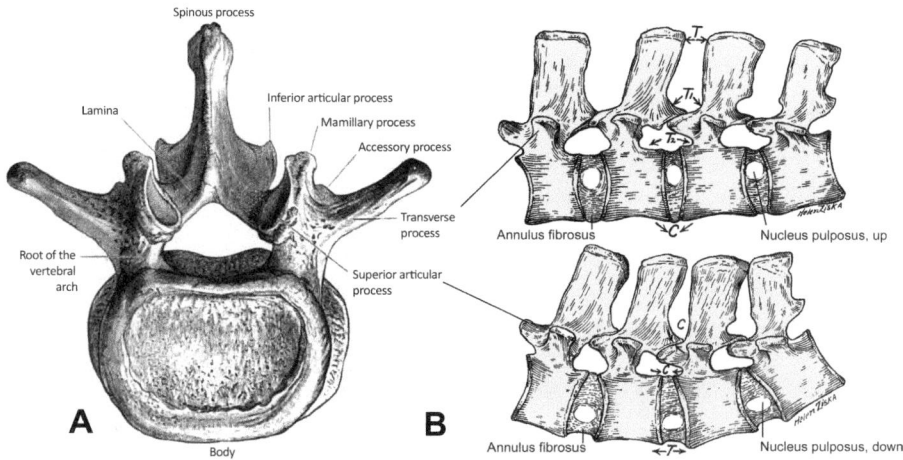

FIGURE 6.1   Mammalian vertebral bodies consist of vertebral centra and neural arches. (A) Frontal view of a human lumbar vertebral body to show the centrum (labelled as Body) and neural arch, which is extended by various neural spines, labelled as processes (Spinous process, Inferior articular process and so forth). Only bone tissue is shown. (B) Lateral view of four generalized mammalian cervical vertebrae, each composed of a centrum, neural arch and neural spine. Also shown are three intervertebral joints (intervertebral disks) consisting of the annulus fibrosus and the persisting notochord (nucleus pulposus). *In vivo*, the nucleus pulposus inside the annulus fibrosus would not be visible from the outside. Upward bending of the cervical spine shifts the nucleus pulposus downwards (Nucleus pulposus, up, Nucleus pulposus, down). The letter C indicates the positions of compression points related to the bending of the vertebral column. The letter *T* indicates the insertion points of ligaments. T, T_1 and T_2 show the insertion point for the intraspinous, subclavian and interarticular and posterior longitudinal ligament, respectively. (A Modified after Thomson, A., 1918. Osteology. The Skeleton. In: Robinson, A. (ed.) *Cunningham's Text-Book of Anatomy*. Fifth Edition, Willam Wood and Company, New York. B Modified after Rockwell, H., et al., 1938. *J Morph* 63: 87–117).

been argued to be essential) is possible without strict resegmentation. Teleost fish are an example. According to Schaeffer (1967), there is no evidence of resegmentation in teleost fish but simply an intrasegmental subdivision of the perichordal tube and the notochord. Wake and Lawson (1973) reached the same conclusion for urodele amphibians. Thus, the term "leaky resegmentation" has a double meaning: "Resegmentation" refers to the intersegmental position of vertebral bodies and "leaky" means that the cells that form the bone around the mineralized notochord sheath derive not only from two adjacent somites but also from many somites.

B. The Notochord and Vertebral Body Development

The centra of vertebral bodies in amniotes are initially made from cartilage. According to the resegmentation hypothesis, cartilage derives from the sclerotome that arises from anterior and posterior somite halves, respectively (Figure 6.2). The amniote notochord becomes completely surrounded by a solid mass of hyaline cartilage that includes

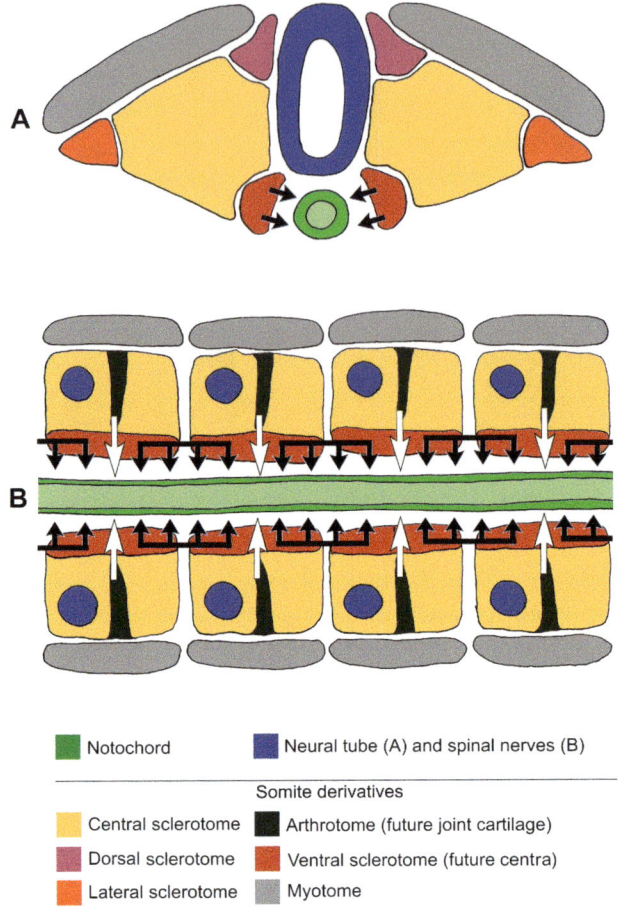

FIGURE 6.2 Resegmentation model for the development of amniote vertebral centra, in large parts based on studies of vertebral column development in embryonic chickens (*Gallus domesticus*). Schematic illustrations of an early stage of sclerotome differentiation, shown as a transverse section across the body (A) and as a longitudinal section (B), both with dorsal at the top. The somites (orange in A) give rise to the different sclerotome derivatives and to the myotome (future muscle) shown in B. Only cells contributed by the ventral sclerotome (red in A and B; black arrows in A) will surround the notochord (autocentrum) and give rise to vertebral centra. Each vertebral centrum receives cells from the posterior and anterior halves of two adjacent ventral sclerotomes (red), shown as black arrow bars in B. Cartilage of the invertebrate joints in birds and the annulus fibrosus in mammals derives from the arthrotome (a term for the joint-forming compartment of avian somites; the former somitocoele, black in A and B), the derivation shown by the white arrows in B. As a consequence, positions of muscle segments (myotome derivatives, gray in A and B) and vertebral centra (ventral sclerotome derivatives, red in A and B) are shifted by half a segment relative to the skeletal derivatives. The arthrotome (black in B) co-locates with von Ebner's fissure that subdivides the central (but not the ventral) sclerotome into anterior and posterior halves. Signaling from the notochord induces arthrotome mesenchyme to chondrify as joint cartilage. Central and dorsal sclerotome give rise to neural arches and neural spines; lateral sclerotome gives rise to ribs. ([(A) Inspired by schemes in Scaal, M., 2016. *Sem Cell Dev Biol* 49: 83–91. (B) Inspired by schemes in Christ et al. 2004 *Anat Embryol* 208: 333–350, both of which should be consulted for further details.])

the neural arches. In non-amniotes, cartilaginous elements that represent the bases of neural and hemal arches can attach to the notochord, but, differing from amniotes, the cartilage may not surround the notochord. Cartilaginous elements that represent the basis of neural and hemal arches are viewed as ancestral parts of a later (later in evolution) complete vertebral centrum (Ota et al. 2011). In either scenario of vertebral body development, the notochord is a placeholder for the developing vertebral bodies but is not involved in the formation of the vertebral centra. In other words, *vertebral bodies, consisting of vertebral centra and arches, are entirely derived from sclerotomes.*

Richard Owen had a different view on the development of vertebral bodies. He emphasized that vertebral bodies derive entirely from the notochord. In his book *On the Anatomy of Vertebrates*, Owen (1866) introduced vertebral body development:

> A few words are requisite to the development of vertebrae… the centrum or 'bodies of the vertebrae', are developed in and from the notochord. The bases of the other elements are laid down in fibrous bands, deriving from the notochord.
>
> *(Owen 1866 p. 31–32)*

One could think that Owen's ideas about vertebral body development are based on a misconception or the restrictions of microscope technology in 1866 and earlier. This is likely not the case. Seven years earlier, independently of each other, Thomas H. Huxley and Albert Kölliker had established the central role of the notochord in vertebral body formation (Huxley 1859; Kölliker 1859).

Huxley described the process in sticklebacks (*Gasterosteus aculeatus*), an advanced teleost species. In his published lecture series "Observations on the development of some parts of the skeleton of fishes," Huxley informs us that vertebral centra in the stickleback have no cartilaginous anlage:

> "In the greater part of its extent it (the notochord) was enclosed neither in cartilage nor in bone—though bony rings, the rudiments (*anlagen*) of the centra of the vertebrae, were developed in the wall of the notochord throughout the rest of the body.
>
> *(Huxley 1859, p. 39) (Figure 6.3)*

Kölliker (1859) described the same process for eel larvae; eels are basal teleosts. Kölliker further suggested that vertebral centra formation is based on notochord sheath mineralization in all "osseous fishes". Huxley's and Kölliker's observations were subsequently backed up by Cartier (1875), Goette (1879) and Balfour and Parker (1882). Moreover, in his later contributions on vertebral development in cartilaginous fishes, Kölliker (1863) communicated that vertebral body anlagen in chondrichthyans are formed by cartilage cells in the notochord sheath.

The observations listed above may provide the background for Richard Owen's statement that "vertebrae are developed in and from the notochord." Indeed, this is the case for teleosts, the largest vertebrate group but it occurs also outside the teleosts in more basal osteichthyans. In teleosts, vertebral body anlagen arise *from* the notochord by segmented mineralization of the notochord sheath and not from sclerotome-derived cells that deposit cartilage or bone *onto* the notochord. Huxley (1859) and others, including Mookerjee et al. (1940), use the term "bony rings" for the vertebral body anlagen that develop in the notochord sheath. Given that the notochord sheath has the characters of a cartilage matrix (Chapter 8), the process resembles more the

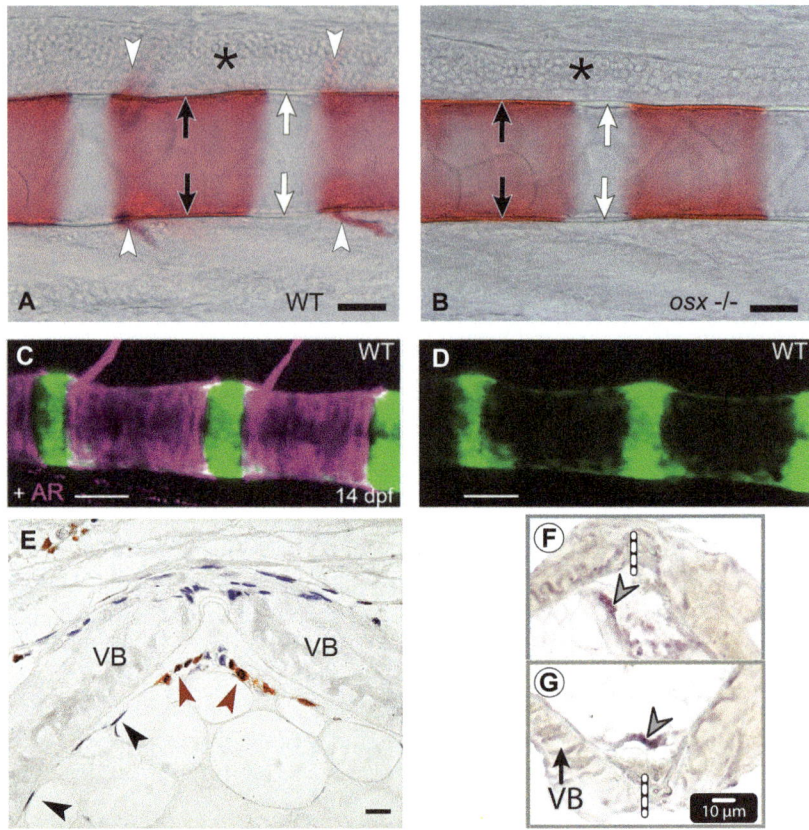

FIGURE 6.3 Initiation of vertebral centra and intervertebral joints in teleost fish. (A) Early anlagen of vertebral centra in wild type (WT) Japanese medaka (*Oryzias latipes*) at 15 days post-fertilization, using calcium staining to demonstrate features initially described by Huxley (1859) for sticklebacks and Kölliker (1859) for eels. Teleost vertebral centra have no cartilaginous precursor. The early centra anlagen consist of segmented mineralization of the notochord sheath (red, black arrows), initiated by the notochord. Sclerotome-derived osteo-cytes (see text) give rise to the neural and hemal arches (white arrowheads). Asterisk, spinal cord. Whole-mount Alizarin red S (red) staining for calcium, scale bar = 25 μm. (B) Shows that notochord sheath mineralization is independent of the sclerotome. The *osterix* (*sp7*) mutant Japanese medaka (*osx-/-*) lacks functional, sclerotome-derived, osteoblasts (Yu et al. 2017) and so neural and hemal arches fail to develop. Nevertheless, mineralization of the notochord sheath (black arrows) is unaffected; compare A and B. White arrows, non-miner-alized notochord sheath; asterisk, spinal cord. Whole mount Alizarin S (red) staining for calcium, scale bar = 25 μm. (C, D) Formation of intervertebral joints in wild type (WT) zebrafish embryos at 14 days post-fertilization. Production of retinoic acid by the notochord accounts for the segmental mineralization of the notochord (purple in C). In the spaces between vertebrae, chordoblasts switch on expression of the gene *cyp26b1* (green). Cyp26 enzymes prevent retinoic acid from spreading into prospective intervertebral joints thus pre-venting mineralization of the notochord sheath in the position of those joints (D). (C) Combined Alizarin red Cyp26 staining. (D) Single staining for the expression of Cyp26.

(Continued)

FIGURE 6.3 (*CONTINUED*) Scale bars in C, D = 50 μm. (E, F, G) Intervertebral spaces of adult Japanese medaka; see Figures 7.6 and 7.7 for anatomical details. (E) The notochord epithelium in the dorsal intervertebral space continues to proliferate, indicated by positive PCNA staining with antibodies (red arrowheads). In contrast, the flattened cells of the notochord epithelium in intravertebral positions are PCNA negative (black arrowheads). Scale bar = 20 μm. (F, G) show the dorsal and ventral intervertebral space, respectively. Cells of the notochord epithelium located in the intervertebral space (grey arrowheads) are positive for sclerostin (*in situ* hybridization), which is a potent inhibitor of bone formation (see text). This indicates that the notochord epithelium continues to prevent ossification of the intervertebral space. Dashed vertical lines mark the border between two vertebral centra. VB, bone of the vertebral body endplate. (A, B Images provided by PEW from a collaborative study with Christoph Winkler. C, D With permission from Pogoda, H.-M., et al. 2018. *Development* 145: dev159418. doi: 10.1242/dev.159418. E Image provided by PEW from a collaborative study with Doris W. Au. F and G from Ofer et al. 2019. *PLoS Biol* 17 (2). doi.org/10.1371/journal.pbio.3000140, used with permission from Ron Shahar).

mineralization of cartilage than it resembles bone formation. Arratia et al. (2001) designate the teleost vertebral centrum anlage as a *chordacentrum*. As a second step, discussed below, sclerotome-derived bone is deposited around teleost chordacentra (Fleming et al. 2004; Nordvik et al. 2005; Grotmol et al. 2003, 2005; Inohaya et al. 2007; Witten and Huysseune 2009; Bensimon-Brito et al. 2012a; Willems et al. 2012; Fleming et al. 2015; Arratia et al. 2001).

6.2   THE CHORDACENTRUM

Despite many studies that confirmed Huxley's and Kölliker's observations, and notwithstanding the fact that expert ichthyologists were always aware of the process of vertebral body formation in teleosts (François 1966; Schaeffer 1967; Laerm 1976, 1982; Arratia et al. 2001), large parts of the scientific community continued and continue to assume that vertebral body formation in teleosts follows the chicken and mouse model. This assumption can perhaps partly be attributed to the fact that studies with chicken and mice serve as the examples of vertebral body development in textbooks. Another source of bias may have been Edwin S. Goodrich's excellent and influential book *Studies on the Structure and Development of Vertebrates* (Goodrich 1930). Knowledge and drawings from Goodrich's book have made their way into essentially all biology textbooks ever since.

Concerning teleost vertebral body development Goodrich relied on a comprehensive review by Gadow and Abbott (1895). This treatise, however, does not take much notice of the observations from Huxley (1859), Kölliker (1859), Cartier (1875), Goette (1879) and Balfour and Parker (1882) concerning the early anlage of vertebral bodies in the notochord sheath. Gadow and Abbott (1895) know about Kölliker's studies but point out that the formation of vertebral body anlagen in the notochord sheath is essentially the same as perichondral formation of vertebral bodies. Furthermore, centra that are formed from the arches (cells that surround the notochord) are assigned to actinopterygians and tetrapods. Chordacentra (cells that derive from the arches and invade the notochord sheath) are assigned to dipnoans and chondrichthyans. The firm statement "We have to bear in mind the following considerations: No centra are formed either in the chordal sheath, or outside it, in the skeletogenous layer, without the previous existence of cartilage" (Gadow and Abbott 1895, p. 191) is definitely wrong for the largest group of vertebrates, the teleosts. This statement, however, shaped the view of large parts of the scientific community on vertebral column development (the non-involvement of the notochord) for more than a century up to the present day. Gadow and Abbott (1895) further emphasize that the bulk of the teleost vertebral centrum is formed from bone that is sclerotome-derived. This is certainly true. Mineralization of the notochord sheath only establishes the vertebral body anlage. Establishing a vertebral body anlage is, however, an important event in vertebral development. Recognizing this process clarifies our knowledge of vertebral development and may (and should) influence our views of the origin and evolution of the vertebral column (Figure 6.4).

François (1958, 1966; confirmed by Grotmol et al. (2003) described chordacentra formation in salmon, explaining that formation of the chordacentrum has been overlooked because, during development, chordacentra are quickly integrated into the bone that forms around the mineralized notochord sheath. Another reason for overlooking mineralization of the notochord sheath likely results from routine demineralization of the notochord sheath and bone around the notochord prior to detailed histological analyses: protocols usually include a step of demineralization. Also non-buffered formalin-based fixatives or Bouin's fixative remove minerals from the notochord sheath. Notochord sheath mineralization is also removed during routine whole-mount staining protocols. Standard double staining methods for bone and cartilage start with a highly acid Alcian blue solution for cartilage visualization. Minerals disappear easily from the notochord and only the bone that is formed around the notochord retains staining for minerals. The consequence is false negative staining of the mineralized notochord sheath (Bruneel and Witten 2015; Bensimon-Brito et al. 2016). The whole-mount staining-related demineralization problem has been addressed by Vandewalle et al. (1998) and also recognized by Springer and Johnson (2000) and Bird and Mabee (2003). A safe protocol is *separate* staining for bone and for cartilage (Vandewalle et al. 1998). Even *in vivo* staining of notochord sheath mineralization is now possible (Bensimon-Brito et al. 2016). Walker and Kimmel (2007) proposed an acid-free double staining protocol for the early stages of zebrafish skeletal development. Eventually, the publications of Angeleen Fleming (Fleming et al. 2001, 2004, 2015), publications in the wake of Kari Nordvik's Ph.D. thesis (Grotmol et al. 2003, 2005, 2006; Nordvik et al. 2005) and the increasing use of the Walker and Kimmel (2007) protocol reinstated the knowledge outside the ichthyological community that the first step of vertebral body centrum formation in teleosts is *segmented mineralization of the notochord sheath*.

The history of viewing vertebral column development in teleost fish justifies a discussion of whether mineralization of the notochord sheath is restricted to teleosts of whether it occurs in other vertebrate groups. Such a discussion leads us to the question of whether mineralization of the notochord sheath is a basal osteichthyan or an advanced teleost character. We address this issue in the next sections. First, we take a closer and detailed look at early vertebral body development in teleosts, a prime example of notochord involvement in vertebral body development (Section 6.3). Subsequently we look at the role of the notochord in other vertebrate groups to find that the role of the notochord in the development of vertebral bodies is larger and more extensive than usually assumed (Section 6.4).

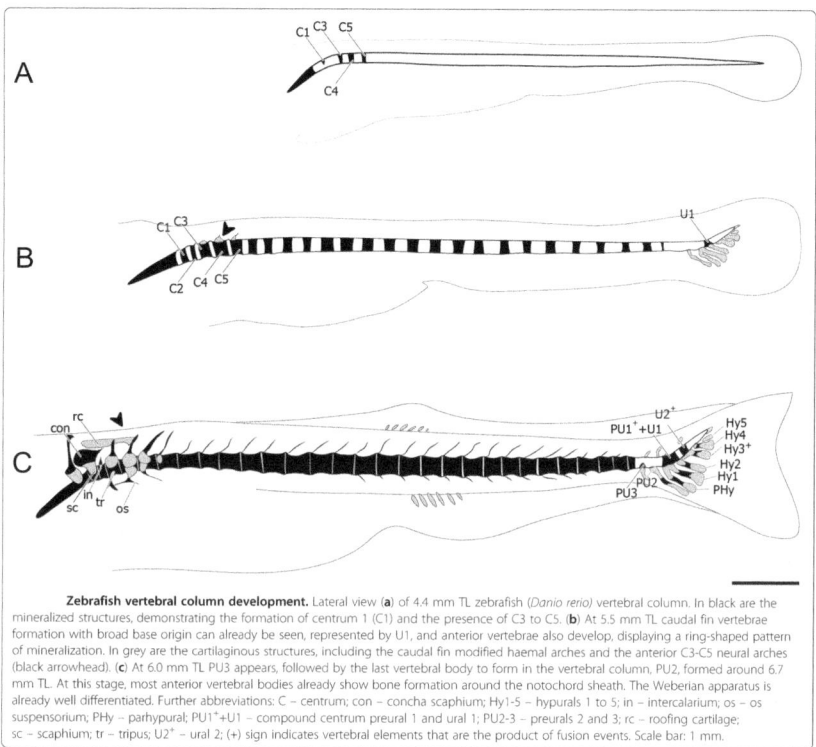

Zebrafish vertebral column development. Lateral view (**a**) of 4.4 mm TL zebrafish (*Danio rerio*) vertebral column. In black are the mineralized structures, demonstrating the formation of centrum 1 (C1) and the presence of C3 to C5. (**b**) At 5.5 mm TL caudal fin vertebrae formation with broad base origin can already be seen, represented by U1, and anterior vertebrae also develop, displaying a ring-shaped pattern of mineralization. In grey are the cartilaginous structures, including the caudal fin modified haemal arches and the anterior C3-C5 neural arches (black arrowhead). (**c**) At 6.0 mm TL PU3 appears, followed by the last vertebral body to form in the vertebral column, PU2, formed around 6.7 mm TL. At this stage, most anterior vertebral bodies already show bone formation around the notochord sheath. The Weberian apparatus is already well differentiated. Further abbreviations: C – centrum; con – concha scaphium; Hy1-5 – hypurals 1 to 5; in – intercalarium; os – os suspensorium; PHy – parhypural; PU1⁺+U1 – compound centrum preural 1 and ural 1; PU2-3 – preurals 2 and 3; rc – roofing cartilage; sc – scaphium; tr – tripus; U2⁺ – ural 2; (+) sign indicates vertebral elements that are the product of fusion events. Scale bar: 1 mm.

FIGURE 6.4 Development of the vertebral column and caudal fin endoskeleton in zebrafish as seen in lateral views, anterior to the left. (A) Segmented mineralization of the notochord sheath (black) at 4.4 mm total-length generates mineralized rings of centra one and three to five (C1, C3–C5). Notice the early mineralization of the anteriormost part of the notochord sheath (black) anterior to C1. (B) At 5.5 mm total-length mineralization of the notochord sheath delineates almost all centra along the body axis. The most anterior centra (C1–C5) show the double-cone shape evident when bone is deposited. The first element of the caudal fin skeleton, ural 1 (U1) has started to mineralize. (C) At 6 mm total-length, hemal and neural arches are present and almost all centra show the double-cone shape of mature vertebrae. Cartilage gray; notochord, white. Scale bar = 1 mm. The text in the figure contains additional information. For details of other skeletal elements see Bensimon-Brito et al. (2012a). (Reproduced with permission from Bensimon-Brito, A., et al. *BMC Dev Biol* 12: 28.)

6.3 IN HALF OF ALL VERTEBRATE SPECIES, NOTOCHORD SHEATH MINERALIZATION AND NOT CHONDROGENESIS CREATES THE ANLAGE OF THE VERTEBRAL CENTRUM

In large parts of the primary and secondary literature nowadays, teleost fish are represented by zebrafish (*Danio rerio*) and the Japanese medaka (*Oryzias latipes*). These species have increasingly been used as model organisms for vertebrate genetics, developmental biology and biomedical research, partly replacing the chick and mouse embryos used in previous decades (Witten et al. 2017).

Teleost fish hatch early from the egg and complete embryonic development outside the egg (Balon 2003).[1] Typical for vertebrate embryos, the notochord is a prominent structure at hatching. As in all vertebrates, the teleost notochord takes part in the development of the vertebral column but in a way that differs from birds and mammals. In teleost fish, vertebral centra lack a cartilaginous anlage. Instead, the notochord sheath mineralizes in intersegmental positions (Figures 6.3 and 6.4). In zebrafish, the first traces of minerals are detected close to the segment boundary. The intersegmental position of the mineralized vertebral body anlage is achieved by anterior and posterior extension of mineralization (Bensimon-Brito et al. 2012a). Mineralized sections of the notochord sheath and not the autocentrum nor sclerotome-derived cartilage represent the first vertebral centra in teleosts. Only in the second step of vertebral body development is sclerotome-derived bone deposited onto the mineralized notochord sheath, producing what are known as autocentra, and which are discussed in Section 6.3C below. Bone formation is direct (intramembranous); autocentra have no cartilaginous anlagen. This bone eventually forms the biconoid-shaped vertebral bodies (François 1966; Arratia et al. 2001).

In contrast to vertebral centra, sclerotome-derived hemal and neural arches are preformed in cartilage. Where this cartilage attaches to the vertebral centrum, additional bone can arise from the cartilage through endochondral bone formation. In species with small individuals such as zebrafish and Japanese medaka, vertebral arches form directly as bone and without a cartilaginous precursor. Only the base of each arch may articulate with the vertebral centra via cartilage vestiges. The lack of cartilage in zebrafish and medaka is most likely related to miniaturization (Witten et al. 2017). In species with small individuals, the contribution of endochondral bone formation to the vertebral centrum is reduced.

Studies on salmon vertebral body development have shown that rearrangement of the cells in the notochord epithelium occurs shortly before the first traces of notochord sheath mineralization can be detected. Cells in the prospective intervertebral joints elongate and arrange themselves perpendicular to the cranio-caudal axis. The cells also express *col11a2* (Wang et al. 2014). Cells in the region of the prospective vertebral body anlage (notochord sheath mineralization) are cubical and lack perpendicular orientation (Grotmol et al. 2003). Reorientation of notochord sheath cells prior to sheath mineralization has also been observed in Japanese medaka (Figure 6.5). Eventually the chordoblasts become highly proliferative, as confirmed by PCNA positive staining (Bensimon-Brito et al. 2012a) (Figure 6.3). By determining the location of the intervertebral joints and through segmented mineralization of the notochord sheath, development and patterning of the anlage of the vertebral column are completed.

Subsequently, not only in teleosts but in all groups of vertebrates — although only postcranially in birds (Piiper 1928) — notochord cells in the spaces between future vertebrae continue to proliferate. They further increase notochord sheath matrix production that contributes to the establishment of the intervertebral joints. In contrast, notochord sheath cells in the region of the teleost vertebral centra become less active. A gradient from high notochord cell activity in intervertebral spaces (future joints)

---

[1] Teleost embryos outside the egg are commonly but erroneously addressed as a series of larval stages. Calling a "freshly hatched" embryo a larva is clearly wrong; see Balon (1990, 2003) for a discussion.

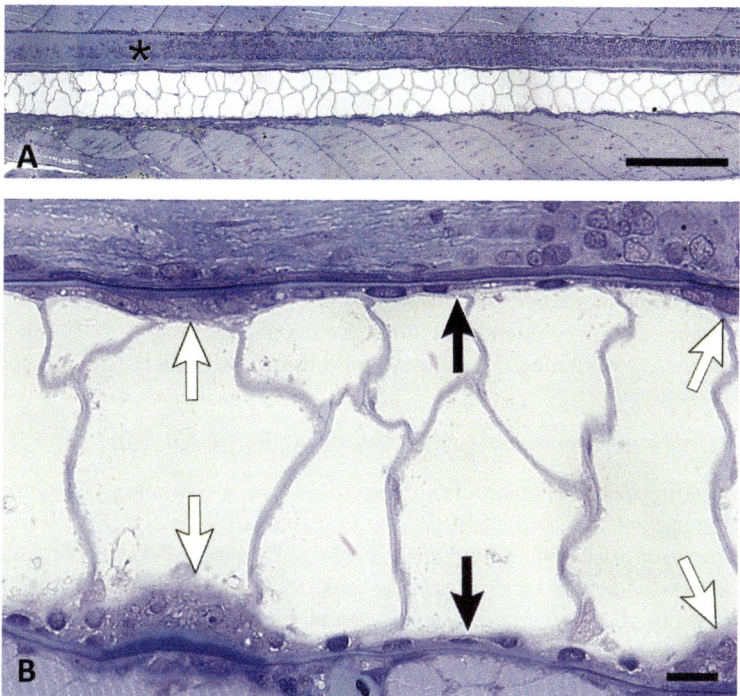

FIGURE 6.5   (A) Early stage of vertebral column development in Japanese medaka with the notochord as the prominent structure of the axial skeleton. Vertebral body formation is initiated by the segmented differentiation of the notochord epithelium. (B) Cells in the location of prospective intervertebral spaces proliferate, increase matrix production and prevent the mineralization of the newly formed matrix (white arrows). Cells in the location of prospective vertebral bodies retain a flat shape and do not increase the production of notochord sheath matrix (Black arrows). The cells facilitate the mineralization of the existing notochord sheath matrix with the expression of retinoic acid (see also Figure 6.3). Black asterisk, spinal cord. Medaka stage 40 (about 4.5 mm total length) according to Iwamatsu (2004). Scale bar in A = 100 μm. Scale bar in B = 20 μm. Epon embedding, Toluidine blue staining. (Figures provided by PEW.)

to low cell activity in the center of the vertebral body contributes to the formation of hourglass-shaped vertebral bodies (Figure 6.5); chordacentra and autocentra follow the shape of the notochord. The bone of teleost autocentra that eventually form the main part of the vertebral centra (cartilage in other vertebrate groups) is somite-derived. Again, hourglass-shaped vertebrae and autocentra, as the main part of the vertebral body, are common to all groups of osteichthyans including amniotes (Remane 1936; Williams 1959; Wintrich et al. 2020).

A. Are Notochord Signals Involved?

The existence of mineralized vertebral centra anlagen that develop in the notochord sheath raises the question of whether this process depends on signals from outside the notochord? Signals could come from somites or from sclerotome-derived scleroblasts.

Alternatively, the notochord could possess an inherent patterning mechanism; that arches and centra are developmental modules (Schaeffer 1967) supports the idea of an intrinsic notochord patterning mechanism. That said, cross talk between notochord and somites is well documented, including the requirement of *shh* signaling from the notochord to induce segmentation of presomitic mesoderm, illustrated in Figure 4.1. Likewise, and as summarized in Figure 4.2, somites signal to the notochord (Smits and Lefebvre 2003). Early signaling from the somites could indeed instruct segmented cell rearrangement and mineralization inside the notochord. Still, the notochord is a closed tube surrounded by a rigid multilayered (usually) acellular wall. Moreover, the notochord is under high internal pressure. This raises the further questions of how signaling molecules from outside the notochord could efficiently reach the notochord epithelium and by what mechanism they could influence notochord segmentation

B. How the Notochord Generates Vertebral Body Anlagen

Notwithstanding possible instruction from the somites, a number of arguments can be raised in support of the hypothesis that the *generation of vertebral body anlagen in the notochord must not depend on sclerotome derived cells*. Recent studies have elucidated the molecular mechanisms by which the notochord establishes vertebral body anlagen by the segmented mineralization of its own sheath.

Cells that form mineralized tissues in vertebrates produce in the first place an extracellular matrix that has the potential to mineralize (Witten et al. 2019; Cotti et al. 2020), mostly a collagen type 1-based matrix for bone and dentine matrix, mostly a collagen type 2-based matrix for cartilage and notochord. Removal of the mineralization inhibitor pyrophosphate facilitates both bone and notochord sheath mineralization. The well-documented alkaline phosphatase activity of osteoblasts is also a pyrophosphatase activity (Apschner et al. 2014). Like osteoblasts, the cells of the notochord epithelium also express alkaline phosphatase (Grotmol et al. 2005). In Atlantic salmon (*Salmo salar* L.), segmental expression of alkaline phosphatase activity by the notochord epithelium coincides with early notochord sheath mineralization. A corresponding expression of alkaline phosphatase and of osteocalcin isoform-1 has been shown in zebrafish (Bensimon-Brito et al. 2012a). Grotmol et al. (2005) concluded that the early expression of alkaline phosphatase is evidence for an instructive role of the notochord for the segmental patterning of the chordacentra: "chordoblasts initiate a segmental differentiation of the notochord sheath into chordacentra and intervertebral regions" (p. 427).

Pogoda et al. (2018) further showed that segmentally expressed retinoic acid can account for the metameric mineralization of the developing chordacentra. In the spaces between vertebrae — where mineralization does not occur — chordoblasts switch on expression of *cyp26b1*, which is a target and regulator of retinoic acid. In positions of notochord sheath mineralization and centra formation, chordoblasts down-regulate collagen 2a1a gene expression but do not inhibit retinoic acid, facilitating notochord sheath mineralization. There is evidence, again from Pogoda et al. (2018), that inhibition by Cyp26 enzymes prevents retinoic acid from spreading into prospective intervertebral joint regions is an evolutionary conserved mechanism (Figure 6.3). As has been shown for the Japanese medaka, the chordoblasts in intervertebral regions remain active in later stages. The cells express sclerostin (*sost*), a glycoprotein known to be secreted by osteocytes in response to mechanical loading. Sclerostin downregulates bone formation through antagonizing the Wnt/β-catenin signaling pathway (Ofer et al. 2019) and likely fulfils the same function in the intervertebral space (Figure 6.3).

To mineralize the notochord sheath chordoblasts segmentally express alkaline phosphatase that can remove the mineralization inhibitor pyrophosphate. Additionally, secretion of pyrophosphate itself is reduced by the segmented expression of ectonucleoside triphosphate/diphospho-hydrolase 5 (entpd5), a mineralization promoter that antagonizes ectonucleoside pyrophosphatase/phosphodiesterase 1 (enpp1) (Pogoda et al. 2018; Lleras-Forero et al. 2018).

Additional evidence for the autonomous action of chordoblasts in setting up the metameric mineralization of the notochord sheath comes from studies on zebrafish and medaka with ablated osterix/sp7-positive osteoblasts or with a genetic loss of Sp7 function.[2] In these animals, the development of autocentra and vertebral arches is compromised, but the chordacentra develop and mineralize (Figure 6.3, and see Willems et al. 2012; Yu et al. 2017; Kague et al. 2021). Similarly, in *fss/tbx24 mutant* embryos, somite borders do not form with the consequence that neural and hemal arches are malformed or missing. In these mutants, patterning and separation of hemal and neural arches is severely disturbed. At the same time, anlage and separation of vertebral centra remains unaffected. The phenotype of *fused somite* mutant zebrafish further supports the idea of a notochord-related patterning of vertebral body anlagen independent from somites (van Eeden et al. 1996; Holley 2007). Fleming et al. (2004) reached the same conclusion in their studies on zebrafish and showed that *in vivo* and *in vitro*, the mineralizing matrix of the notochord sheath is (i) the product of the notochord, and (ii) does not derive from osteoblasts outside the notochord. They further demonstrated that targeted laser ablation of notochord cells prevents centrum formation, i.e., prevents mineralization of the notochord sheath. So, in conclusion we finally understand the mechanisms that led to Huxley's and Kölliker's observations about the development of vertebral body anlagen in the teleost notochord sheath (Huxley 1859; Kölliker 1859).

C. VARIATIONS OF VERTEBRAL BODY CENTRA ANLAGEN FORMATION

The studies on salmon, zebrafish and medaka provide prime examples for notochord-based development of the vertebral centrum anlagen. Before these studies, chicken and mice were chosen as the models for vertebral development, assumptions being that basic processes of vertebral column development in these tetrapod models apply to almost all vertebrates. If salmon or zebrafish had been chosen as common models in developmental biology we would now have a different (textbook) view of early vertebral column development. Like tetrapods, teleosts are representatives of a highly derived branch of vertebrates. They are neither primitive nor basal vertebrates. Still, below, we argue that early steps of vertebral centrum formation in teleosts and in amniotes are more similar than commonly assumed.

---

[2] Sp7 is a zinc finger transcription factor essential for osteoblast differentiation.

Arratia et al. (2001), who reviewed the origin of the tissues that contribute to the formation of vertebral centra in gnathostomes, identified four basic types of vertebral body centra anlagen (1–4):

1. *Chordacentra*: Vertebral centrum anlagen develop in the notochord sheath, as discussed above. In chondrichthyans, mineralization not chondrification of the notochord sheath establishes chordacentra.
2. *Autocentra*: Bone forms directly around the notochord and without a cartilaginous template. This is the second step of vertebral body formation in teleost fish but is also considered as a general process of centrum formation in vertebrates (Owen 1866).
3. *Arcocentra*: Arches contribute cells to the development of the vertebral body that forms a cartilage around the notochord. The arcualia hypothesis views this type of vertebral centrum formation as ancestral. Arcocentra, however, are highly disputed, see below and Chapter 9.
4. *Holocentra*: Vertebral centra derive from cartilage that develops outside the notochord sheath to form complete templates of vertebral bodies that include centra and arches. A holocentrum is the vertebral body anlage in birds and mammals.

The concept "arcocentra" (3) is today highly disputed but remains presented in textbooks and is still used by parts of the scientific community (Ota et al. 2014; Cumplido et al. 2020). Arcocentra were introduced by Hans Gadow as part of his arcualia hypothesis (Gadow 1896). The hypothesis was rejected by Remane (1936) who criticized the lack of developmental evidence that would support the hypothesis; there is no evidence that the bases of arches contribute to the formation of vertebral centrum anlagen. Since then the arcualia hypothesis has been dismissed by most scientists (Danto et al. 2017) for osteichthyans and for tetrapods (Williams 1959; Schaeffer 1967; Andrews and Westoll 1970; Carroll 1988; Fleming et al. 2015). Ernest E. Williams was quite outspoken in his criticism:

> It is necessary to make clear, once and for all, that Gadow's hypotheses as applied to tetrapods are very remote from the facts, so remote that even a superficial adherence to his terminology complicates with false simplicities a story that, sadly enough, cannot, given the modern evidence, ever again have the tidiness seemingly given it by Gadow's four pairs of arcualia.
>
> *(Williams 1959, p. 2)*

Those who reject the arcualia hypothesis emphasize that vertebral body centra formation starts either with (i) a chordacentrum and the notochord sheath proper, or (ii) with an autocentrum as sclerotome-derived bone or cartilage that directly forms around the notochord sheath (in both scenarios independent from the arches).

The autocentrum as the basic process of vertebral centrum formation in vertebrates was proposed long ago by Gegenbaur (1862), Hasse (1892), Hay (1895) and by Müller (1938). The autocentrum can (i) consist of cartilage that is later transformed into bone, (ii) directly forms as bone without a cartilaginous precursor or

(iii) consist of both bone and cartilage (Williams 1959). Cartilage in the autocentrum of teleosts was possibly lost as a consequence of the early hatching of teleost embryos. In such a scenario, active swimming of the embryo (Balon 1990) requires and induces rapid notochord sheath mineralization (see Fiaz et al. 2012) to develop functional vertebral bodies as early in ontogeny as possible, followed by deposition of the bone of the autocentrum onto the mineralized notochord sheath. Welsch and Storch (1971) reached this conclusion to explain the presence of mineralized rings that develop in late embryonic stages around the notochord in the Koh Tao Island caecilian *Ichthyophis kohtaoensis* (Amphibia, Gymnophiona). A fascinating example that also underscores the hypothesis of Welsch and Storch (1971) comes to us from the fossil record. It illustrates the advantages to be had when an informed paleontologist incorporates a developmental approach in her/his analysis. Aystopods are a group of extinct, limbless, elongate reptiles with up to many hundreds of vertebrae. Aystopods and Palaeozoic tetrapods share two patterns of vertebral development, both involving the notochord. Vertebrae developed either from a perichondral tube around the notochord or from a medial notochord fibrous sheath, a mechanism that allowed more rapid ossification than does perichondral ossification (Carroll 1989).

Whether vertebral centra initially materialize as cartilage, bone or mineralized notochord sheath may in large part be a function of the speed of development. Wake and Wake (2000) compared vertebral body development between gymnophiones and mammals and concluded that the observed major differences can be attributed to vastly different developmental rates. Furthermore, large parts of the vertebral centrum develop in response to mechanical load which reflects a generalized adaptive response of sclerotome tissue to functional demands for vertebral consolidation (Laerm 1979a, b).

6.4 HOW COMMON IS NOTOCHORD SHEATH MINERALIZATION?

For technical reasons discussed above, and because mineralization of the notochord sheath is a developmental process and cladistic character that is hardly ever considered outside the ichthyological community, the literature about notochord mineralization in tetrapods is scarce. We have a report of mineralization of the notochord sheath in mammals from Williams (1908) who studied late notochord development in pigs. According to Williams, formation of vertebral centra starts with the mineralization of cartilaginous anlagen, while at the same time the notochord sheath mineralizes in embryos of 32 mm crown-rump length. A complete mineralized notochord has been identified in sauropod dinosaurs (brachiosaurus) of the Permo-Carboniferous. Werneburg (2008) concluded that, as shown in extant salamanders by Wake and Lawson (1973), the brachiosaur notochord became cartilaginous prior to complete mineralization (Werneburg 2008).

While early mineralization of the notochord sheath appears to be the rule for teleosts (half of all vertebrate species on the planet) it is also a pre-teleostean feature (Patterson 1968). We know that the notochord sheath can mineralize in cartilaginous fishes (chimeras, François 1966), and that notochord sheath mineralization is a neoselachien character, also known from the fossil record of this group (Daniel 1934; Maisey 2008). Mineralization of the notochord occurs in primitive neopterygian ray-finned fish such as Stylodontidae (von Zittel 1911, p. 106) and notochord sheath mineralization was present in Palaeozoic tetrapods (see Chapter 7 for further examples from stem-ward osteichthyans).

Indeed, notochord sheath mineralization has occurred and occurs outside the teleosts, respectively outside the actinopterygians. Could notochord sheath mineralization even be a tetrapod character? Gymnophiona from the genus *Ichthyophis* have ring-shaped mineralizations of the notochord sheath at a time when vertebral bodies are still entirely cartilaginous (Welsch and Storch 1971; Barteczko and Jacob 2002). The earliest recorded actinopterygian (ray-finned fish) vertebral centra in *Haplolepis* from the Carboniferous were made from mineralized notochord sheath (Patterson 1968; Gardiner 1984). Gardiner (1983) provides an account of Triassic actinopterygian groups with members (Palaeoniscids, Pholidopleurids and Halecostomi) that had a mineralized notochord sheath. *Polypterus senegalus*, placed by Bob Carroll at the base of the Palaeoniscoidea, starts vertebral centrum development with notochord sheath mineralization (Carroll 1988; Bartsch and Gemballa 1992). Notochord mineralization can also take place as partially or completely separated half rings (Patterson 1968; Schaeffer 1967), although this pattern is no longer a character found in adult teleosts. Separate dorsal and ventral mineralization of the notochord sheath are only observed during development, before ring-shaped notochord sheath mineralization is complete, after which the mineralized notochord sheath is encircled by intramembranous bone (Bensimon-Brito et al. 2012a).

A functional, but not developmental equivalent to "bony rings" that develop within the notochord sheath are bony rings that surround the notochord sheath (autocentra). In lobe-finned fishes (rhipidistians), such as *Rhizodopsis*, *Megalichthys*, *Ectosteohachis* and *Strepsodus*, vertebral centra are made from membrane bone around the notochord (Andrews and Westoll 1970; Gardiner 1984). If centra formation went through the same developmental steps as in teleosts, there would have been a faint mineralized notochord sheath as a substrate for that bone. Such bone rings are seen in the notochord of extant urodeles, in the northern two-lined salamander *Eurycea bislineata* (Wake and Lawson 1973) and in the common smooth newt *Triton vulgaris* (Mookerjee 1930). The published figures in Mookerjee (1930) are, however, difficult to interpret, although as Danto et al. (2019) report mineralization of the notochord sheath in urodeles, it may be that in *Triton vulgaris* mineralized notochord sheath rather than bone is present around the notochord. Schaeffer (1967) points out the difficulty of recognizing chordal centra in fossils once notochord mineralizations are completely covered by perichordal bone.

Our view about the significance of a mineralized notochord may have to be revised. Obviously, the notochord sheath of vertebrates has the capacity to mineralize but only few studies suggest that mineralization of the notochord sheath occurs in tetrapods and even in mammals. Few studies suggest that the notochord sheath can also chondrify (reviewed by Danto et al. 2019). In salamanders, caecilians and early tetrapods, vertebral centrum formation is induced by the notochord. In salamanders and caecilians, cartilage forms within the notochord sheath and this cartilage can ossify. Until we have more comparative data about notochord mineralization from vertebrates other than teleosts, one must conclude that *notochord sheath mineralization is chiefly a teleost character*. However, as in the mineralization of bone, mechanical load (tension) could initiate the process (Weinans and Prendergast 1996). The capacity of the notochord sheath to mineralize in response to load could be a shared gnathostome character and explain notochord mineralization outside teleosts.

6.5 DEVELOPMENT OF CARTILAGINOUS VERTEBRAL BODIES FROM THE NOTOCHORD

Chondrichthyans share with teleosts a chordacentrum in which the anlage of the vertebral centrum develops within the notochord sheath. Differing from teleosts, however, it is not mineralization but chondrification of the notochord sheath that establishes the anlage of the vertebral centrum, as illustrated in Figures 5.5 and 5.6.

Chondrichthyan vertebral bodies are cartilaginous; later in life the cartilage can mineralize but without undergoing endochondral bone formation. This results in heavily mineralized cartilaginous vertebrae with structures, properties and shapes similar to the vertebral bodies in bony vertebrates (Ridewood 1921). The connective tissue that surrounds the cartilage also can mineralize. Kölliker (1863) is very explicit about this connective tissue. To him, it is mineralized connective tissue, not bone. Other authors (Peignoux-Deville et al. 1982, 1985) have considered the possibility that this tissue has characters of bone; see Hall (1982), Debiais-Thibaud (2019) and Berio et al. (2021) for reviews of this topic. In some sharks, the lumen of the notochord can mineralize, as observed in the kitefin shark *Dalatias licha* (Bonnaterre 1788) listed by Kölliker (1859) as *Scymnus lichia*.

The evolutionary origin of the cells that develop inside the chondrichthyan notochord sheath into the anlage of the vertebral centra are the subject of a controversy that started prior to 1900. In the wake of the discovery of a notochord in amphioxus and in tunicates (as illustrated in Figures 2.1 and 2.3), the chondrichthyan notochord in the process of vertebral body formation received much attention. Because the cells that form the cartilage of chondrichthyan vertebral body anlagen are clearly located below the outer elastin layer of the notochord, Kölliker (1859) concluded that the cells must be notochord derived. He supports his view about notochord-derived cartilage with the fact that in skates (Rajidae) the anterior end of the notochord continues as cartilage. This cartilage (*chordaler Knorpel*) derives, according to Kölliker (1863), without doubt from the notochord sheath, see also Kryvi et al. (2020) for the cranial and caudal extension of the notochord in Atlantic salmon. Concerning the origin of the cartilage below the chondrichthyan notochord sheath, Klaatsch (1893a, b) and Gadow and Abbott (1895) reached different conclusions. They observed that cells (chondroblasts) from the bases of the cartilaginous neural and hemal arch anlagen acquire a flattened morphology and move into the notochord sheath. In this process, the outer elastic layer of the notochord sheath becomes fenestrated. The invasion of these cells into the notochord sheath has indeed been observed in examination of the three-month-old embryos of the small-spotted catshark, *Scyliorhinus canicula* and is discussed in Chapter 5 and illustrated in Figures 5.5 and 5.6.

Goodrich (1930) provides a well know description of the development of shark vertebral body anlagen. Since vertebrate chondrocytes have no cell processes and are motionless, the "chondrocytes" that move into the notochord sheath must be undifferentiated or dedifferentiate in order to be able to invade. Chondrocyte dedifferetiation and subsequent re-differentiation is not uncommon in vertebrates; it is a regular part of endochondral bone formation and it facilitates the separation of cartilaginous radials during teleost pectoral fin development (Roach 1997; Dewit et al. 2011; Yang et al. 2014; Park et al. 2015; Hall and Witten 2019).

Since the cells that transform the notochord sheath into cartilage appear to derive from the bases of the neural and hemal arches and not from the notochord itself, shark vertebral bodies can easily be taken as an example of arcocentra. This is, however, not the case. Even the proponent of the osteichthyan arcocentra hypothesis, Hans Gadow (Gadow 1896, p. 194), explicitly stated that

> The formation of chordacentra being independent of the arcualia explains how and why the number of centra does not necessarily agree with that of the arcualia or with that of trunk segments, e.g. *Hexanchus*, and tail of most other elasmobranchs.
>
> *(Gadow and Abbott 1895, p. 194)*

Šećerov (1911) provides an impressive account of the range of non-matching numbers of centra, arches and myosepta in sharks from the genera *Scyliorhinus* (catsharks), *Galeus* (catsharks), *Chiloscyllium* (bamboosharks), *Squatina* (angelsharks), *Heterodontus* (bullhead sharks), *Sphyrna* (hammerhead sharks), *Mustelus* (smoothhound sharks), *Hexanchus* (sixgill sharks), and the broadnose sevengill shark *Notorynchus cepedianus*.

Ridewood (1921) describes the process of elasmobranch vertebral centra formation inside the notochord sheath in detail:

> The notochord in the course of its development becomes invested by an envelope or sheath, produced by the chordal cells. This envelope in the case of Elasmobranch fishes undergoes differentiation into an external cuticular sheath, the membrana elastica externa, and an inner fibrous sheath, which is considerably thicker. On the differentiation of the sclerotomes from the myotomes the skeletogenous tissue applies itself to the sides of the notochordal sheath, but the right and left tracts of skeletogenous tissue are not continuous above and below the notochord. Skeletogenous cells now invade the fibrous sheath of the notochord through fenestration in the membrana elastica externa, particularly from the arch-bases, and produce a thickness of cartilage between the outer membrane (membrana elastica externa) and the structureless layer which is now recognisable as the membrana elastica interna. The rings thicken and lengthen, and develop into the vertebral bodies or centra, the chordal sheath in the intervening regions passing into a fibro-cartilaginous and ultimately into a fibrous condition, and persisting as the intervertebral ligaments.
>
> *(Ridewood 1921, p. 323)*

What Ridewood describes is an autocentrum; sclerotome-derived connective tissue forms around the notochord and contributes to centrum formation independent of the arches. In his words: "The neural and hemal arches of the vertebrae are developed independently of the centra. At a later period the arch-cartilages begin to differentiate in the upper and lower parts of the tracts, while the middle part of each tract (along the axis) becomes reduced, and thinned out into a layer of about two cells in thickness" (Ridewood 1921, p. 324). The contribution of the notochord to the vertebral body anlagen (chordacentra) is described by Klaatsch (1895) who points out that the cells that move into the notochord sheath do not produce cartilage matrix. They move into a matrix that is already present and was produced by the notochord epithelium. The result is a mixed cartilage with sclerotome-derived chondrocytes in a notochord-derived cartilage matrix; chordacentrum plus autocentrum.

The movement of somite-derived cells into the notochord sheath matrix is not restricted to elasmobranchs. Ridewood points out that this also happens but not to same extent in lungfish and holocephalians (chimeras, and the second extant group of chondrichthyans). Gloria Arratia and colleagues describe this similar but less pronounced process for the South American lungfish *Lepidosiren paradoxa* (Figure 5.6 and see Ridewood 1921; Arratia et al. 2001). Lungfish, however, do not develop vertebral centra and whether all cartilage cells in the notochord sheath are indeed sclerotome-derived remains to be elucidated. Interestingly, chondrocyte movement into a notochord generated cartilaginous matrix also takes place in chicken embryos, as discussed in Hall (1977): The chicken notochord actively produces extracellular matrix before sclerotomal cells begin to migrate. Therefore, the migrating cells are moving into an extracellular environment rich in collagen fibers and glycosaminoglycans (sulfated and unsulfated) produced by the notochord (Kosher and Lash 1975). Some of this matrix may indeed be integrated into the matrix that later surrounds the sclerotomal cells. Further studies on chicken and mice showed that collagen type II produced by the notochord not only hosts the sclerotome-derived cells but is required for their migration, survival, and chondrogenic differentiation (von der Mark 1980; Oettinger et al. 1985; Cheah et al. 1991; Sandell 1994; Ng et al. 1993). *In vitro*, somitic cells only differentiate into cartilage on a collagen type II-containing substrate (Minor et al. 1975) and *in vivo* the cells do not survive if the collagen type II-producing notochord is removed (Teillet and Le Douarin 1983). Even if the notochord is grafted dorsal to the somites it induces somitic cells to differentiate into cartilage (Pourquié et al. 1993).

In addition to providing the matrix, by the early 1970s it was known that the chicken notochord induces sclerotomal mesenchyme to chondrify as vertebral cartilage. Glycosaminoglycan-rich extracellular matrix accumulates around the notochord of chick embryos before chondrogenesis of the sclerotomal mesenchyme is initiated, leading to the proposal that these notochord extracellular matrix products progressively become the matrix of the vertebral cartilage (Frederickson and Low 1971; Strudel 1971; Corsin 1974). Such a proposal equates notochord with vertebral cartilage as a partial transformational series: notochord extracellular matrix becomes vertebral cartilage extracellular matrix but notochord cells do not become cartilage cells (chondrocytes). *The issue here is whether notochord (in this case the sheath) contributes to another tissue (vertebral cartilage), which is a different matter from whether notochord should be regarded as a cartilaginous tissue*, a conclusion that requires comparison of notochord cells as well as notochord extracellular matrix (taken up below and in Chapter 8).

6.6   A NEW START OF VERTEBRAL COLUMN DEVELOPMENT

From an evolutionary-developmental biology perspective, the question of the homology of vertebral bodies in different classes of vertebrates is linked to developmental processes and to what we consider as the onset of vertebral body development (Hall 2003). Gardiner (1983) emphasized that consideration about the vertebral centrum must begin with the origin and structure of the notochord and its sheath, which in form and function is identical and thus homologous in all vertebrates. The arcualia hypothesis considered the formation of cartilage outside the notochord as the primitive state of vertebral body development. Likewise, classical embryology considers condensation of somite-derived chondrogenic cells around the notochord as the start of vertebral body development (Figure 6.2). As the notochord is now recognized as active during vertebral column development, both earlier and later than previously acknowledged, it is possible to define more events as the manifestations of a start of vertebral column development. For example:

   i. Notochord signaling subdivides the presomitic mesoderm into somites.
   ii. Subdivision of the notochord into intra- and intervertebral regions by expansion of the notochord in the prospective intervertebral joints, a process that can be observed in all gnathostome groups.
   iii. Notochord signaling by the product of the gene *Brachyury* induces the differentiation of skeletal precursor cells in the sclerotome-derived tissue (autocentrum) that forms around the notochord in all gnathostomes.
   iv. Segmented (intersegmental) mineralization of the notochord sheath (chordacentrum) and or the segmented perinotochordal mineralization (autocentrum).
   v. Cartilaginous condensation of vertebral arch anlagen.
   vi. Formation of cartilage (later bone or mineralized cartilage) or of bone around the notochord.

Defining intervertebral spaces (future intervertebral joints) and mineralization of the notochord sheath are two notochord-based key events that pattern the vertebral column before cartilaginous or bony vertebral bodies develop. Even after ablation of the somitic clock the notochord is capable of fulfilling this role (Chapters 4 and 9). As vertebrates from all classes follow the above listed first three steps (i–iii) of vertebral body development these steps must then be considered as homologous. In terms of evolutionary change, patterning of the vertebral column to define the positions of vertebral centra and intervertebral joints, is an important a process as is the mode of vertebral centra formation.

After the pattern has been set, vertebral centra formation continues around the notochord sheath in essentially one of two ways: (i) by direct formation of membranous bone or (ii) by the formation of cartilage that can subsequently mineralize (chondrichthyans) or be replaced by bone.

It is interesting that *Col10a* is co-expressed with *osterix* by the osteoblast population that produces the bone around the notochord sheath in medaka (Renn et al. 2013; Seemann et al. 2015). In mammals, *Col10a* is typically expressed in mineralizing cartilage but absent from osteoblasts. *Col10a* expression in this osteoblast population may reflect the dual nature of these cells that can produce bone or cartilage that later mineralizes. An analysis by Debiais-Thibaud et al. (2019) demonstrates the ancestral association of *Col10a* with the mineralization of dermal and endoskeletal elements and further provides evidence that *Col10a* evolved prior to the divergence of osteichthyans and chondrichthyans. If so, differences concerning the involvement of cartilage or bone in vertebral centrum formation are the consequence of descent with modifications and not characters that evolved independent in different lines of vertebrates.

6.7 SUMMARY

Several unsubstantiated and therefore incorrect conclusions concerning the notochord are addressed in this chapter. The first, despite much evidence to the contrary, is that the notochord is a transitory embryonic structure, merely a placeholder for the future vertebral column. In the majority of vertebrate species, the notochord remains as a continuous structure along the length of the vertebral column throughout the life of the individual. The second is the assumption that there is a one-to-one relationship between somites and vertebral bodies. On the contrary, a process of resegmentation or leaky resegmentation results in the cells that form an individual vertebra arising from two or more adjacent somites. The third, and widespread misunderstanding, is that vertebral centra and vertebral arches are entirely derived from somite-derived sclerotomal cells. This was known not to be true over 150 years ago. Why? Because in teleosts, which comprise half of all vertebrate species, notochord sheath mineralization and not chondrogenesis creates the anlage of vertebral centra. Much of this chapter deals with the evidence for this statement and documents the existence of four basic types of anlagen for vertebral body centra; in the notochord sheath; as bone around the notochord; as cartilage around the notochord, or as cartilage that also forms the vertebral arches.

7 The Notochord in Adult Vertebrates

CONTENTS

7.1 INTRODUCTION

A common misconception repeated in many textbooks and scholarly articles is to view the notochord only as an embryonic organ, a transient structure that disappears in the course of development. The notochord indeed disappears in tunicates during metamorphosis. In cephalochordates, it turns into muscle tissue (Welsch 1968). In all vertebrates without complete vertebral bodies the notochord maintains its original function, which is to provide axial support as a non-constricted hydroskeleton (Schmitz 1998a; Koob and Long 2000). In vertebrates that develop complete vertebral bodies, a notochord is maintained in juveniles of most groups and often also in adults. In reptiles (mostly squamates), the notochord becomes constricted first in the intervertebral regions and remains relatively wide in the middle of the centrum well into post-natal life. *Sphenodon* and the geckos maintain a prominent notochord also as adults (Winchester and Bellairs 1977). In birds, the notochord is an embryonic structure only (Figure 7.1). No notochord remains are present in adult birds. In actinopterygians (with the exception of *Lepidosteus*, see Figure 4.1) and in mammals, the notochord and/or notochord-derived cartilaginous tissue becomes a functional component of the intervertebral joints. The notochord remains fluid-filled and functions in the intervertebral joint as a hydroskeleton (Nordvik et al. 2005; Ellis et al. 2013). Discussion of the fate of the notochord in adult vertebrates and of the different cellular mechanisms associated with notochord retention/modification in those adults is the topic of this chapter.

    A vertebral column is a flexible system that consists of vertebral bodies, intervertebral joints and ligaments (Symmons 1979). The importance of the intervertebral

DOI: 10.1201/9781315155975-10

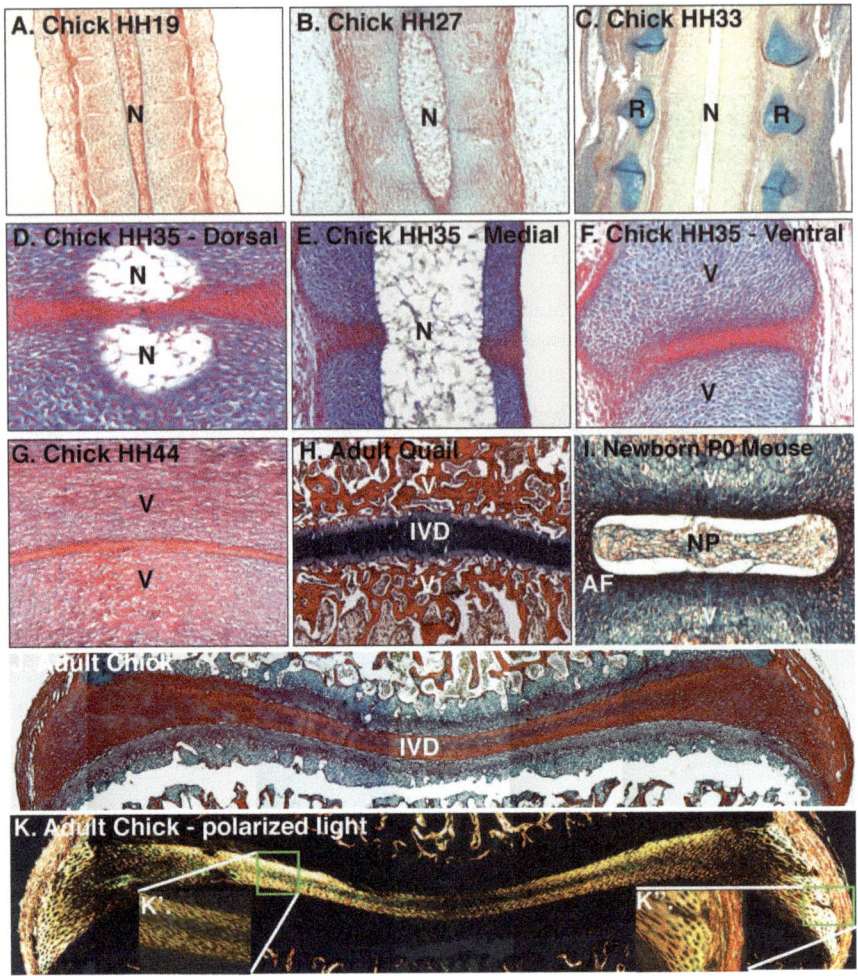

FIGURE 7.1 Analysis of bird intervertebral disk formation that shows the replacement of the notochord by intervertebral cartilage according to Bruggeman et al. (2012). Picrosirius red and Alcian blue were used to determine the structure of the chicken intervertebral disk beginning at Hamburger and Hamilton stage (HH) 19 through adulthood (Hamburger and Hamilton 1951). (A–F) At HH19 (A), HH27 (B), and HH33 (C) no disk structures were observed. At HH35 formation of a "ring" of collagen was found in the intervertebral space between adjacent vertebrae, shown as sections through the same disk in D–F. (D–F) Sections through the same disk. (E) The notochord was present in the middle of the ring of collagen. (G) At HH44, a thin layer of collagen was located between adjacent vertebrae. (H) In adult quail, the intervertebral disk was composed of glycosaminoglycans (GAG). (I) Intervertebral disks from newborn mice contained a nucleus pulposus surrounded by an annulus fibrosus. (J) Adult chickens did not contain a nucleus pulposus. (K) An analysis of collagen fibers using polarized light revealed that the middle region of the adult chicken disk was enriched for collagen III (K′). Collagen I/II was found at the ends of the disc and surrounding collagen III (K″). N, notochord; R, rib; V, vertebrae; IVD, intervertebral disk; NP, nucleus pulposus; AF, annulus fibrosus. (Figure and figure legend text from Bruggeman, et al., 2012. *Dev Dyn* 241: 675–683.)

joints for the function of the system is all too well known to humans who have suffered a hernia, which is the disruption and protrusion of the notochord out of the annulus fibrosus and which exerts pressure on the spinal nerve roots (Lotz et al. 1998; Hunter et al. 2003). Humans who suffer from a hernia receive a painful reminder of the existence of their own notochord and of the 400 million years of history they share with their gnathostome ancestor.

Among the diapsid reptiles that arose in the late Permian 250 MYA and include crocodilians, lizards, snakes, turtles and birds, birds are exceptional in that they do not preserve a notochord or notochord-derived structures in postnatal life stages. There is no nucleus pulposus and no intervertebral and intravertebral preservation of the notochord (Figure 7.1). In those bird species that have been studied in detail, the notochord is lost in a process in which cartilage precursor cells invade intervertebral spaces (future vertebral joints) and thus replace the notochord. The same process can be observed in amphibians and reptiles, but in birds, the invading cartilage completely replaces the notochord (Figure 7.2G and H). The precursor cells of this cartilage derive (according to a study by Bruggeman et al. 2012) from the rostral half of each sclerotome. According to Mittapalli et al. (2005), the cells can be traced back to somitocoel cells. They are located in the posterior part of the sclerotome close to the intervertebral fissure (von Ebner's fissure). The part of the sclerotome that gives rise to the articular cartilage is designated as the arthrotome (Senthinathan et al. 2012) (Figure 6.2).

Independent from the discussion about the exact origin of the cells within the sclerotome, also in birds, the signaling factors that attract the cartilage precursor cells and the early extracellular matrix for this cartilage are products of the notochord (see Chapter 8). A second misconception, admittedly also nursed in parts of this book, is that a notochord always consists of large vacuolated cells. Only a minority of vertebrate species maintains a differentiated notochord with typical vacuolated cells as adults. Differentiated notochords with large vacuolated cells are typical for agnathans and for early and some basal osteichthyans that have, for other reasons, no vertebral centra, incomplete centra or non-constricted notochords (Hay 1895) (Figure 7.2A and B). Past this stage, if vertebral centra are present, vacuolated notochord cells further differentiate. In most species, the vacuolated notochord cells differentiate either into cartilage cells, cartilage-like cells (nucleus pulposus) and/or the cells condense into a fibrous connective tissue which provides space for extracellular fluid-filled vacuoles (Romer 1956; Figure 7.2).

7.2 NON-CONSTRICTED NOTOCHORDS IN ADULT VERTEBRATES

A non-constricted notochord occurs in those vertebrates that (i) lack vertebral centra altogether, (ii) have only partly developed centra, or (iii) have centra that are simply ring-shaped (Figure 7.2A and B).

Extant jawless vertebrates (lampreys and hagfishes) possess lifelong a non-constricted notochord as the principal axial skeleton. The first description of a notochord in jawless vertebrates comes from Guilelmus Rondelet. In 1554/1555, he reported that the sea lamprey, *Petromyzon marinus*, back then classified as an elongated slimy cartilaginous fish, has a continuous chorda instead of a vertebral column

The Fate of the Notochord as Part of the Vertebral Column in Vertebrates

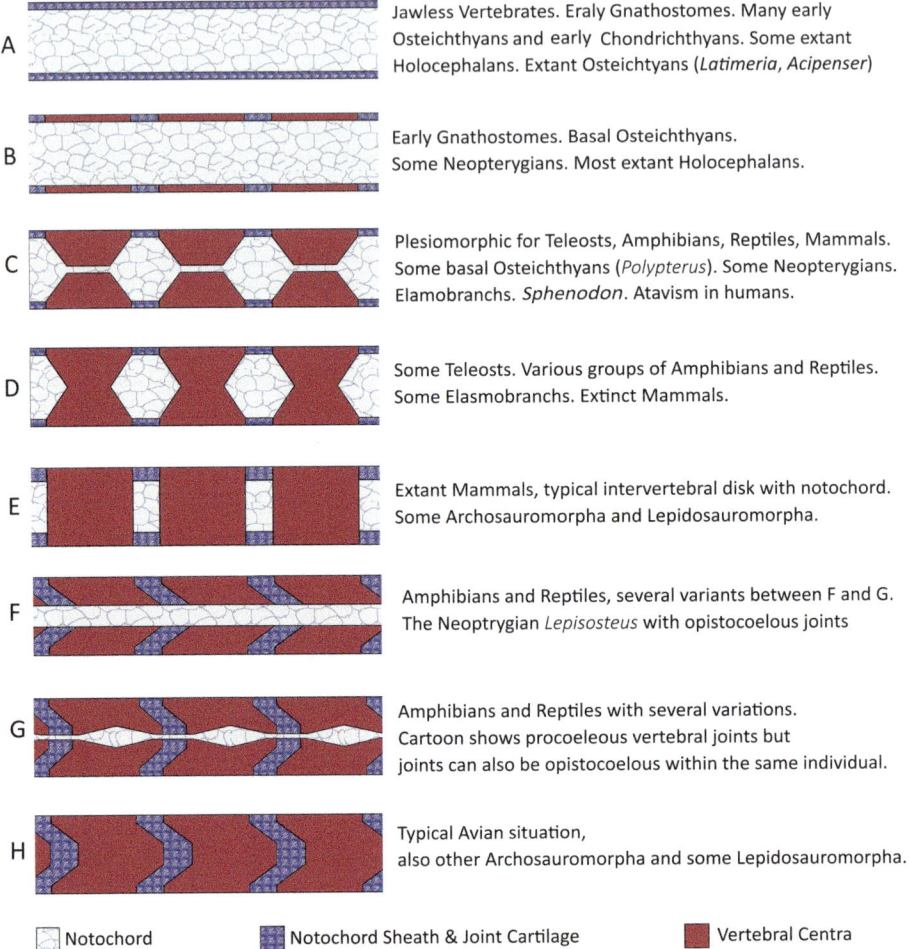

A — Jawless Vertebrates. Eraly Gnathostomes. Many early Osteichthyans and early Chondrichthyans. Some extant Holocephalans. Extant Osteichtyans (*Latimeria, Acipenser*)

B — Early Gnathostomes. Basal Osteichthyans. Some Neopterygians. Most extant Holocephalans.

C — Plesiomorphic for Teleosts, Amphibians, Reptiles, Mammals. Some basal Osteichthyans (*Polypterus*). Some Neopterygians. Elamobranchs. *Sphenodon*. Atavism in humans.

D — Some Teleosts. Various groups of Amphibians and Reptiles. Some Elasmobranchs. Extinct Mammals.

E — Extant Mammals, typical intervertebral disk with notochord. Some Archosauromorpha and Lepidosauromorpha.

F — Amphibians and Reptiles, several variants between F and G. The Neoptrygian *Lepisosteus* with opistocoelous joints

G — Amphibians and Reptiles with several variations. Cartoon shows procoeleous vertebral joints but joints can also be opistocoelous within the same individual.

H — Typical Avian situation, also other Archosauromorpha and some Lepidosauromorpha.

Notochord   Notochord Sheath & Joint Cartilage   Vertebral Centra

FIGURE 7.2   This set of diagrams shows, in longitudinal sections, the fate of the notochord as part of the vertebral column in postnatal and adult vertebrates. Groups in which particular patterns are found are shown beside each of images A–H (see text for cross references to pattern A–H). Blue structures represent both the notochord sheath and the tissues of the intervertebral joints. Joint tissue can derive from the notochord sheath, from notochord-derived cartilage, from somite-derived cartilage or from both notochord-derived and somite-derived cartilages. Joints are absent if centra fuse as part of normal development. Examples are the sacrum (fused sacral vertebrae in mammals), the notarium (fused vertebrae of the shoulder girdle in birds and some pterosaurs) and the synarcual (fused anterior vertebral bodies in batoids). Red structures represent vertebral centra which can consist of the mineralized notochord sheath, of cartilage (also notochord-derived cartilage) and/or of bone that develops from the autocentrum around the notochord (somite-derived). The notochord itself can be hollow, contain vacuolated chordocytes, contain cartilaginous cells (true cartilage or cartilage-like tissue, or can contain condensed fibers (that may mineralize) combined with extracellular vacuoles. (Figure by PEW.)

(Rondelet 1554/1555, pp. 398–399). In contrast to other vertebrate groups where science chiefly focuses on characters of early notochord development, studies have been dedicated to the notochord of adult jawless vertebrates (Schwarz 1961; Welsch et al. 1998).

Members of several stemward gnathostome groups such as holocephalans (ratfishes), dipnoans (lungfishes), coelacanths (crossopterygians), and sturgeons (chondrosteans) have a continuous non-constricted notochord (Maisey 2000; Arratia et al. 2001; Greenwood 1988; Schmitz 1998a; Leprévost et al. 2017).

Gnathostomes without vertebral bodies typically possess cartilaginous basal elements of the neural and hemal arches (basidorsals and basihemals). These elements abut the notochord sheath but do not grow around the notochord to form complete vertebral centra (Lauder 1980). Vertebrate groups with this design include species with large individuals. The holocephalan *Chimera monstrosa* (rabbit fish, rat fish) can grow to 1.25 m length (Froese and Pauly 2019). *Latimeria chalumnae* (coelacanth) reaches 2.0 m and a weight of 95 kg (Froese and Pauly 2019). Smith (1953) describes the notochord of the second coelacanth discovered. This individual had a length of 1.39 m and a massive notochord with a total diameter of 56 mm (internal diameter 40 mm, notochord sheath thickness 8 mm). Meunier et al. (2019) show a 2 mm-long coelacanth with a total notochord diameter of about 64 mm (Figure 7.3). The Australian lungfish *Neoceratodus forsteri* reaches from 1.1 m to a maximum of 1.7 m and a weight of 40 kg (Arratia et al. 2001; Froese and Pauly 2019). The known growth record for the white sturgeon *Acipenser transmontanus* is 6.1 m and the species' weight record is registered as 817 kg (Froese and Pauly 2019). The largest known extinct primitive teleost *Leedsichthys* (Pachycormiformes) from the Middle Jurassic of England reached a length of at least 12 m without solid vertebral bodies (Maisey 1996). Notochords as the main axial skeleton in these specimens reach an impressive size. It has been shown that the central axial structure of the sturgeon notochord, the funiculus, becomes mineralized with age (Figure 7.4). Our knowledge about function, growth and metabolism of large non-constricted notochords is still limited. Definitely, these organs function differently from notochords that we picture as early and transient embryonic structures. Notochords with a low level of constriction and non-constricted sections also can be found in adult actinopterygian deep-sea fish from the order Stomiiformes, which contains over 400 species. This relates to a decreased level of vertebral centrum ossification or to the secondary lack of postcranial vertebral bodies. In the deep-sea dragonfishes from the family Stomiidae, the uncovered and thus flexible postcranial notochord forms a functional head joint (Schnell et al. 2010; Schnell and Johnson 2017; Germain et al. 2019) (Figure 7.3).

A. The Naked Notochord of Jawless Vertebrates

Jawless vertebrates not only lack jaws, a third semicircular canal and a horizontal myoseptum, they also lack vertebral centra. The cellular composition of their non-constricted notochords resembles the composition of a typical gnathostome notochord prior to vertebral centra formation. The notochord sheath has a basal membrane, a thin inner elastin layer, a thick layer without cells, densely packed collagen fibrils and an outer elastin layer (Welsch et al. 1998). A study on the

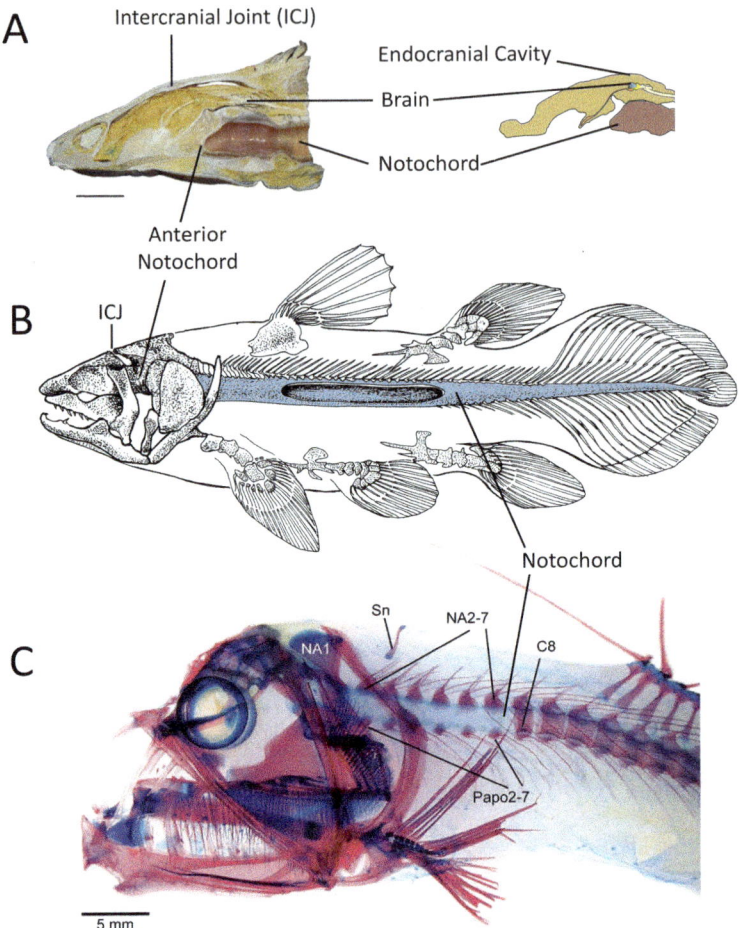

FIGURE 7.3 Notochords without vertebral bodies are found in a limited number of extant osteichthyan taxa. (A, B) The Coelacanth *Latimeria chalumnae*; (C) the deep-sea Sloan's viperfish *Chauliodus sloani*, a teleost from the order Stomiiformes. (A, B) The notochord of *Latimeria* is massive and hollow and functions as the prime axial skeleton. The lumen contains a single-layered notochord epithelium but no chordocytes. Notice the enormous size of the notochord, particularly the anterior notochord in the bisected head (A). The extended anterior notochord stands in contrast to the extremely small size of the brain. In other vertebrates, the anterior most notochord is typically narrow and pointed (see Figure 3.4). (C) *Chauliodus sloani* represents a condition that typifies Stomiid genera of deep-sea ray-finned fishes. The postcranial notochord lacks vertebral centra which increases the mobility of the animal's head. Shown in the cleared and stained whole mount (C) is an example of the loss of vertebral centra; NA2-7 label the remaining neural arches, C8 the most anterior centrum. Complete loss of centra seen in *Latimeria* is discussed in Section 7.2B. Scale bars in *A* and *C* = 5 cm. B, not to scale. Adult coelacanths can reach 2 m in length. ((A) Modified after Dutel, H., et al., 2013. *Naturwissenschaften* 100: 1007–1022 and Dutel, H., et al., 2019 *Nature* 569: 556–559. (B) Modified after P. H. Greenwood, P. H., 1988. *A Living Fossil Fish. The Coelacanth Latimeria chalumnae*. British Museum (Natural History), London. (BMNH) ©The Trustees of the Natural History Museum, London, used with their permission. C, from Schnell, N. K., et al., 2010. *J Morph* 271: 1006–1022, image taken by Ralf Britz)

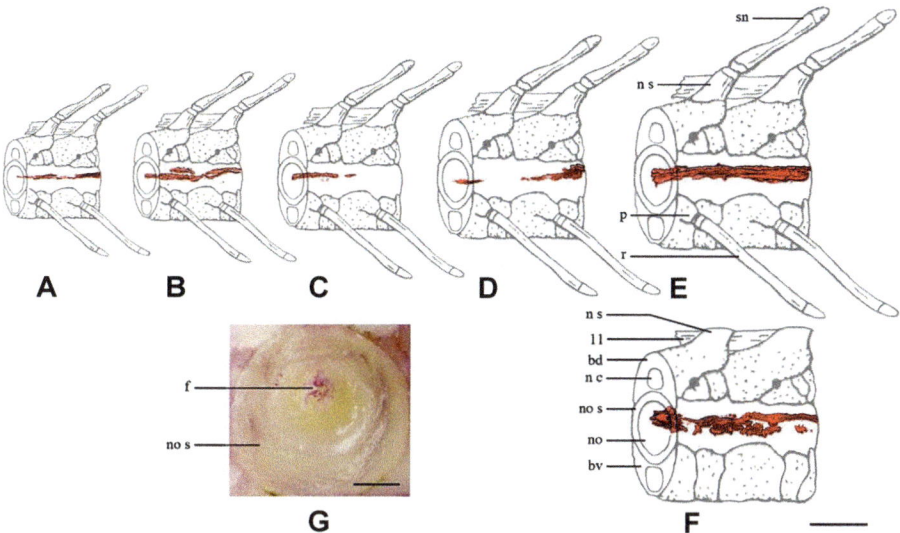

FIGURE 7.4   3D representation in lateral view (A–F) of a pair of vertebrae in a growth series in 2–20-year-old (70–144 cm total length) Siberian sturgeon (*Acipenser baerii*), anterior to the left. Mineralized structures (shown in red) within the notochord were detected by X-ray microtomography A–F. (A) Two-year-old, 70 cm total length (TL). (B) Three-year-old, 80 cm TL. (C) Five-year-old, 88 cm TL. (D) Seven-year-old, 108 cm TL. (E, F) Twenty-year-old, 144 cm TL. (A–E) Abdominal vertebrae; (F) caudal vertebrae. (G) Photograph of a transverse slice through the abdominal regions of the notochord of the 7-year-old specimen depicted in D. Alizarin red staining of this specimen shows the mineralized area of the notochord in the funiculus (f). Abbreviations in E and F: bd, basidorsal; bv, basiventral; ll, longitudinal ligament; nc, neural canal; no, notochord; no s, notochord sheath; ns, neural spine; p, parapophysis; r, rib; sn, supraneural. Scale bars: *A–F* = 2 cm, *G* = 1 cm. (From Leprévost, A., et al., 2017. *J Morph* 278: 1586–1597.)

development of the elastic fibers in ammocoete larvae of lampreys suggests that, as in gnathostomes, elastic fibers develop from oxytalan fibers via elaunin fibers (Schinko et al. 1992). In adults, the vacuolated chordocytes are surrounded by a persisting notochord epithelium. The cells of the notochord epithelium contain a large amount of endoplasmic reticulum, indicative of metabolically active cells. The cells of the notochord epithelium continue to give rise to chordocytes with intracellular vacuoles. The metabolic activity of the vacuolated chordocytes is indicated by the presence of numerous mitochondria. Both cell populations are designated as epithelial cells by Schwarz (1961) chiefly based on the presence of intermediate filaments which contain cytokeratins. Vacuolated chordocytes are firmly connected to each other by numerous desmosomes, indicative for a high amount of mechanical load that the cells have to withstand. Indeed, extant agnathans show extensive coiling of their body and even the formation of knots. The cell population inside the notochord allows these extreme bending movements and withstands compression and stretching. Caveolae — membrane structures that allow extreme cell compression and stretching in gnathostome notochords — are, however, missing (Welsch et al. 1998; Lim et al. 2017).

B. Notochords of Extant Sarcopterygians

The structure and cellular composition of non-constricted notochords in adult sarcopterygians has been compared with embryonic, respectively larval, notochords of other vertebrates. According to Locket (1980) the principal structures of the premetamorphic bullfrog notochord (*Rana catesbeiana*), as described by Bruns and Gross (1970), are present in the adult lungfish notochord. The main difference is that the constituents in lungfish are correspondingly larger. Concerning the structure of the notochord and its sheath, there are no significant differences between the two lungfish genera *Lepidosiren* and *Protopterus* (Arratia et al. 2001). Schmitz (1998a) provides a detailed description of the notochord in adult West African lungfish *Protopterus annectens*. The notochord sheath is massive with an outer elastic membrane and contains cells that have the shape of fibroblasts (see also Figure 5.6). Cells with this shape are also present in the notochord sheath of extant coelacanths (*Latimeria chalumnae*). At its inner side, the notochord sheath of *Latimeria* has a single-layered notochord epithelium. Different from the notochord epithelia of other vertebrates, the cells of the notochord epithelium contain little endoplasmic reticulum and only a few small Golgi vesicles Schmitz (1998a). The notochord lumen of *Protopterus annectens* is filled with vacuolated cells. Arratia et al. (2001) show cartilage inside the caudal notochord of a juvenile (30 cm length) *Protopterus* (see also Chapter 8 for caudal notochord to cartilage transition). In the Devonian lungfish taxa *Dipterus*, *Scaumenacia*, *and Barwickia*, the caudal part of the notochord was mineralized, whereas the anterior part remained non-mineralized. Mineralization started in the region of the second caudal fin and extended to the caudal tip of the notochord (Johanson et al. 2009). The cells inside the notochord of the extant lungfish *Protopterus annectens* are connected with desmosomes, interdigitate and caveolae are present. The very center of the notochord contains a strand of condensed chordoblasts, orientated perpendicular to the dorsoventral axis. Schmitz (1989a) describes this strand as made from collapsed vacuolated cells and identifies the structure as a funiculus (a bundle of fibers) following Schauinsland (1906). According to Hasse (1893), a funiculus is also present in teleosts and elasmobranchs. Leprévost et al. (2017) describe a funiculus in a basal actinopterygian, the Siberian sturgeon *Acipenser baerii*. The funiculus is also called the notochord strand (von Ebner 1896; Schaffer 1930), a common name for this structure in adult teleosts (Figures 7.4, 7.6, and 7.7).

Lungfish share the non-constricted notochord and the absence of vertebral centra with fossil and extant coelacanths, the latter represented by *Latimeria chalumnae* (Figure 7.3). A non-constricted notochord is also found in most early gnathostomes such as acanthodians, placoderms and in early actinopterygians (Laerm 1979a; Forey 1991; Greenwood 1988). Smith (1953) describes the notochord of the second coelacanth specimen found as massive and entirely cartilaginous. This would suggest that the notochord lumen was filled with cartilage.

Cartilage inside the notochord is found in the lungfish *Protopterus*, in adult amphibians and in adult reptiles. Thus, it would perhaps not be surprising to find a cartilaginous notochord in *Latimeria*. The specimen examined by Smith (1953) must have had a cartilaginous notochord or the description refers only to the massive notochord sheath. Curiously, later examinations of *Latimeria*'s notochord did not find cartilage inside the notochord. Forey (1991) portrays the notochord as a thick and relative inflexible rod of constant diameter. The massive notochord sheath has indeed a density that is very similar to the density of cartilage (Schultze and Cloutier 1991) and is surrounded by a thin outer elastic sheath (Schaeffer 1967).

Locket (1980) points out that differing from the notochord in lungfish, the notochord lumen of *Latimeria* is not only deprived from cartilage, it is hollow and completely fluid-filled. Different from the notochord epithelium of *Protopterus*, the cells of *Latimeria*'s monolayered notochord epithelium contain extensive endoplasmic reticulum and pinocytotic vesicles, indicative of active fluid transport across the cell membranes. Cells, others than cells of the notochord epithelium, are only present in the caudal-most part of the notochord (Locket 1980). Although most segments of the notochord are hollow, *Latimeria*'s notochord is internally subdivided by transverse divisions. Schultze and Cloutier (1991) identified the divisions based on Nuclear Magnetic Resonance Imaging and concluded that the divisions represent the convolutions described earlier by Millot and Anthony (1958). The structure of the notochord sheath of *Latimeria* follows that of the other vertebrates with an inner elastic membrane, a massive fibrous sheath made from circumferentially arranged collagen fibers and an outer elastic membrane (Mathews 1975). As in *Protopterus* and *Neoceratodus*, the fibrous sheath of *Latimeria* contains cells that are described as fibroblasts (Locket 1980) (Figure 7.3A and B).

Unique among vertebrates is the enormous extension of the cranial portion of *Latimeria*'s notochord (Dutel et al. 2019) (Figure 7.3). In gnathostomes, the anterior tip of the notochord is typically reduced in size and retracts from the hypophyseal region early in development. Not so in *Latimeria*. Here, the anterior notochord undergoes a proportionally greater degree of expansion than the brain. In fact, the brain of *Latimeria* is extremely small compared to the anterior notochord because the extended notochord fills up the space behind the hypophyseal fossa that, in other gnathostomes including lungfish, is occupied by gray matter. The diameter of the anterior notochord in *Latimeria* is about nine times larger than the diameter of the rhombencephalon. In *Polypterus*, the rhombencephalon has about eight times the diameter of the anterior notochord. Dutel et al. (2019) conclude that the expansion of the notochord probably causes major spatial packing constraint on the brain; *Latimeria* has notochord volume where other vertebrates have brain volume. Furthermore, the anterior expansion of the notochord may underpin the unique crossopterygian division of the neurocranium, the basis for the development of the crossopterygian intracranial joint (Forey 1991) (Figure 7.3).

The composition of *Latimeria*'s viscous or gel-like notochord fluid has been analyzed by Griffith et al. (1975) based on a live caught female specimen. The osmolarity of the notochord fluid (1,058 mOsm) exceeds that of serum (942 mOsm). Compared to serum increased levels of potassium, chloride, urea, trimethylamine oxide and total free amino acids were measured. Lower concentrations, compared to serum, were measured for sodium, magnesium, calcium, bicarbonate, sulfate, total carbohydrates, glucose, lactate, cholesterol, bound phosphate and total proteins. Inorganic phosphorus was essentially identical. *Latimeria*'s notochord fluid also contains a matrix of fibers which are 10 nm in diameter and of variable indefinite lengths. The fibers resemble a sialoglycoprotein that is stabilized by disulfide bonds. In addition, cellular debris was found in the fluid. Unfortunately, *Latimeria*'s notochord fluid has given rise to a dangerous fallacy: the rarity, longevity and survival of the coelacanths for millions of years created a superstition that the notochord fluid will work as a life-prolonging elixir. This superstition has established a black market for *Latimeria*'s notochord fluid (Balon 1991). One wonders if consumers who believe in the power of *Latimeria*'s notochord fluid also consider that the fluid could shrink their brain (Figure 7.3).

C. NON-CONSTRICTED NOTOCHORDS AND VERTEBRAL CENTRUM REGRESSION

According to Arratia et al. (2001), a vertebral column consisting of a persistent notochord and ossified arcocentra is the primitive condition for gnathostomes (see Chapter 9 for a discussion of this concept). George Lauder's scenario for the primitive gnathostome condition is an unrestricted notochord with small accessory perichondral ossifications and notochord calcifications (Lauder 1980). The phylogenetic trend of the loss of vertebral centra and consequently preservation or reappearance of a non-constricted notochord within the osteichthyans is not easy to interpret. Functionally, it is possible for primary aquatic gnathostomes to substitute a vertebral centra-based vertebral column with a notochord-based vertebral column (Witten et al. 2019; Cotti et al. 2020). This could have facilitated the loss of vertebral centra.

Osteichthyan species that we assign close to the osteichthyan stem can have fully ossified vertebral centra and a pronounced constricted notochord or may lack vertebral centra and have a non-constricted notochord. At the level of chondrostean organization *Polypterus* and *Calamoichthyes* (previously assigned to the sarcopterygians), have complete and heavily mineralized vertebral bodies. In contrast, mineralized vertebral bodies are lacking in sturgeons and paddlefish. For the palaeontologist Bobb Schaeffer, sturgeons and paddlefish may never have had ossified vertebral centra (Schaeffer 1967). Another leading palaeontologist, Alfred Romer, had a different view and suggested that sturgeons and paddlefish likely show extreme reduction of a mineralized skeleton. Romer further speculates that in the course of evolution members of these groups will lose all bones and turn into "cartilaginous fishes" (Romer 1974).

At the holostean level, the bowfin (*Amia calva*) and gars (*Lepisosteus spp.*) develop completely ossified vertebral centra that are lacking in other basal actinopterygian taxa. Within the extinct Jurassic holostean order *Amiiformes*, one can find species with and species without vertebral centra (Romer 1974). A possible explanation, particularly from a cladistic point of view, would be that vertebral centra evolved several times independently in different vertebrate lineages (Arratia et al. 2001). Loss of vertebral centra is another explanation (see above, loss of vertebral centra in sturgeons). For the sarcopterygian *Latimeria*, it has been argued that vertebral centra have been lost. The utilization of lobed fins in forward movement is in sharp contrast to the sinusoidal movement of actinopterygians and may not require further enforcement of the notochord with vertebral bodies (Griffith et al. 1975).

The presence of ossified vertebral centra in Devonian lobe-finned fish such as *Osteolepis* and *Eusthenopteron* and other coelacanths from the rhipidistian clade

(Arratia et al. 2001; Romer 1974; Laerm 1979a) argues indeed for a loss of bony vertebral centra in *Latimeria*. Andrews and Westoll (1970) provide data about the notochord constriction by developed bony vertebral bodies in nine rhipidistian taxa. Notochord constrictions range from slight (wide notochord canal) to strong (narrow notochord canal), from ring-shaped to typical hourglass-shaped vertebral body centra. The loss of vertebral centra in *Latimeria* could also relate to the general tendency of deep-sea fish species to reduce the lung/swim bladder organ and the skeleton. This may have contributed to the loss of coelacanth vertebral centra (Griffith et al. 1975). The regression of vertebral bodies is exemplified by the loss of postcranial vertebrae in stomiiform deep-sea fishes (Schnell and Bernstein 2007) (Figure 7.3). Wake (1970) points out that the development of bone is plastic and that bone can be produced where required for functional reasons and without regards to homologies. Mechanical load is indeed a driving factor for the development of skeletal elements (Murray 1936; Hall 1967, 1968; and see reviews by Hall 2015; Witten and Hall 2015; Hall and Witten 2019). Witten and Hall (2021) address the problem of absence and presence of vertebrae in different osteichthyan lineages from a developmental perspective and argue, in agreement with Fleming et al. (2015), for a deep homology of vertebral centra. Emphasis is placed on notochord-based early patterning mechanisms that establish the positions of the intervertebral joints, a process that is similar in all vertebrates. The mechanism of vertebral column patterning could be more important than the presence or absence of bony and cartilaginous vertebral centra (see Chapter 9).

7.3 CONSTRICTED NOTOCHORDS IN ADULT VERTEBRATES

Notochord constriction occurs if vertebral centra develop around the notochord. The term *"constriction"* was used by Goodrich (1909) and remains widely used. In many cases, however, the notochord is not 'choked'. Instead, the notochord actively expands into the location of the prospective intervertebral joints while its diameter does not further expand into the location of the future vertebral centra (Figure 7.2C).

In osteichthyans, bone and cartilage development initially follows the shape of the notochord. This results in the typical hourglass shape of vertebral centra that give the impression of intervertebral notochord constriction While constriction usually refers to the narrow notochord in the middle of vertebral centra, actual notochord constriction can occur in the region of the intervertebral joints. Examples are frogs of the genera *Rana* and *Bufo* and gars in the actinopterygian genus *Lepisosteus* (Figure 4.3). Here, constriction is caused by sclerotome-derived cartilage that forms around the notochord sheath and subsequently grows inward (Gardiner 1982). As shown by Moffat (1973) in the New Zealand basal frog genus *Leiopelma*, the notochord persists in the adults. It is expanded in intravertebral regions and becomes constricted by cartilage in intervertebral regions. In urodele amphibians, this cartilage can fuse with notochord-derived cartilage (Wake 1970).

Within the actinopterygians and the sarcopterygians (amphibians and amniotes respectively) two types of notochord constriction occur. Ancestral for actinopterygians and sarcopterygians is expansion of the notochord in the intervertebral joints (Figure 7.2C), giving the impression of a notochord that is constricted intravertebrally. We see this type of notochord constriction in most teleost species. It is also the ancestral, plesiomorphic condition for amphibians, reptiles and mammals. Complete replacement of the intravertebral notochord leads to the intervertebral disk of extant mammals (Figure 7.2C–E). Constriction of the notochord in the spaces between vertebrae by invading cartilage cells of sclerotomal origin is the second type of constriction, found in various groups of amphibians and reptiles and some actinopterygians (Figure 7.2F and G). This type of constriction is most extreme in birds. The complete replacement of the intervertebral notochord by fibrous tissue (Figure 7.2H) facilitates the connection of vertebral centra in extant birds (Williston 1925; Wintrich et al. 2020).

A. Notochord Constriction in Non-Mammalian Tetrapods

All types of notochord constriction described above can be found in amphibians and reptiles. The plesiomorphic condition for both groups, and also for mammals, is amphicoelous vertebral bodies with a continuous notochord. As Romer (1974) points out, early reptiles and some of the earliest amphibians are so similar in their skeletons that it is almost impossible to distinguish the two groups. For example, adult giant labyrinthodont amphibians in the genus *Eryops* (Temnospondylii) from the Early Permian had heavily mineralized vertebral centra with a well-defined notochord canal (Rockwell et al. 1938), which provides clear evidence for the presence of a continuous notochord.

Variation of adult notochord design is large within the vertebrate clades. Earlier in this chapter, we discussed the conundrum of presence and absence of vertebral centra even within closely related groups of stemward osteichthyans. Also, the design of vertebral centra joints and thus the design and function of the notochord can vary to a large degree between members of the same clade. Most remarkable is the variation in the design of cervical vertebrae in turtles (Anapsida, Testudines), reviewed by Williston (1925). The earliest turtle had amphicoelous vertebrae throughout the column, but most others have an extraordinary combination of all types—amphicoelous, procoelous, opisthocoelous, plano-concave, plano-convex, and even biconvex—otherwise known in only the first caudal vertebra of the procoelian crocodiles. *Platypeltis spinifera*, a living soft-shelled river-turtle, has opisthocoelous cervical vertebrae, while certain pleurodiral turtles have saddle-shaped articulations (Williston 1925). Wintrich et al. (2020) provide an overview of the variation of vertebral body and notochord design in different clades of reptiles.

The notochord has a prominent role in the function of the vertebral column in adult extant amphibians, for many groups of reptiles but not for birds. The notochord is involved in the formation of intervertebral joints, either by attracting cartilage to this region or by contributing cartilage cells to the connection between vertebral centra. Amphibians but also squamate reptiles, develop cartilage inside the notochord independent from the joint cartilage (Figure 7.2F and G). Lawson (1966) and Jonasson et al. (2012) show that notochord cartilage truly derives from the notochord and is not generated by cells that invade. In ambystomatids (salamanders), intravertebral cartilage remains as hyaline cartilage and as fibrocartilage. The same applies to extant Gymnophiona (*Apoda,* caecilians). Lawson (1966) studied in detail and up to adulthood, the development of vertebral bodies and the contribution of the notochord in the Frigate Island caecilian *Hypogeophis rostratus* (Gymnophiona). His analysis confirms the observations of Gegenbaur (1862), Ebner (1896) and Klaatsch (1895) which is that the cartilage inside the vertebral bodies is directly derived from the notochord. Lawson (1966) concludes from his analysis that there is no indication for

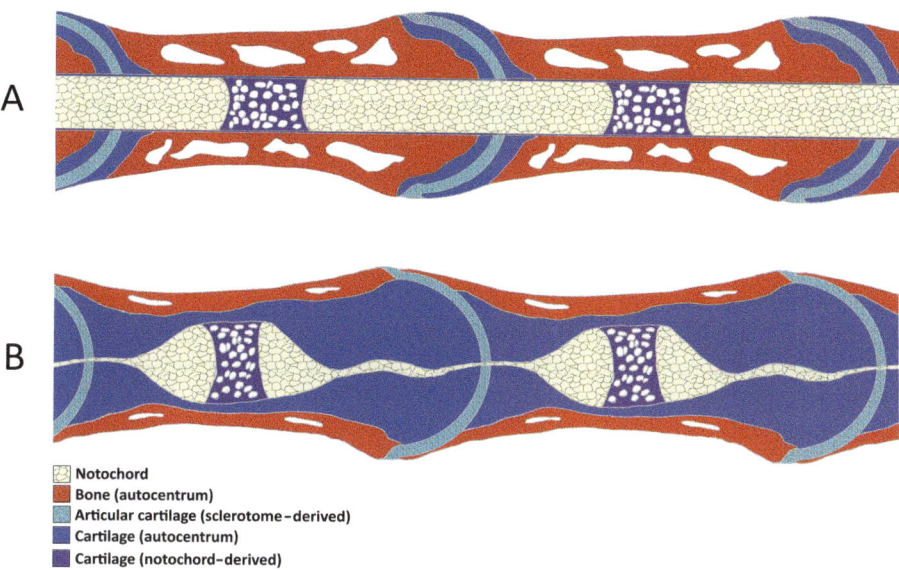

FIGURE 7.5   Variation of the notochord and surrounding tissues of the vertebral column as illustrated diagrammatically for two species of squamates (scaled reptiles). (A) Depicts the pattern in the adult thick-tailed gecko *Phyllurus milii* (Holder 1960) while (B) depicts the pattern in a young legless lizard *Anguis fragilis* (Winchester and Bellairs 1977). See also F and G in Figure 7.2. Variations of these two types of vertebral centra with preservation of the notochord intravertebrally, development of intravertebral cartilage and constriction of the notochord in the intervertebral joints is seen in many amphibian and reptilian taxa that develop centra with pro- or opisthocoelous joints. Typical is development of cartilage and bone around the notochord *without* intravertebral notochord replacement, seen in A and B. The notochord of many amphibians and reptiles differentiates into cartilage that is generated inside the notochord. This cartilage can remain or can undergo endochondral bone formation together with the cartilage that develops outside the notochord. Unlike chondrichthyans, osteichthyans and mammals (Figure 7.2. C–E), in amphibians and reptiles the notochord becomes constricted in the intervertebral spaces by sclerotome-derived cells that invade the joint area and form the articular cartilage. Complete disappearance of the notochord from pattern B would represent the situation in birds. See Lawson (1966), Wake (1970), and Wake and Lawson (1973) for the adult amphibian notochords. See Jonasson et al. (2012) for further details about notochord-derived cartilage in geckos. (Image prepared by PEW.)

any migration of cells through the notochord sheath, prior to the formation of the notochord cartilage. The cartilage inside the adult vertebral bodies is the product of the notochord cells (Figure 7.5, and see Chapter 8).

We conclude that notochord-derived cartilage can be found in two places: (i) In the section of the notochord that is located in the middle of the vertebral centra and (ii) In the region of the intervertebral joints. In amphibians (as in other vertebrates), cartilage contributes to the formation of intervertebral joints and can further transform into fibrocartilage. The notochord cartilage that develops in the middle of the vertebral centra eventually even contributes to the bone of the vertebral centrum via endochondral bone formation. Intravertebral notochord cartilage typifies the amphibian notochord; only in anurans is this cartilage restricted to the postcranial vertebral bodies (Wake 1970). A third type of possibly notochord-derived cartilage is its caudal cartilaginous extension, discussed in the next chapter.

For salamanders, Wake (1970) emphasizes that the notochord is much more than a remnant; it comprises a very large part of the centrum. In *Ambystoma* and more derived groups within salamanders and plethodontids, cartilage production inside the notochord starts before vertebral centra are well established. At the same time little bony pieces develop around the notochord and serve as attachment for trunk muscle tendons. The second phase of notochord cartilage development is related to the maturation of the intervertebral joint when notochord-derived cartilage is joined by cartilage that develops in this location outside the elastic membrane of the notochord. The elastic membrane eventually becomes fenestrated, a process also observed in dogfish and lungfish (Figures 5.5 and 5.6), and which unites the two populations of cartilage, the notochord-derived cartilage and the sclerotome-derived cartilage. At the periphery of the intervertebral joint cartilage blood vessels can invade and further penetrate toward the notochord cartilage that resides in the middle of the vertebral centrum, which, at this stage, is already surrounded by a solid perichondral bone collar. The vascularization initiates endochondral bone formation within this cartilage. The notochord cartilage is remodeled into bone and joins with the bone collar. The result is a bony vertebral centrum made from bone that is derived from notochord cartilage and from sclerotomal cartilage. Differentiated cartilage with a function in the joint between vertebral centra is also notochord- and sclerotome-derived.

The situation is different in frogs that have a vertebral centrum entirely derived from perichordal tissues. Vertebral centra have a cartilaginous anlage and the cartilage is remodeled into bone. No notochord cartilage contributes to the adult vertebral centrum (Lawson 1966) and there is no remodeling of notochord cartilage into bone in adult urodeles.

Notochord constriction by sclerotome-derived cartilage that forms around the notochord sheath and expands into the intervertebral space is most pronounced in birds. During development postcranial vertebral body anlagen in birds go through typical stages with intervertebral notochord expansion, intervertebral ligaments and intravertebral notochord constriction by surrounding cartilage (Piiper 1928). Eventually cartilage completely replaces the notochord at the sites of the intervertebral joints. In squamate reptiles, sclerotome-derived cartilage can form around the notochord sheath and populate the intervertebral space but unlike birds, this process must not replace the notochord. Lynette Holder described three different types of intervertebral joints for Australian geckos (Figure 7.5). All are characterized by sclerotome-derived joint cartilage that develops around the notochord while notochord cartilage is present in the middle of the vertebral centrum (Holder 1960):

> The notochord in the amphicoelous joint of adult broad-tailed geckos *Phyllurus platurus* is expanded despite being surrounded by cartilage in this position.
>
> Adults of thick-tailed geckos *Phyllurus milii* have procoelous joints (cranial facing socket, caudal facing ball) the notochord is not expanded but remains non-constricted but surrounded by sclerotome-derived cartilage.
>
> In the procoelous joints of adult Parker's least gecko *Sphaerodactylus parkeri* the notochord is constricted but remains continuous.

Intravertebrally, geckos preserve the notochord which becomes filled with notochord-derived cartilage that can also mineralize. Jonasson et al. (2012) provide a detailed account of the development of notochord-derived cartilage in geckos based on a study of the common leopard gecko *Eublepharis macularius* and six other species. True cartilage in geckos develops from pre-existing chordoid tissue during late embryogenesis. The cells are embedded in a gel-like metachromatic colloidal matrix that is rich in glycosaminoglycans. They have the same characters as chondrocytes and chondroblasts present elsewhere in the skeleton, for example being located within lacunae that do not interconnect (i.e., no canaliculi). Moreover, the tissue is avascular, it lacks nerves and lymphatic vessels. In the tuatara *Sphenodon punctatus* (Rhynchocephalia), we find amphicoelous vertebrae with a persisting notochord but the notochord is interrupted mid-vertebrally by bone and intervertebrally by fibrous cartilage (Wettstein 1931; Seligmann et al. 2008). The authors provide no information on whether the intervertebral cartilage is entirely of sclerotomal origin or has contributions from notochord cartilage as it is the case in amphicoelous amphibian vertebrae.

Birds, as far as we know, do not retain a notochord as adults but the matrix of the early intervertebral spaces into which somite-derived cells migrate may well be notochord-derived. Later, intervertebral spaces develop into synovial joints encapsulated by a synovial membrane (Shapiro 1992). During development, the bird notochord soon becomes constricted. Midvertebrally, at the location of the prospective vertebral centra, the notochord is removed during endochondral bone formation together with the massive cartilage that surrounds the notochord (Linsenmayer et al. 1986) (Figure 7.2H).

B. Notochord Expansion Forms the Actinopterygian Intervertebral Disk

The largest group of actinopterygian fishes and also the largest group of vertebrates on the planet are the modern bony fish, teleosts. Teleost diversity increased substantially at the end of the Cretaceous period and speciation gained further speed during the early Tertiary, about 60 million years ago (Carroll 1988; Maisey 2000). With about 30,000 species, teleosts conquered virtually all aquatic environments. The two most successful teleost groups are the Ostariophysi that dominate the freshwaters and the Acanthomorphs that dominate marine environments but also reinvaded the freshwater systems (Maisey 2000).

The enormous number of teleost species naturally restricts any general statements about the adult teleost notochord. The animals' size, adaptations to extreme habitats or other functional constraints influence development and design of the skeleton (Martini et al. 2021), and consequently, the structure and design of the adult notochord. Examples provided here represent the small number of teleost species that have been studied. This must be considered when the text refers to the "teleost notochord". A must read for those interested in the development, evolution and function of the actinopterygian vertebral column are Joshua Learm's, Bobb Schaeffer's and Gloria Arratia's treatises that cover chondrostean and neopterygian vertebral centra and the design of teleost vertebrae (Schaeffer 1967; Laerm 1976, 1979b, 1982; Arratia et al. 2001). For the function and biomechanics of the adult teleost notochord, there is Siada Symmons' excellent and comprehensive review of the elastic components of the teleost axial skeleton (Symmons 1979).

Basal actinopterygians can have fully ossified vertebral bodies with a constricted notochord or a non-constricted notochord and no vertebral centra (Figures 7.2A–C). A non-constricted notochord can be a primary or secondary condition. In many basal actinopterygian species, the notochord is neither surrounded nor replaced by mineralized centra. In contrast, teleosts typically have fully mineralized vertebral centra (Schaeffer 1967). Teleost vertebral centrum development starts with the mineralization of the notochord sheath, a process regulated by the notochord. The centra are designated as chordacentra (Fleming et al. 2015) (Figure 6.3). It has been proposed that chordacentra were absent in early teleosts and only evolved within the teleost lineage (Peskin et al. 2020). Chordacentra are, however, already present in palaeoniscoidian genera (Silurian — Cretaceous) such as *Turseodus*, *Haplolepis*, and *Pygopterus*. Members of pholidopleuriform chondrostean genera such as *Pleuropholis* and *Australosomus* also have chordacentra (Gardiner 1983; Gardiner and Schaeffer 1989). Moreover, the fully ossified vertebral bodies of the Senegal bichir *Polypterus senegalus* start their development as chordacentra (Bartsch and Gemballa 1992). Within the non-teleost neopterygian pachycormids, the extant Bowfin, *Amia calva* and its extinct relative *Huletta americana* have chordacentra. Clearly, notochord sheath mineralization evolved earlier than teleosts and perhaps even outside the actinopterygians (see Chapter 9).

Actinopterygian vertebral centra cause different degrees of notochord constriction. *Amia*'s vertebral centra are complete, ring-shaped, and notochord constriction is slight. The caudal region of the vertebral column is diplospondylous with two vertebrae per segment (Hay 1895; Schultze and Arratia 1986). Following John G. Maisey's account (Maisey 2000), the earliest teleosts that we know from the fossil record, the pholidophorids, had a non-constricted notochord, albeit with complete vertebral bodies in the shape of hollow cylinders that surround the notochord (see also Schaffer and Patterson 1984; Schultze and Arratia 1986, 1988) (Figure 7.1B). Given that mineralization of the notochord sheath (chordacentrum) was present already in early chondrosteans and elasmobranchs it is possible that pholodorphorid ring-shaped vertebrae initially represent notochord sheath mineralization.

Parchicomids, another extinct group of early teleosts, lacked solid vertebral centra. The up to 6 m-long ichthyoldelids, a third primitive teleost clade, had solid bony vertebral centra that resemble those of extant teleosts. The notochord was constricted midvertebrally. Extant members of the next basal teleost group, the osteoglossomorphs, *Arapaima agassizii* and *Arapaima gigas*, also have fully ossified vertebral bodies with the typical hourglass shape one finds in modern teleosts (Carroll 1988; Maisey 2000). Their notochord is constricted in the middle of the vertebral centra and expanded in the intervertebral spaces (Hilton et al. 2007; Scadeng et al. 2020).

In most neopterygians, the chordacentra are covered and eventually replaced by a perichordal cylinders of membrane bone (Gardiner and Schaeffer 1989). Depending on the degree of vertebral centrum development, the notochord remains as a cylinder of constant diameter inside a ring-shaped vertebral centrum or becomes constricted to various degrees. Full constriction occurs if a double cone (hourglass-shaped) centrum with amphicoelous joints develops, which is typical of teleost vertebral centra (Figure 7.6).

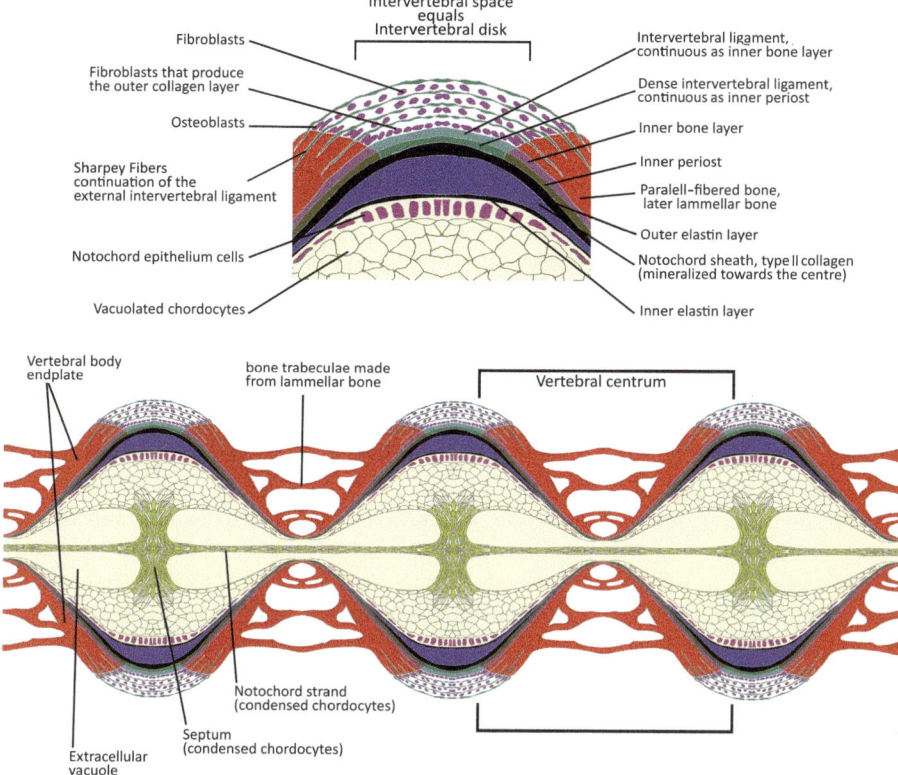

FIGURE 7.6 The contribution of the notochord to the design of teleost vertebral centra and to the composition of the teleost intervertebral disk is shown in these schematic drawings that show a longitudinal section of two complete vertebral centra. The dorsal part of the intervertebral space with the intervertebral ligament is magnified on top. Extension of the notochord in the location of future intervertebral spaces is a general osteichthyian and tetrapod character, while in teleosts, intravertebral notochord extension continues into adulthood and shapes the vertebral centra. All layers of the initial notochord sheath are preserved in the intervertebral space and become part of the intervertebral ligament. The notochord sheath layers constitute the inner part of the intervertebral ligament. The outer part of the intervertebral ligament is composed of collagen type I fiber bundles derived from the autocentrum and the surrounding mesenchyme. The outermost collagen type I fiber bundles connect the bone of the vertebral endplates and continue inside the bone as Sharpey fibers. A similar design is realized in the mammalian intervertebral disk with notochord-derived inner parts of the annulus fibrosus and mesenchyme-derived outer parts (see Figure 7.8). The vacuolated cells inside the notochord condense into a septum and notochord strand, structures that can keratinize and mineralize. Within this process, intracellular vacuoles of chordocytes are replaced by extracellular vacuolar spaces. Functionally, the notochord remains between the vertebral centra as a fluid-filled cushion. Schaeffer (1967) designates the intervertebral notochord and the surrounding ligaments in adult teleosts as the intervertebral disk. (Figure by PEW. References used for figure labelling: von Ebner, V., 1896. *Sitzungsber Kais Akad Wiss Wien* 105 (39): 123–16; Kvellestad, A., et al., 2000. *Dis Aquat Organ* 39: 97–108; Nordvik, K., et al., 2005. *J Anat* 206: 103–114; and P. E. Witten, P. E., et al., 2019. *J Exp Biol* 222: jeb188763.)

The collagen type I-based bone matrix (autocentrum) that becomes the membrane bone around the notochord is deposited onto the collagen type II-based mineralized notochord sheath in the second phase of teleost centrum development. The shape of the autocentrum follows the shape of the notochord. Schaeffer (1967) summarizes the situation for actinopterygians as follows: (i) at the chondrostean level, ossification around the notochord is rare; (ii) at the holostean level there is maximal experimentation with patterns of central ossification; and (iii) at the teleostean level there is relative stabilization in central ossification.

Gardiner (1983 p. 13) emphasized the importance of the collagen matrix that is deposited independent from its degree of mineralization onto the mineralized notochord sheath, the autocentrum: from early on, a perichordal ring of fibrous tissue encloses the notochord. This skeletogenic material forms a tube and appears identical in form and relationships in all vertebrate groups; it is difficult to see why it should not be considered homologous.

The adult teleost notochord that expands between the vertebral centra functions as an intervertebral disk (Schaeffer 1967; Symmons 1979) despite the fact that it remains continuous in most species (Figures 7.2 and 7.6). Sharpey fibers insert into the periosteal bone around the margin of the bony vertebral centrum endplates, which constitute a joint capsule according to Wintrich et al. (2020). Indeed, the adult teleost intervertebral disk (notochord) shares a number of characters with the mammalian intervertebral disk. Its core acts as a fluid-filled cushion, as does the mammalian nucleus pulposus (Figure 7.8). The core is surrounded and contained by ligaments composed from type II collagen and type I collagen (inside to outside), similar to the mammalian annulus fibrosus. There are however important differences. As teleosts mature, the vacuolated chordocytes of the continuous notochord give rise to two structures: a transverse septum in the intervertebral space and a horizontally orientated notochord strand. The transverse intervertebral septum connects the dorsal and the ventral intervertebral ligament (Figure 7.6). The notochord strand connects septa of neighboring intervertebral disks. The literature also refers to the teleost notochord strand as the axial strand or fascicle. The latter term infers its composition from large collagen fiber bundles (Kvellestad et al. 2000; Kryvi et al. 2017). The strand also becomes keratinized (Kague et al. 2021). An unsurpassed detailed description of the notochord as part of the intervertebral joint in the freshwater pike (*Esox*) is provided by von Ebner (1896) and shown in Figure 7.7.

As vacuolated cells turn into dense, non-vacuolated chordocytes extracellular vacuoles arise. These vacuoles are lined by dense filamentous chordocytes and maintain pressure inside the notochord. Functionally, the extracellular vacuoles replace the intercellular vacuoles of the early chordocytes. Kryvi et al. (2017) describe the process of transformation from vacuolated notochord cells to condensed notochord cells for juvenile Atlantic salmon (*Salmo salar*) as devacuolization and intracellular filament accumulation. The notochord strand is under tension that keeps the connected vertical septa in the center of the intervertebral joints in position (Symmons 1979). The degree of condensation into a septum and a strand can be different in different species. Moreover, a ring of non-condensed vacuolated notochord cells is usually present below the intervertebral ligament, again of different extent in different species.

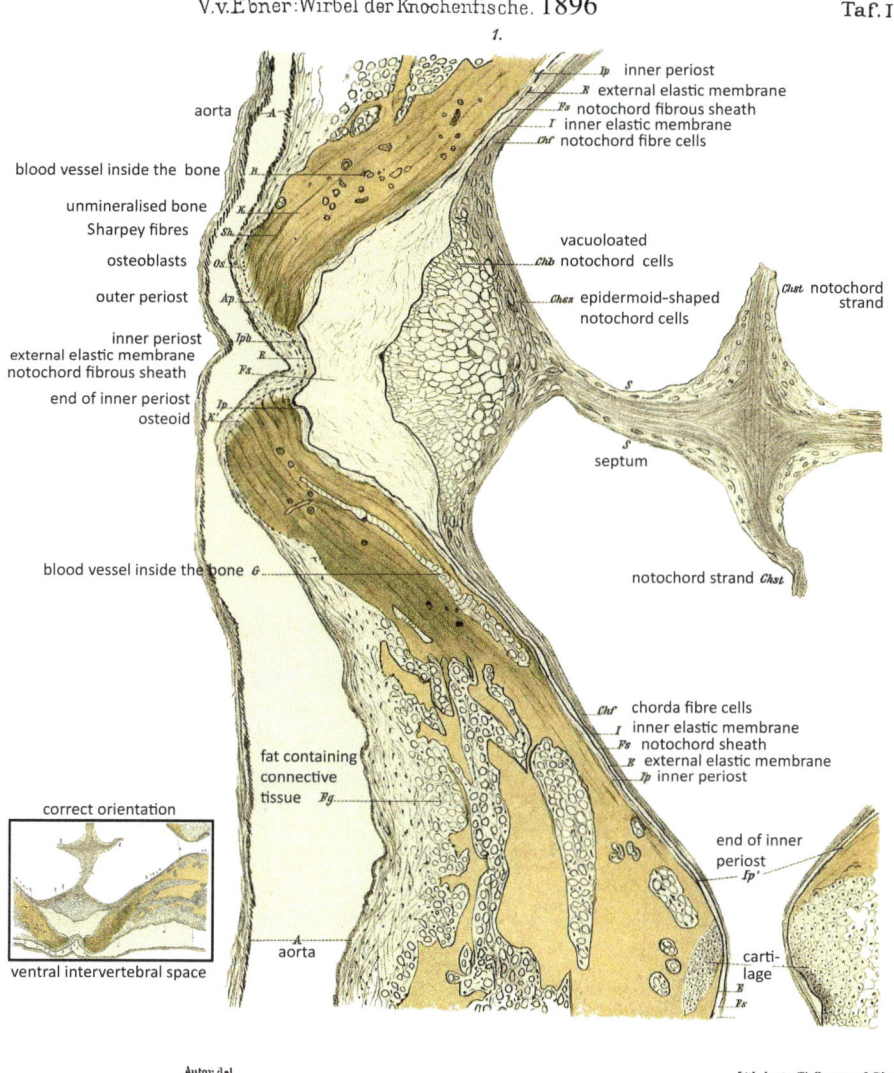

FIGURE 7.7 The ventral intervertebral joint region of the pike (*Esox sp.*). von Ebner's unmatched detailed analysis of the notochord structure and the tissues of the intervertebral joint in an adult teleost produced 126 years ago. (Compare with Figure 7.6. von Ebner, 1896. *Sitzungsberder Kais Akad Wiss Wien* 105 (39): 123–161. Labels translated by PEW.)

C. A Neopterygian Notochord that Follows the Tetrapod Scheme

Within the neopteryginas, there is at least one exception concerning the design of the notochord in the intervertebral spaces. Lepisosteiformes (gars) are placed at the basis of the neopterygian group (Carroll 1988). According to Joshua Laerm's

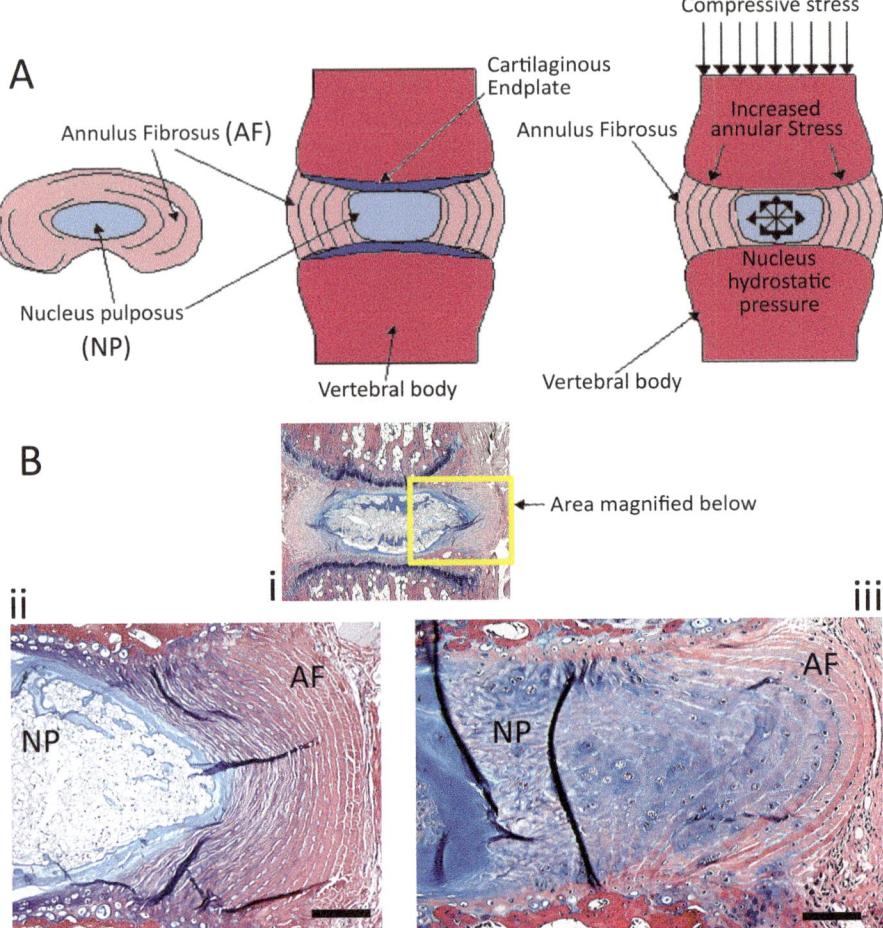

FIGURE 7.8 Mammalian Intervertebral discs. (A) Three schemes of a human interverte-bral disk viewed in the transverse plane (first scheme, left) and in the sagittal plane (2nd and 3rd schemes, middle and right). These three schemes show the extent and shape of the two notochord derivatives, the nucleus pulposus (NP, pale blue) and the annulus fibrosus (AF, pink) in relation to the bone of the vertebral centra (red). The third scheme indicates the trajectories of mechanical load in response to compressive stress. Compressive load (arrows) increases stress on the fibrocartilage of the annulus fibrosus and increases hydrostatic pres-sure on the nucleus pulposus (horizontal double-headed arrow). (B) Histology of a normal (ii) and a degenerated (iii) murine tail intervertebral disk, from the area identified by the box in a normal disk (i). After compressive loading and recovery, notochord cell death is accompanied by proteoglycan production (blue) and the inner architecture of the nucleus pulposus (NP) is lost; compare the area of the NP in the normal (ii) with the degenerated (iii) disks. Metaplasia of fibroblasts generates chondrocytes that fill the nucleus pulposus in the degenerated disk (iii), extending into the annulus fibrosus (AF), seen by comparing images ii and iii. Scale bars = 500 μm. ((A) Modified with labels corrected after C. J. Hunter, et al., 2003. *Tissue Eng* 9 (4): 667–677. (B) From J. C. Lotz, et al., 2002. *Biochem Soc Trans* 30 (6): 853–858, with permission from J. C. Lotz who kindly provided the color images.)

documentation, in *Lepisosteus* vertebral centrum development and notochord maturation follows the tetrapod scheme (Laerm 1982) and thus is very different from other neopterygians and from teleosts; a distinct modification from the ancient actinopterygian pattern according to Schaeffer (1967). Starck (1979) compares the vertebrae of *Lepisosteus* with vertebrae of urodeles (Figure 4.3 and 7.2F). As in urodele amphibians, the notochord is continuous and becomes constricted at the site of the intervertebral joint by somite-derived cartilage. Intravertebrally, the notochord remains expanded. Also, as in amphibians, vertebral centra articulate with procoelous joints with a socket and ball design, (Schultze and Arratia 1986) and not with amphicoelous joints. According to Schaeffer (1967), in amphicoelous intervertebral joints of teleosts and other actinopterygians, the notochord constitutes the joint by forming the intervertebral disk. In contrast, in amphibians and in *Lepisosteus*, the intervertebral joint is formed by somite-derived cartilage and the notochord is largely reduced in this position.

D. Constricted and Non-Constricted Chondrichthyan Notochords

John G. Maisey, personal communication: Stem chondrichthyans, but also stem elasmobranchs and stem holocephalans have a non-constricted notochord, condition A in Figure 7.2. In contrast to knowledge about condition A in "early chondrichthyans" there is no evidence of any modern "basal" chondrichthyan with condition A. Most, but not all, modern chimaeroids exhibit condition B, and so do several stem holocephalans going back as far as the Jurassic (*Ischyodus*, *Acanthorhina*, *Myriacanthus*, *Metopacanthus* and *Squaloraja*). Many earlier stem holocephalans (including Paleozoic forms) have non-constricted notochords without vertebral centra (Figure 7.2A), but the Lower Carboniferous *Chondrenchelys* has a series of elements identified as chordacentra forming a cylindrical axial column (Finarelli and Coates 2014). Most extant holocephalans have ring-shaped vertebral centra (Figure 7.2B) but *Callorhynchus* (the elephant fish found in the Atlantic and Pacific oceans off South America) is an example of the lack of vertebral centra in an extant species from this group. The presence of condition B in so many extinct relatives of modern chimaeroids suggests, however, that the absence of condition B is secondary in *Callorhynchus*. Some stem elasmobranchs (hybodonts, for example) have primitively retained a non-constricted notochord (Figure 7.2A). Among the chondrichthyans, a constricted notochord with hourglass-shaped vertebral centra (Figure 7.2C and D) occur only in the modern elasmobranchs, plus in a few extinct taxa that are close to the elasmobranch stem. An example is *Palidiplospinax* (formerly known as *Palaeospinax*), a Lower Jurassic shark with vertebral centra and triple-layered tooth enameloid (another feature otherwise found only in modern sharks). Constricted notochords (Figure 7.2C and D) have not been found in any fossil chondrichthyan outside of the elasmobranch crown and the proximate part of its stem. The best-known "sister group" of the crown elasmobranchs are the hybodont sharks, all of which have non-constricted notochords and not even ring-shaped vertebral centra (Figure 7.2A). Curiously, there are no data about extinct stem elasmobranchs or stem chondrichthyans with condition B which could be seen as an intermediate stage between condition A and C. Ring-shaped vertebral centra anlagen (formation of chordacentra as discussed in Chapter 5 and illustrated in Figures 5.5 and 5.6) that resemble condition B are part of development. Still, as far as adult chondrichthyans are concerned, condition B in adults has not been found as an intermediate stage between condition A and C/D in Figure 7.2. Instead, condition B seems to be apomorphic for chimaeroids and a few related stem holocephalans. Condition C/D seems to be apomorphic for crown elasmobranchs and a few related stem elasmobranchs. The scenario for osteichthyans is different as there are both ontogenetic and phylogenetic data that show condition B-like stages as intermediates between condition A and condition C. Perhaps the chondrichthyan fossil record is missing these "intermediate" forms, or they may never have existed. Many extinct placoderm fishes had perichondrally ossified neural and hemal arches. These were usually separate, but in rhenanids (flattened ray-like placoderms; e.g., *Gemuendina*, *Jagorina*) the neural and hemal arcualia were fused around the notochord although bony centra were probably absent (Denison 1978).

7.4 THE DISCONTINUOUS MAMMALIAN NOTOCHORD

For obvious reasons, we know more about the composition of the adult mammalian/human notochord than we know about the notochord from other vertebrate groups. This is because the intervertebral disks that house our notochord are of utmost importance for the function of the human spine. The book *"The Intervertebral Disc. Molecular and Structural Studies of the Disc in Health and Disease"* (Shapiro and Risbud 2014) describes the human disk in great detail. Mammals expand the notochord in intervertebral joints into a nucleus pulposus. The nucleus pulposus is contained by the annulus fibrosus, a mostly somite-derived fibrocartilage but the inner parts are also notochord-derived (Hall 2015).

The plesiomorphic condition of the mammalian vertebral body is an amphicoelous design with an initially continuous notochord. Discontinuation of the notochord and flattening of the vertebral body endplates led to the typical platycoelous mammalian vertebral bodies connected by notochord-based intervertebral disks (Figure 7.2C–E) (Wintrich et al. 2020). As in other vertebrates, notochord and somite-derived cells contribute to the formation of the intervertebral joint (Wake and Lawson 1973; Jonasson et al. 2012) (Figure 7.8). Primordia of the intervertebral disks in rats appear as condensations at 14 days of gestation, weakly positive for collagen I and II, and positive for both chondroitin-6- and dermatan sulfates (Rufai et al. 1995). The outer wall of the disk, the annulus fibrosus, is composed of flattened cells with collagen I as the main collagen and versican as the main proteoglycan. The inner part of the annulus fibrosus is designated as a transition zone as it also receives contributions from the notochord (Shapiro and Risbud 2014). The transition zone contains a mix of rounded nucleus pulposus cells and flattened annulus fibrosus cells as well as a mix of collagen II and collagen I fibrils. The mammalian notochord eventually becomes discontinuous due to its disappearance inside the bony vertebral centra. How the mammalian notochord eventually disappears is not understood fully. Notochord cells neither undergo apoptosis nor do they proliferate at this stage. This has led to the hypothesis that notochord cells either succumb to pressure from the developing vertebral centrum mesenchyme and are forced into the future nucleus pulposus, or that the cells actively migrate to sites of the future nucleus pulposus (McCann and Séguin 2016). Lawson and Harfe (2015) point out that the two models are not mutually exclusive. Possibly, a combination of pressure exerted by the forming vertebrae coupled with an attraction/repulsive signaling system emanating from discrete regions of the vertebral column are responsible for the formation of nuclei pulposi. The notochord persists midvertebrally for a long time after birth in the base of the skull (inside the odontoid process of the axis) and in the coccyx (Müller and Kölliker 1858). Infrequently, the persistence of a continuous notochord is observed in adult humans (Christopherson et al. 1999). As described by Musgrove (1891, p. 386) human lumbar vertebrae can "reproduce the characters of the osseous fishes". Indeed, the intervertebral disk in teleosts has a similar design. It is a fluid-filled cushion that is surrounded by an inner collagen type II based ligament of notochord origin and an outer collagen type I ligament of somitic origin.

Roughley (2004) and Richardson et al. (2014) provide concise information about the composition of the human notochord represented by the nucleus pulposus, the gelatinous core of the intervertebral disk (Cortes and Elliott 2014). The core of the disk and the inner part of the annulus fibrosus are notochord-derived structures. In a 3-year-old human, the nucleus pulposus cells are large and vacuolated. By the age of 10, the cells have become smaller and have changed to a more chondrocyte-like phenotype (Richardson et al. 2014). In humans, healthy nucleus pulposus cells are typically described as chondrocyte-like (Sive et al. 2002). They occur as cell clusters embedded in a proteoglycan-rich matrix. The cells have a rich endoplasmic reticulum and a defined Golgi system but they contain few, if any mitochondria (Shapiro and Risbud 2014). The main type of collagen in the nucleus pulposus is collagen II. Collagen III, VI, IX and XI occur in smaller quantities. Collagen II is accompanied by a high amount of large aggregating proteoglycans, particularly aggrecan and versican (Roughley 2004; Stemple 2005a) with large numbers of negatively charged glycosaminoglycan side chains. Cations bound to the side chains attract water and increase the osmotic pressure inside the nucleus pulposus. With collagen type II and proteoglycans the principal components of the mammalian nucleus pulposus are the same as in hyaline cartilage. The amount of water is 70%–85%. Other organic components, essentially identical to the known components of cartilage, are aggrecan, keratan sulfate, chondroitin sulfate and hyaluronic acid. Different from hyaline cartilage, the nucleus pulposus contains ten times more proteoglycans in comparison to collagen II. The negatively charged hyaluronic acid side chains are responsible for the accumulation of sodium ions inside the nucleus pulposus. Ion accumulation creates the required osmotic pressure inside the intervertebral disk. There is a loose network of long collagen II fibers inside the nucleus pulposus. The fibers also run across the nucleus pulposus to connect both endplates (Cortes and Elliott 2014). Typical for a notochord the nucleus pulposus contains neither blood vessels nor nerve fibers nor lymphatic vessels. Its core is low in nutrients and oxygen, the pH is low and there is accumulation of lactic acid. Nucleus pulposus and inner annulus fibrosus cell nutrition depend on loading and unloading which facilitates fluid exchange, the excretion of waste products and the intake of nutrients and oxygen. This is also a typical scenario for cartilage nutrition.

The mammalian nucleus pulposus remains active in its notochord function as a crucial signaling center that regulates vertebral body development and most likely intervertebral disk morphology into adulthood (Bruggeman et al. 2012). In all chordates, the developing notochord is a key source of the secreted Bmp antagonist Chordin (Lowe et al. 2015). In mice, Chordin binds to Bmp and inactivates Bmp in the developing intervertebral disk. At the same time, Chordin moves Bmp to the developing vertebral body where BMP, together with the product of the gene *Crossveinless2,* is required for the formation of the cartilage vertebral body anlage. Loss of Chordin or *Crossveinless2* results in small vertebral bodies and increased intervertebral joints (Zakin et al. 2010; Cox and Serra 2014). With ageing, clusters of chondrocytes increasingly appear inside the nucleus pulposus. Furthermore, mechanical overload can turn nucleus pulposus cells into cartilage-like cells (Boss et al. 1997; Lotz et al. 2002) (Figure 7.8).

The annulus fibrosus is the flexible wall that surrounds the adult mammalian notochord and connects the vertebral body endplates. It is composed of cartilage and fibrocartilage to form the annulus fibrosus. *Pax1*-expressing cells that migrate from the ventral sclerotome invade the notochord sheath (see Chapter 9 for notochord sheath invasion by sclerotomal cells in sharks) in the position of the original somite and opposite the myotome. Subsequently, these cells produce the fibrocartilage of the annulus fibrosus (Peters et al. 1999). Cartilage and fibrocartilage in the annulus fibrosus are based on collagen types I and II. The outer wall has higher amounts of collagen type I, the inner wall contains more collagen type II. Collagen fibers from the outer wall directly connect to the bone of the vertebral centra. The collagen fibers of the inner wall connect to the cartilaginous endplates of the vertebral centra (Cortes and Elliott 2014).

The origin of the mammalian nucleus pulposus from the notochord is now established but has long been a matter of debate (Cox and Serra 2014) despite firm early statements about its notochord origin by Albert Kölliker (Kölliker 1858; Müller and Kölliker 1858). The intervertebral ligament of a child of 1 year contains a cavity which is filled with the continuously growing mass of the notochord. This mass of notochord consists of a soft matrix and many cells with characteristic vacuoles arranged in clusters or in a network of strands. In large part of the nucleus pulposus, characteristic fetal notochord cells can be recognized. Kölliker (1861) further elaborates on the increase of the notochord in the position of the intervertebral disk in a 2-month-old human fetus. The gelatinous cell mass inside the annulus fibrosus, which also is present in adults, represents for Kölliker, without the slightest doubt, the notochord.

Being a strong supporter of Darwin and the new idea of evolution, Kölliker emphasized the developmental and structural similarities between the notochord-derived intervertebral joints in humans and the joints in fish and fish-like amphibians. In his words, "One could think the connection of vertebral bodies in fish with the notochord is somehow special but it is not because it is the same in humans and mammals" (Kölliker 1864). Surely, being able to establish the homology of the human intervertebral disk with the notochord was in line with Albert Kölliker's view on evolution when he points out that neither anatomical nor mental features allow him to give humans any exclusive position. Furthermore, Darwin's work must be praised extremely highly because of his most serious and most detailed studies and because of his many new insights about this very important topic (evolution); for these reasons alone his work will be epoch-making for all times. Indeed, the notochord is one of those conserved organs that bear witness to the evolutionary theory.

Today, the origin of the mammalian nucleus pulposus from the notochord is no longer debated (Cox and Serra 2014). The expression of *Brachyury* in the mammalian nucleus pulposus, expressed early in development in the notochord of all chordates (Annona et al. 2015), provides evidence for the notochord origin of the nucleus pulposus. The continuous expression of the homeobox gene *Notochord* (*Noto*) and *Shh* together with cell lineage tracing studies further confirm that the mammalian nucleus pulposus represents the notochord (Shapiro and Risbud 2010; Risbud and Shapiro 2011). Finally, ultrastructural analysis of human nucleus pulposus demonstrates the presence of chondrocyte-like notochord cells at all ages ranging from 26 weeks after fertilization to 91-year-old adults (Kölliker 1861; Williams 1908; Leeson and Leeson 1958; Choi et al. 2008; Shapiro and Risbud 2010; Bruggeman et al. 2012; Cox and Serra 2014).

7.5 DISCONTINUOUS NOTOCHORDS IN NON-MAMMALIAN VERTEBRATES

The previous section portrayed the mammalian notochord as typically discontinuous. A discontinuous notochord canal is, however, not restricted to mammals. Moreover, the paedomorphic trait of a continuous notochord canal can occur also in mammals (humans) as a type of atavism (Figure 7.2). The most radical interruption and eventually disappearance of the notochord is observed in birds.

We conclude from studies on chicken, ducks and quails that birds are apparently the only vertebrate group that does not maintain a notochord in juvenile or adult individuals. The perinotochordal space in the chicken is initially free of cells but filled with extracellular matrix. The perinotochordal space becomes invaded by sclerotomal cells which use radially oriented fibers as a substratum. The cells will give rise to the cartilaginous (later bone) vertebral bodies and to the intervertebral disks (Christ and Ordahl 1995). Linsenmayer et al. (1986) showed that the notochord inside the cartilaginous vertebral bodies, together with the surrounding cartilage, expresses type X collagen, which is a typical expression pattern for hypertrophied cartilage. Subsequently the notochord, together with the surrounding cartilage, is remodeled into bone.

At the location of the intervertebral joints, the notochord shows a strong expression of type II collagen. Different from mammals where the notochord also expresses type II collagen, chicken do not maintain notochord cells in the intervertebral joints. Instead, the type II collagen-based intervertebral joint matrix becomes populated by sclerotome-derived cartilage. A nucleus pulposus that would represent the notochord is eventually lacking. After 17 days of development post-hatching, the chicken notochord has largely disappeared; the only remains are inside the cartilaginous vertebral bodies (Shapiro 1992). As long as it is present the chicken notochord is of course not without function for the development of the avian vertebral column. Similar to early development when the notochord induces segmentation of the presomitic mesoderm, ablation experiments have clearly shown that the notochord induces the development of the cartilage that forms around the notochord (Ruggeri 1972). Collagen type II in the region of the prospective intervertebral joints is expressed much earlier by the notochord than by sclerotome-derived chondroblasts outside the notochord. Ablation of the notochord prevents the formation of cartilage around the notochord, i.e., it prevents the development of cartilaginous vertebral body anlagen.

Different from birds, other vertebrates maintain their notochord or notochord-derived tissue lifelong as an important component of the intervertebral joints. In mammals, a narrow continuous notochord canal can remain inside the vertebral centra, a paedomorphic trait according to Eastman et al. (2014). The mammalian notochord typically disappears inside the vertebral bodies, although as noted by (Williams 1908) a canal, representing its former position, traverses each vertebral body for a considerable time. In the intervertebral regions, the notochord persists as large flat disks, forming the pulpy nuclei of the fibrocartilages. This is the nucleus pulposus surrounded by the annulus fibrosus, two structures that continue to expand during development. The notochord canal inside the vertebral bodies disappears when the cartilage of the vertebral body anlagen is eventually remodeled into bone during what is endochondral bone formation. The notochord canal is remodeled into bone together with the cartilage. According to Christopherson et al. (1999), human notochord cells regress before endochondral bone formation, at the stage when the vertebral body centrum is still cartilaginous. The notochord sheath disappears later during endochondral bone formation as described by Williams (1908).

Many vertebrates maintain a small canal that contains the notochord inside the vertebral bodies, sometimes named the *notochord canal*. The rare persistence of the notochord canal in humans usually is asymptotic. It is occasionally discovered as an incidental finding with imaging (Christopherson et al. 1999). A continuous notochord canal in teleost fish is often described as becoming increasingly constricted as the animals grow. Based on a study of Antarctic notothenioid teleosts, Eastman et al. (2014) described the canal as being larger in smaller individuals and to decrease in size with increasing body length. This is, however, only an apparent constriction because the size of the notochord canal is reduced in relation to the diameter of the vertebral body. Eastman et al. (2014) do not provide data to show whether the absolute diameter of the notochord canal decreases with age. As long as the notochord canal is maintained in teleosts its diameter remains as small as it was during early vertebral body development. According to Maisey (2000), eventually the notochord canal can disappear from inside the vertebral bodies in teleosts. Further invasion of the notochord by the bone of the vertebral centrum obliterates the notochord in most teleosts and also in other neopterygian fishes such as *Amia calva* (Maisey 2000). Amphibians and reptiles can form mineralized rings around the notochord that also restrict the diameter of the notochord inside the vertebral bodies. A similar constriction of the notochord is observed in sharks in which the notochord is surrounded by a ring of hyaline cartilage that forms in the notochord sheath followed by a ring of mineralized fibrocartilage. Proliferation of the hyaline cartilage in the inner ring can constrict the notochord up to obliteration in the blunt-nose sixgill shark genus *Hexanchus*, and in the bramble shark genus *Echinorhinus* (Ridewood 1921).

7.6 SUMMARY

This chapter has explored the enormously diverse nature of the notochord in adult vertebrates and the cellular mechanisms that regulate retention or modification of the notochord.

In all vertebrates *without* complete vertebral bodies the notochord maintains its embryonic function, which is axial support as a non-constricted hydroskeleton. In those vertebrates in which *vertebral bodies develop*, the notochord (or notochord-derived cartilage) becomes a functional component of the intervertebral joints. Only a minority of vertebrate species maintains a notochord composed of the vacuolated cells described in the previous chapter.

A *non-constricted notochord* is present in those vertebrates that lack vertebral centra, have only partly developed centra, or have simple, ring-shaped centra. The non-constricted notochords of extant jawless vertebrates (hagfishes and lampreys) and extant sarcopterygians (lungfish, *Latimeria*) are examples.

Constricted notochords develop in those vertebrates in which centra develop around the notochord in a characteristic hourglass shape (Figure 7.1). There is an initial pattern of notochord expansion in the intervertebral spaces. In amphibians and amniotes, the notochord becomes restricted intervertebrally but can remain expanded in intravertebral spaces; in extant mammals, intervertebral notochord forms the intervertebral disks, while in non-mammalian tetrapods (amphibians and reptiles, including birds) a variety of mechanisms result in notochord constriction and the development of cartilage derived from the notochord or from sclerotomal cells. An intervertebrally expanded notochord forms the intervertebral disks in bony fish (actinopterygians), using a considerable diversity of different tissues discussed in this chapter. In mammals, the notochord is discontinuous and contributes the cells that form both the nucleus pulposus and the inner annulus fibrosus, which are, the gelatinous core and the outer portion of the intervertebral disks, respectively. As part of the complexity of notochord evolution, discontinuous notochords are found not only in mammals but also in several other groups of vertebrates.

Section IV

Relationships: Notochord-Cartilage, Notochord-Vertebrae, Notochord-Vertebral Column

Two major themes and questions tackled in this book are summarized in this section. One is the relationship between notochord and cartilage (Chapter 8), the other is the relationship between the notochord, vertebrae and the vertebral column (Chapter 9). Both require understanding whether and/or the degree to which notochord and cartilage (on the one hand) and notochord and vertebrae (on the other) share development, structure, function and evolution and whether such sharing would indicate homology and degree of evolutionary relatedness. Would sharing any (only one?) or all these four features indicate common evolutionary history through shared ancestry or convergence among more distantly related organisms? Would sharing a single feature be sufficient to infer homology and relatedness or must all four features be shared; i.e., how conserved, evolvable and/or independent are development, structure, and function? Chapters 8 and 9 discuss how these questions have been answered since 1859 when Kölliker described the gelatinous substance of the notochord as comprised of "soft cartilage cells" (p. 215) and when both Kölliker (1859) and Huxley (1859) concluded that teleost vertebral body anlagen arise as segmented mineralizations of the notochord sheath to produce what are now known as chordacentra (Huxley 1859; Kölliker 1859). Subsequently vertebrae grow through chondrogenesis (sometimes osteogenesis) of sclerotomal cells that surround but do not invade the notochord sheath. The discovery of clock genes and the discovery of genes expressed

DOI: 10.1201/9781315155975-11

by the notochord that independently pattern the vertebral column provide further evidence for the concept that centra and arches are developmental modules. This fuels a discussion about the origin and homology of vertebral bodies, a discussion that we carryout from a notochord perspective.

8 Relationships between Notochord and Chondrogenic Cells and Tissues
Transformational Series

CONTENTS

DOI: 10.1201/9781315155975-12

8.1   INTRODUCTION

Although it is the first supporting skeletal structure to develop in all vertebrate embryos, the notochord is recognized and categorized as a tissue and structure distinct from the skeletal tissues cartilage and bone that replace it in many but not all taxa: The notochord forms the permanent functional adult axial skeleton in many extant basal (see Chapter 7) chondrichthyans and osteichthyans, taking the place of and performing the function undertaken by cartilage or bone in other vertebrates. Clearly, the relationship(s) between notochord and cartilage is a close one and in the past, notochord has been regarded as a subtype of cartilage (see below).

Evidence supporting possible relationships between cartilage and notochord, especially between notochord and chondrogenic cells, is summarized in this chapter. Questions we ask include:

1. Should notochord be identified as a form of cartilage and included as a member of a cartilage superfamily?
2. Should cartilage and notochord be regarded as distinct members of a connective tissue family with many features in common; or should they be classified as distinctive and separate tissues?
3. Do the features shared by notochord and cartilage represent an ancient shared evolutionary history that may predate the vertebrates, or are they the result of convergence of cells and extracellular matrix (ECM) products?
4. How do we relate notochord and (vertebrate) cartilages to tissues that are intermediate between cartilage and notochord — chordoid and chondroid in normal development and repair; chordomas and chondromas as neoplastic/cancerous/malignant tissues — especially taking into consideration that notochord cells can and do form cartilage during vertebral formation in teleosts, urodeles and reptiles (below and see Chapter 7)?
5. How do we relate notochord and (vertebrate) cartilages to the many known invertebrate cartilages, especially considering that (i) many invertebrates evolved cartilage but none evolved a notochord, and that (ii) notochord evolved before cartilage in chordates?
6. Given these many and varies relationships, can we identify one or more transformational series that connect notochord to cartilage or to specific types of cartilage?

8.2   SHARED MORPHOLOGICAL FEATURES OF NOTOCHORD AND CHONDROGENIC CELLS

Shared features of notochord and chondrogenic cells were discussed in Chapters 1 and 2. Box 8.1 contains a summary of similarities and differences between notochord and cartilaginous cells — chordocytes and chondrocytes respectively, — and their ECMs, some of which are highlighted below.

The early notochord, which is an epithelial organ derived from mesoderm, consists of enlarged vacuolated cells (chordoblasts) connected by desmosomes and surrounded by a notochord sheath secreted by the chordoblasts. Cartilage, which is a mesenchymal tissue/organ derived from mesoderm or neural crest, consists of chondrocytes, often enlarged and hypertrophic, embedded in the ECM they secrete.

Essential *similarities* between notochord and cartilage are their mode of development as a "stack of coins," shared components in the notochord sheath and cartilaginous ECM (both of which can mineralize), a shared ability to induce cartilage from mesoderm, and absence of nerve fibers, blood vessels and lymphatic vessels (Box 8.1 and see Figure 1.5).

BOX 8.1   STRUCTURAL SIMILARITIES AND DIFFERENCES BETWEEN NOTOCHORD AND VERTEBRATE CARTILAGE

SIMILARITIES

Notochord and cartilage are both cellular tissues, composed of chordoblasts (notochord) and chondroblasts (cartilage).

Notochord cells have cell processes as do chondroblasts early in their differentiation. As chondroblasts mature into chondrocytes and deposit the ECM, cellular processes are less evident.

The gelatinous substance of the notochord was described over 160 years ago as consisting of "soft cartilage cells" (Kölliker 1859, p. 215).

A very similar "stack of coins" arrangement is found in the early stages of the development of the notochord, and of rod-like invertebrate and vertebrate cartilage, indicating an ancient shared mechanism of morphogenesis (Section 1.4A).

The inner layer of the notochord sheath shares molecules with cartilage ECM, notably collagen type II, proteoglycans and glycosaminoglycans (GAGs; Table 5.1). Indeed, the composition of the notochord sheath is essentially identical to the composition of vertebrate cartilage ECM.

Elastin is a matrix component, present in the notochord sheath and in the ECM of many vertebrate cartilages. The notochord epithelium produces both a layer of elastin fibers that surrounds the notochord sheath and a thin inner elastin layer of the sheath.

Notochord and cartilage share unique properties that distinguish them from all other connective tissues — neither notochord nor cartilage contains nerve fibers, blood vessels or lymphatic vessels.

Both the notochord sheath and cartilage ECM can mineralize.

Both notochord and cartilage can induce cartilage formation from somitic (sclerotomal), somatic and lateral plate mesoderm.

The notochord can directly produce cartilage as part of normal development. Injuries or amputation can transform the notochord into cartilage. Notochord tumors (chordomas) share features of cartilaginous and notochord cells (Section 8.6B).

Notochord and cartilage both participate in joint formation, including diarthrodial joints, intervertebral disks and the nucleus pulposus (Chapters 5 and 7)

DIFFERENCES

Notochord cells contain vacuoles, while cartilage cells only rarely contain vacuoles (vesicular cartilage).

Many chondrocytes become hypertrophic when terminally differentiated. Notochord cells, which are filled with vacuoles, give only the impression of hypertrophy.

The notochord sheath contains laminin and type IV collagen, both of which characterize basement membranes and neither of which are present in cartilage ECM (Table 5.1).

Peripheral chordocytes are connected by cell-to-cell connections (desmosomes), which are typical of *epithelial tissue organization*. Desmosomes are not found in mature chondrocytes, in which separation of cells by the ECM is typical of *mesenchymal cellular organization*. The notochord sheath, which is an elaborate basement membrane, and the epithelial organization of the peripheral notochord cells, both stand in contrast to the ECM and mesenchymal organization of cartilage cells.

The notochord can both induce and produce cartilage, but cartilage has *not* been shown to induce notochord formation nor have chondrocytes been shown to transform into notochord cells.

Differences are the presence of vacuoles within chordocytes and of desmosomes between chordocytes (shown in Figures 1.2 and 6.5), and the absence of both vacuoles and desmosomes in cartilage. The notochord sheath, which is an elaborate basement membrane typical of epithelial organization, contrasts with the ECM of cartilage, which is of mesenchymal cell organization. The sheath of the zebrafish notochord contains three isoforms of laminin, a family of ~400 to ~900 kDa basement membrane proteins essential for the early differentiation of the notochord and not found in cartilages (Parsons et al. 2002; Pollard et al. 2006). Apart from the laminin isoforms, the matrix components of notochord sheath and cartilage are essentially identical.

Both notochord and cartilage can mineralize the extracellular sheath or the ECM. Mineralization of cartilage is primarily only seen as a prelude to replacement of cartilage in osteichthyans; cartilage mineralization can be permanent in chondrichthyans. The extensive discussion in Chapter 6 of notochord sheath mineralization during vertebral body formation highlights the skeletal potential of the notochord and the similarities in cellular activity and function between notochord and chondrogenic cells. Are these features consistent with:

   i. Notochord as a type of cartilage;
   ii. Notochord as a "primitive" early type of cartilage;
  iii. Notochord as an epithelium;
   iv. Notochord as a member of a transformational series (notochord-cartilage; notochord-chordoid; notochord-chordoid-chondroid-cartilage);
   v. Cartilage as a type of notochord, and/or
   vi. Convergent evolution of cartilage and notochord?

The available evidence outlined in the book and summarized below is consistent with either the *notochord as an early type of cartilage*, or a *very close cartilage-notochord relationship* but is not consistent with cartilage as a type of notochord or convergent evolution of cartilage and notochord (v and vi above).

A century and a half ago, Kovalevsky (1866a, b, 1867, 1871a, b, 1877 and see Chapter 2) drew attention to the similarity of early notochord development in amphioxus, ascidians and vertebrates. As discussed in Section 1.4, the first indication of notochord cells is as a single column of cells perpendicular to the length of the notochord in what has been termed a stack of coins pattern (see Figure 1.5). This may be a universal arrangement among chordates, having been described in *Branchiostoma lanceolatum* (amphioxus), lampreys, chondrichthyans, teleosts and urodele amphibians (see Chapter 1 for references). This morphogenetic mechanism may be even older and more primitive as the fundamental process for forming rod-like cartilages in vertebrates and in invertebrates (Huyssseune 1989, 2000; Bird and Mabee 2003; Zhang and Cohen 2006; Cole 2011).

col2a1 plays a role in the ability of notochord cells to initiate the stack-of-coins arrangement. A mutant zebrafish, *cyclops*, in which the floor plate is disrupted, shows abnormal notochord expression of *col2a1* and fails to initiate the stack-of-coins morphogenetic pattern required for notochord development (Yan et al. 1995). *Cyclops* encodes the protein nodal-related2 (ndr2), a member of the Tgfβ superfamily of growth factors.

A further similarity, early in development, is that notochord and cartilage both can induce cartilage formation from somitic (sclerotomal), somatic and lateral plate mesoderm (Lash 1963; Cooper 1965). Chondrocytes are only inductively active when undergoing hypertrophy and before deposition of type X collagen into the cartilage matrix. Similarly, fully vacuolated notochord cells with type X collagen in the sheath are no longer able to elicit chondrogenesis (Hall 1977). Cartilage has not been shown to induce notochord formation nor have chondrocytes been shown to transform into chordoblasts, although notochord tumors (chordomas) share features of cartilaginous and notochord cells, and notochord cells can and do transform into chondrocytes during vertebral development (see Figures 7.2 and 7.5 and Section 8.6C).

8.3 SIMILAR GLYCOSAMINOGLYCANS IN THE NOTOCHORD SHEATH AND CARTILAGE EXTRACELLULAR MATRIX

Both the inner layer of the notochord sheath and cartilaginous extracellular matrices contain a closely similar range of glycosaminoglycans (GAGs) and proteoglycans, including aggrecan, chondroitin sulfate and keratan sulfate (see Table 5.1). Collagen types II, IX and X (discussed in Section 8.4) also are found in the matrices of both tissues (Ruggeri 1972; Mathews 1975; Linsenmayer et al. 1986; Schmitz 1995; Cole and Hall 2004a, b; Stemple 2005a). A short evaluation of glycosaminoglycans and collagens is provided in the following two sections.

A. NOTOCHORD SECRETES THE GLYCOSAMINOGLYCANS OF THE NOTOCHORD SHEATH

The first experimental evidence that notochord cells deposit the products of the notochord sheath came from studies on chick embryos initiated in the 1960s. ^{35}S-labeling revealed that sulfated glycosaminoglycans (GAGs) appear around the notochord as early as 20 hours of incubation (H.H. stage 10; Franco-Browder et al. 1963). Importantly for the interpretation of this research, 20 hours of incubation is before any sclerotomal cells are present around the notochord and ventral neural tube; i.e., the GAGs could not have been deposited by sclerotomal (mesodermal) cells but were deposited by noto-chord cells. See Hall (1977, 2015) for reviews of these older studies.

[35]S-labeling revealed the presence of sulfated GAGs but not the type(s) of GAG being deposited. Transmission electron microscopical and histochemical analyses by Toole (1972) demonstrate chondroitin sulfate and hyaluronan associated with the notochord in chick embryos as young as 60 hours of incubation (H.H. stage 16). This stage is immediately before the appearance of a basement membrane around the notochord and the colonization of this notochord matrix by sclerotomal cells at H.H. stage 17 (reviewed in Hall 1977).

These GAGs were shown by Mathews (1971, 1975) to be chondroitin sulfate–protein complexes (also known as proteoglycans). Among the small proteoglycans produced by the notochord are complexes previously thought to be cartilage-type proteoglycans and only found in cartilage (see Vasan 1987 for a summary). Later demonstrations, such as the presence of chondroitin sulfate in the hagfish notochord (Ueoka et al. 1999) indicated that chondroitin sulfate in the notochord sheath was neither a late evolutionary event nor confined to jawed vertebrates.

B. Aggrecan

Using a monoclonal antibody (S103L) to the core protein of the chick chondroitin sulfate proteoglycan, the proteoglycan aggrecan was shown to be present throughout the length of the notochord from its initiation in chick embryos (Domowicz et al. 1995). With continued development, expression of aggrecan is restricted to the notochord sheath in a segmental pattern that parallels the patterning of the somites (Bundy et al. 1998).

Aggrecan is also synthesized by cartilage cells. Notochord and cartilage aggre-cans differ at the post-translational level; notochord aggrecan lacks the keratan sul-fate chains that are prominent in chicken cartilage aggrecan. Notochord aggrecan is rich in the HNK1 epitope (involved in cell-to-cell adhesion). Cartilage aggrecan lacks the HNK1 epitope. The aggrecans in both tissues are the product of the same gene, the differences illustrating the importance of tissue-specific post-translational modification in individual extracellular matrices (Domowicz et al. 1995).

The cells inside the mammalian nucleus pulposus are notochord-derived cells. They resemble articular chondrocytes in that they are spherical and occur in clusters of four to six cells. These cells express genes and proteins characteristic of articular cartilage, including aggrecan, collagen type II, and vimentin, as do other notochord and cartilage cells (Hunter et al. 2003, and see Figure 7.8).

C. Chondromodulin-1

Chondromodulin-1, a 25-kDa glycoprotein product of the gene *chm1*, is an inhibi-tor of the growth of the endothelial cells that line blood vessels (Dietz et al. 1999). Found in cartilage and chondrogenic precursors, chondromodulin promotes the growth and later stages of differentiation of chondrocytes and inhibits vascular invasion of cartilage by inhibiting endothelial cell growth. Inactivation of chondromodulin when chondrocytes hypertrophy allows cartilage to be invaded by blood vessels (Shukunami et al. 1999). Cloning of the *chm1* gene was used by Sachdeva et al. (2000) to demonstrate expression of chondromodulin-1 in the notochord and in developing cartilage of zebrafish.

8.4 SIMILAR COLLAGENS IN THE NOTOCHORD SHEATH AND CARTILAGE EXTRACELLULAR MATRIX

Studies in chick embryos on the presence of collagen in early vertebral development led to the identification of a family of collagen molecules in tissues of mesenchymal origin, many with a tissue-specific distribution.

Collagen types II, IX and X are present in the notochord sheath (Table 5.1). Collagen type II (the product of the *Col2a1* gene) is of special interest: (i) because it is commonly referred to as cartilage-type collagen, and (ii) because invertebrate cartilages lack type II collagen. Collagen type X also is of special interest in the context of its role in tissue resorption associated with change in cell fate.

A. COLLAGEN TYPE II

From the early 1960s onward, the suggestion slowly emerged that the notochord and neural tube might have the ability to synthesize and export collagen. As with the demonstration of GAGs around the notochord, transmission electron microscopy demonstrated perinotochordal fibrils, perhaps of sclerotomal but perhaps of notochord origin. These microfibrils were attached to GAGs and either sensitive to removal by collagenase (and so identified as collagen), or sensitive to removal by hyaluronidase and amylase but not by collagenase and so identified as hyaluronate; see O'Connell and Low (1970), Frederickson and Low (1971), Minor (1973), Jurand (1974) and Bancroft and Bellairs (1976) for these perinotochordal fibrils.

Were these microfibrils sclerotomal or notochord in origin? Sensitivity to collagenase and co-localization with regions of uptake of ^{3}H-proline — which is transformed to hydroxyproline during collagenogenesis — led Bazin and Strudel (1972, 1973) and Ruggeri (1972) to identify the fibrils as collagen. Because collagen was regarded as a molecule only produced by mesenchymal cells, Bazin and Strudel concluded that the fibrils were incorporated into the matrix of the sclerotome-derived cartilage of the developing vertebrae. However, Cohen and Hay (1971) demonstrated that neural tubes isolated from chick embryos could secrete collagen fibrils that were deposited in association with the basement membrane. *Importantly, collagen is not restricted to mesenchymal cells.*

Separate studies showed that the notochord could reform a basement membrane after the original one had been removed, and that fibrils with the 51 nm axial periodicity of collagen were deposited in the basement membranes that formed *in vitro*. Using antibodies, these perinotochordal fibrils were shown to be type II collagen (Linsenmayer et al. 1973; Trelstad et al. 1973; H. von der Mark et al. 1976; K. von der Mark et al. 1976; von der Mark 1980). Two decades later, notochords from chick embryos were shown by Hayashi et al. (1992) to begin to secrete type IX collagen immediately before initiation of vertebral chondrogenesis in the adjacent sclerotomal cells (for which see Carlson et al. 1974).

From the mid-1980s, type II collagen was shown to be present in the *basement membranes of epithelia* in chick, mouse and *Xenopus* embryos (Thorogood et al. 1986; Kosher and Solursh 1989; Cheah et al. 1991). Sometimes the association was with epithelia — notochord, brain vesicles such as the optic cups — known to be required to initiate chondrogenesis in adjacent mesenchyme (Thorogood et al. 1986; Mina et al. 1991; Seufert et al. 1994; Hall 2015) but sometimes not. The *developmental significance* of the discovery that notochord and neural tube, which promote chondrogenesis in sclerotomal cells, produce the same collagen type (type II) as the tissue they induce, was the possibility that type II collagen may play an inductive role in chondrogenesis.

The conclusion from these and other studies was that *cartilage-type collagen might be predominant in cartilage but it is not limited in its distribution to cartilage* (Figure 8.1).

This conclusion from developmental studies has important implications for the question (discussed below) of the tissue in which type II collagen first evolved. Vertebrate cartilage and notochord are the two candidate tissues; the collagen in cartilage and in the notochord sheath in the shovelnose sturgeon *Scaphirhynchus platorynchus* is type II collagen. Miller and Mathews (1974) concluded that during development, collagen type II is first synthesized by the notochord before it appears in vertebral cartilage.

The gene for collagen type II, *col2a1*, was cloned from zebrafish by Yan et al. (1995) and shown to be expressed in the notochord, floor plate and hypochord, with a pattern of expression very similar to that of the helix-loop-helix gene *Twist*. Based on *in situ* hybridization, immunostaining and RT-PCR, Zhang and Cohn (2006) demonstrated that hagfish and lampreys express *col2a1* in the notochord, floor plate and hypochord. The expression of *col2a1* in *some* cartilage elements in these jawless vertebrates changed the view that *col2a*-based cartilage evolved first in jawed vertebrates. We now know that *col2a*-based cartilage is a general vertebrate character. However, whereas all cartilage elements in jawed vertebrates contain type II collagen, only some cartilages in hagfish and lampreys contain type II collagen. The coexistence of

FIGURE 8.1 Lateral view of collagen type II (Col II) immunostaining in the axial skeleton of a zebrafish of 5.9 mm total length. (A) Collagen type II is present in all cartilaginous preformed structures, including the Weberian apparatus (left arrow) and the hypurals of the caudal fin endoskeleton (right arrow). The thickening notochord sheath in the intervertebral spaces is also positive for collagen type II (white arrowheads). (B) Although not as strong as in cartilage, labeling for collagen type II in the thickening notochord sheath can be clearly seen (white arrowheads). (Reproduced with permission from Bensimon-Brito et al. 2012a, *BMC Dev Biol* 12: 28.)

cartilages with and without type II collagen in extant agnathans provides an indication of the evolution of type II collagen-based cartilage in gnathostomes. This could suggest that not only during development (Miller and Mathews 1974) but also during evolution, type II collagen appeared first in the notochord before it became the main collagen type of cartilage in the axial skeleton.

In summary, by the early 1970s, it was known that GAG-rich ECM accumulates around the notochord of chick embryos before initiation of chondrogenesis of the sclerotomal mesenchyme. It was concluded that notochord ECM products progressively become the matrix of the vertebral cartilage (Frederickson and Low 1971; reviewed by Hall 1977). Such a proposal essentially equates notochord with vertebral cartilage as a partial developmental transformational series: Notochord ECM that becomes vertebral cartilage ECM appears to be common but also the transformation of notochord cells (chordoblasts) into cartilage cells (chondrocytes) is a common event in amphibians and reptiles (see Chapter 7) and is also a reaction of the notochord to injury or mechanical overload. The mixed cartilage that develops within the notochord sheath of sharks is a good example of notochord ECM that becomes vertebral cartilage ECM — somite-derived chondrocytes are located in a notochord-derived cartilage matrix, as illustrated in Figures 5.5 and 5.6. Direct cartilage formation from the notochord contributes to intravertebral cartilage in lissamphibians — the modern amphibians — and in reptiles with the exception of birds (see below, Chapter 7 and Figure 8.2).

B. Other Fibrillar Collagens

The family of fibrillar (fiber-forming) collagens is an ancient one, regulated by an equally ancient gene network under the control of Sox genes (Wada 2010). Col I, II and III evolved as fibrillar collagens in vertebrates. Other collagen types are non-fibrillar.

Collagen type II, IX and X (discussed below) are found in both cartilage and notochord matrices (Table 5.1 and see Ruggeri 1972; Mathews 1975; Linsenmayer et al. 1986; Schmitz 1995; Cole and Hall 2004a, b; Stemple 2005a, b). Collagen type VII is expressed in the notochord and in early cartilage in zebrafish and in mice (Christiansen et al. 2009). Collagen type XV is a component of the extracellular sheath of the zebrafish notochord, is required to maintain the sheath — perhaps through attachment of the basement membrane to the notochord cells — and, through interaction with shha, is involved in signaling muscle development in the adjacent mesoderm (Pagnon-Minot et al. 2008). Jagged-1–Notch signaling from the notochord also acts via *Shh* to integrate muscle and notochord development (Yamamoto et al. 2010).

C. Secretion of Type X Collagen

Type X collagen is expressed in chordoblasts and hypertrophic chondrocytes when both are undergoing dissolution, resorption or transformation (see Hall 2015 for details and references). Recent studies have shown type X collagen expression in early zebrafish and medaka osteoblasts (Spoorendonk et al. 2008; Yu et al. 2017), which may indeed indicate the dual nature of these cells as cartilage or bone precursor cells (Debiais-Thibaud et al. 2019). During vertebral centra development in the chick, the notochord within the intervertebral space between the future vertebral centra deposits collagen type X. Expression of collagens II and X is segmental, with collagen type X deposited in prospective intervertebral areas from which the notochord is later removed: This is a feature of avian notochord. Other vertebrates maintain the notochord in the intervertebral joints. Collagen type X is not expressed in future avian vertebral centra but collagen type II is (Linsenmayer et al. 1986).

8.5 SHARED GENES AND TRANSCRIPTION FACTORS

Genes that characterize the notochord were discussed in Chapters 1 and 2. Here, we emphasize four transcription factors with important developmental roles in notochord and cartilages, listed along with other molecules in Table 5.1. The four are *Brachyury*, *Sox5*, *Sox6* and *Sox9, each discussed below.*

A. BRACHYURY

Brachyury — also called "no tail" and often abbreviated as T because of its membership in the T-box gene family — is a canonical notochord gene that was introduced in Chapters 2 and 4.

Expression of *Brachyury* characterizes the notochord of all chordates including the most basal chordates. *Brachyury* is expressed in the annelid axochord but not in the hemichordate stomochord (see Figure 3.4), consistent with the non-homology of notochord and stomochord, but also a good example that the expression of a single gene is insufficient to draw conclusions about homology or convergence.

Brachyury is required and sufficient to initiate notochord development, i.e., *Brachyury* acts at the onset of notochord formation (Capellini et al. 2008; Satoh et al. 2012; Annona et al. 2015). *Brachyury* also acts later in notochord development. *Brachyury* is expressed in cells of the nucleus pulposus in adult mammals (Risbud and Shapiro 2011), providing additional evidence for the notochord origin of the nucleus pulposus. Chordomas, which are notochord tumors (see below), specifically express *Brachyury*.

Brachyury also is a cartilage gene capable of initiating chondrogenic differentiation of C3H10T1/2 cells *in vitro* and *in vivo* (Hoffmann et al. 2002; Hall 2015). *Brachyury*-bound enhancers are an important mechanism of site-specific regulation of *Brachyury* (Lolas et al. 2014).

B. SOX5 AND SOX6

Sox5 and *Sox6*, which are high mobility group (HMG) transcription factors, are required for (i) survival of notochord cells, (ii) formation of the notochord sheath, and (iii) development of the nucleus pulposus within the intervertebral disks in mammals (Smits and Lefebvre 2003). Both genes are activated after *Brachyury* initiates notochord formation (Chapter 4) and so are high up in the cascade of notochord genes, as shown in Figure 4.2.

Both *Sox5* and *Sox6* also are important transcriptional regulators of chondrogenesis. Upregulated by *Sox9*, both *Sox5* and *Sox6* are active from the condensation stage onward, regulating the expression of *Col2a1* in chondrocytes (Akiyama et al. 2002).

Sox5 and *Sox6* upregulate *Bmp6* and down-regulate *Ihh*, *Fgfr3* and *Runx2* in growth-plate chondrocytes (Smits et al. 2004).

Sox5 (*LSox5*), *Sox6* and *Sox9* are expressed at sites of chondrogenesis in mice. By binding to a 48-base-pair chondrocyte-specific enhancer (chondrocyte-specific enhancer-binding protein), each transcription factor can activate *Col2a1* in C3H10T1/2 and murine MC615 cells (Ng et al. 1997; Lefebvre and de Crombrugghe 1998; Zhao et al. 1997). *Sox6* is a downstream regulator of *Bmp2*-induced chondrogenesis of C3H10T1/2 cells (Fernández-Lloris et al. 2003). Interactions between *Sox6*, *Sox9* and *LSox-5* and between *Sox8*, *-9* and *-10* and interactions with *Bmps* initiate chondrogenesis (Chimal-Monroy et al. 2003).

C. *Sox9*

Sox9, which is the major transcription factor for cartilage development, also is required for notochord development; *Sox9* and *Col2a1* are co-expressed in the notochord (Ng et al. 1997; Zhao et al. 1997). Unlike *Sox5* and *Sox6*, which are required to initiate the differentiation of notochord cells (see Figure 4.2), *Sox9* is required to regulate the differentiation of prechondrocytes into chondrocytes (see Figure 4.1); absence of *Sox9* in mouse embryos results in severe malformations of the vertebrae after a notochord develops but disintegrates, *Sox9* being required for segmentation of the ventral sclerotome and for chondrogenesis (Zhao et al. 1997; Smits and Lefebvre 2003; Barrionuevo et al. 2006) (Table 5.1).

In comparing notochord and cartilage (see Figure 1.5), one can see that *Sox9* plays the role in cartilage (transforming prechondroblasts to chondroblasts) that *Sox5* and *Sox6* play in the equivalent development stage in the notochord, transforming prechordoblasts to chordoblasts. For chondrocytes to go on to become hypertrophic, *Sox5* and *Sox6* genes have to be turned off (see Figure 4.2).

8.6 TRANSFORMATIONS BETWEEN NOTOCHORD AND CHONDROGENIC CELLS

Although cartilage is found in many invertebrate phyla, in contrast to vertebrate cartilages, invertebrate cartilage cells maintain cell-to-cell connections and lack type II collagen (Chapter 1). Notochord and the cartilages of all jawed vertebrates possess type II collagen; hagfish and lampreys express type II collagen in the notochord with limited expression in cartilage. Depending on the *nature of the chordate ancestor and whether it possessed cartilage* (Chapter 2), the ability to synthesize type II collagen may have:

  i. originated in a common chordate cellular precursor of notochord and cartilage, or
  ii. originated in cartilage or notochord and been co-opted by the other tissue (Cole 2011; Hall and Kerney 2012).

As notochord is phylogenetically older than vertebrate cartilage, *the notochord may have acquired the ability to synthesize and deposit type II collagen before cartilage* (ii), in which case *"cartilage-type collagen"* should be named *"notochord-type collagen."* Either way, a close relationship is evident.

The intriguing similarity in ECM products and shared transcription factors between the notochord and vertebrate cartilage, coupled with the restriction of type II-collagen-containing cartilage to vertebrates, are consistent with seeking a close relationship (perhaps even a common origin) of notochord and vertebrate cartilage (Hall 1998a, 2015; Cole and Hall 2004a, b, 2009; Cole 2011). While proposing this close evolutionary relationship, we have to be mindful of the possibility of convergence, i.e., that notochord and cartilage independently obtained their range of similar features early in vertebrate evolution, including the ability to deposit type II collagen into their extracellular environments. Neural crest (ectodermal) cartilage in the gill arches and a mesodermal cartilage from the notochord is an obvious indication that features can exist in two different mesenchymal cell populations — neural crest and mesodermal — perhaps through co-option of gene pathways from one to the other (Cole 2011; Hall and Kerney 2012; Hall and Gillis 2013; Hall 2014; Huysseune et al. 2022a).

It is in these contexts that we introduce and discuss two tissues — *chordoid* and *chondroid* — that share a range of features with notochord and cartilage. Research into these tissues is consistent with *chordoid as a tissue allied to notochord and chondroid as a cellular transformation from chordoid, allied to cartilage.*

A. Chordoid and Chondroid

Chordoid (or chordoid tissue) is composed of large vacuolated cells known as *chordocytes* that secrete limited amounts of ECM. Chondroid (or chondroid tissue) is distinguished from chordoid by smaller chondrocytes that lack vacuoles and that deposit a more extensive ECM.

In his classic discussion of types of cartilage, Schaffer (1930) proposed notochord as a subclass of cartilage. Based on histology and a perceived evolutionary sequence, Schaffer identified notochord, chordoid, chondroid and cartilage as increasingly differentiated types of cartilage. Importantly, this sequence exists, independently of whether we regard the tissues as types of cartilage or as a transformation series between closely related cells or tissues (Hall 1971, 2015; Beresford 1981, 1983). For analyses of these tissues from various perspectives, see Moss and Moss-Salentijn (1983), Benjamin (1989, 1990), Hall and Witten (2007) and Witten et al. (2010).

Supporting the tissue hierarchy established by Schaffer (1930), a study of vertebral column development in geckos showed that chordoid appears early in embryogenesis and that *chondroid both forms late in embryogenesis and develops from chordoid* (Jonasson et al. 2012). Notochord persists throughout life in geckos as alternating bands of chordoid and chondroid along the length of the vertebral column.

Cartilage development inside the notochord, even endochondral bone formation from notochord cartilage, is not uncommon in extant amphibians and non-avian reptiles (Gegenbaur 1862; Holder 1960; Lawson 1966; Wake 1970; Winchester and Bellairs 1977; Jonasson et al. 2012) (Figure 8.2). Chondrification of the notochord speaks to segmented organization; chordoid develops at intervertebral articulations, chondroid in mid-vertebral locations where the diameter of the notochord is reduced (see Figure 5.8E). Both chordoid and chondroid develop from notochord cells.

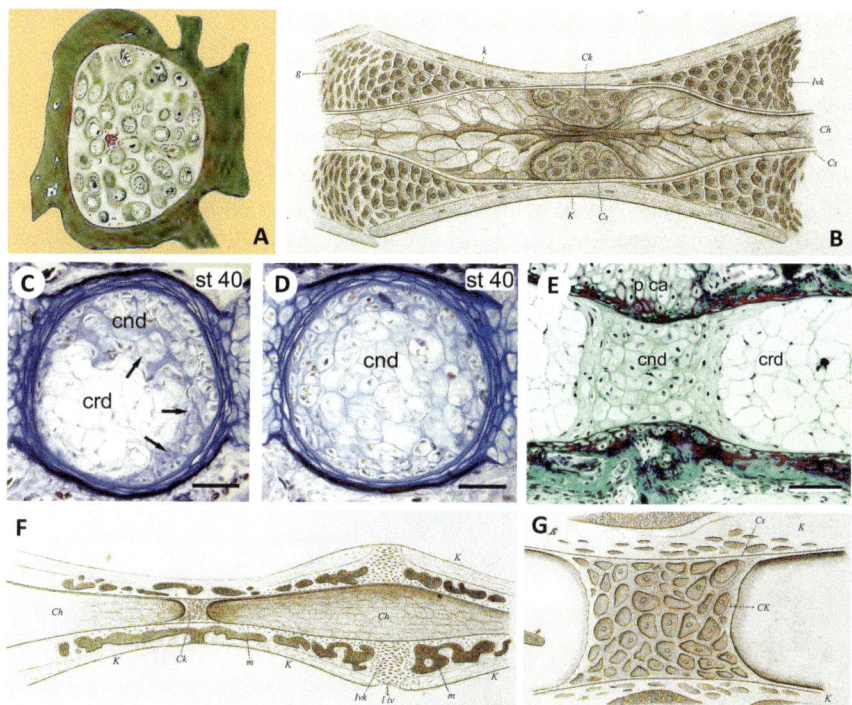

FIGURE 8.2 Cartilage is produced by the notochord as part of normal vertebral column development and maturation in amphibians and reptiles.

(A) Cross section through a vertebral body (green) of a 3.6-cm-long Mexican axolotl (*Ambystoma mexicanum*). The notochord is completely filled with cartilage cells. Krauss (1909) described two possible mechanisms for the differentiation of such notochord cartilage: Cartilage is derived from the notochord epithelium or cartilage forms by metaplasia from chordocytes.

(B) Intravertebral cartilage completely surrounded by the notochord sheath in a great crested newt *Triturus cristatus*. The specimen is close to the completion of metamorphosis. See panels F and G for labels.

(C, D) Transverse sections through the embryonic tail of Bibron's thick-toed gecko *Chondrodactylus bibroni* at stage 40 (st 40) shows the chordocytes (crd) of the notochord and notochord-derived intravertebral cartilage (cnd), Scale bars = 50 μm.

(E) Longitudinal section of the tail of the leopard gecko *Eublepharis macularius* at the intravertebral fracture (autotomy) plane shows alternating chondroid (cnd) and chordoid (crd), providing an excellent example of the differences in histology between the two (see the text). The notochord is encircled by perichordal cartilage (p ca). Scale bar = 100 μm.

(F, G) Sagittal sections throught the vertebral column of the Australian velvet gecko *Amalosia lesueurii* show the location of the notochord cartilage (Ck) in relation to the other parts of the notochord (Ch), to bone and to the cartilage that constitutes the intervertebral joint. Ch, notochord (Chorda); Ck, notochord cartilage (*Chorda Knorpel*); Cs, notochord sheath (*Chorda Scheide*); g, joint surface (*Gelenkfläche*); Ivk, intervertebral cartilage (*Intervertebralknorpel*); m, bone marrow (*Mark*).

(A) From Krauss, 1909. *Arch Mikrosk Anat Entwicklungsgesch* 73: 69–116; (B, F, G) from Gegenbaur, 1862. *Untersuchungen zur Vergleichenden Anatomie der Wirbelsäule bei Amphibien und Reptilien*, Leipzig; (C, D) from Jonasson, et al., 2012. *J Morphol* 273 (6): 596–603; (E) from Hall, 2015. *Bones and Cartilage*, Elsevier Academic Press, San Diego. Image provided to BKH by Matt Vickaryous.)

If lost, gecko tails can regenerate at a specialized autotomy plane. Zones of chondroid in the tail align with fracture (autotomy) planes, along which the vertebrae rupture when the tail is "dropped" or removed experimentally, providing further indication of regional specialization of the fate of the notochord in these lizards (Jonasson et al. 2012). Although lineage tracing was not used in this study, the evidence is consistent with the vacuolated chordocytes that form chordoid arising directly from vacuolated notochord cells (chordoblasts) and with the ability of chordoblasts to deposit an ECM, linking notochord → chordoid → chondroid → cartilage as a *transformation tissue series* and chordoblasts → chordocytes → chondroid cells → chondroblasts as a *transformational cell series*.

As pointed out by Stemple (2005a), the ultimate fate of the notochord emphasizes another close relationship between notochord and cartilage. Most of the notochord in mammals transforms into the nucleus pulposus. Although Choi et al. (2008) demonstrated that the nucleus pulposus is entirely notochord-derived in mice, not all notochord cells necessarily transform into the nucleus pulposus. With aging or as a reaction to increased mechanical load, the cells inside the nucleus pulposus acquire the phenotype of (transform into) chondrocytes, including the expression of cartilage-specific genes and proteins (Hunter et al. 2003 and see Figure 7.8).

B. Chordomas and Chondromas

Chordomas (notochord tumors) are rare tumors that form from notochord cells in adult mammals. Notochord remnants found in the spinal column of mice are suspected to be a source of late chordoma onset (Tamplin 2009). Reports about chordomas also exist from rats, dogs and cats (Hunter et al. 2003). In humans, the incidence is 0.08/100,000 individuals in the USA (McMaster et al. 2001). The low incidence has been said to reflect resistance of the notochord to invasion by malignant tumors (Schroyens et al. 1991) but most if not all chordomas *arise from transformation of notochord cells themselves*. Chordomas have features of cartilage and notochord (hence the name) and so represent what have been called *intermediate tissues* (Hall 2015). Indeed, a gradient of tissues extends from connective tissue to cartilage and many cartilage types consist of large chondrocytes with little or almost no ECM (Beresford 1981; Benjamin 1989, 1990; Hall and Witten 2007; Witten et al. 2010). Chordomas are a neoplastic transition between notochord and cartilage and as such represent a transformation series.

The presence of large cells with an ECM rich in mucin and glycogen have long been the histological criteria used to identify a chordoma. Indeed, these features were used over 150 years ago to propose that chordomas arose from the notochord. Connective tissue separates the lobules of chordomas. In adult humans, notochord remnants (perhaps only a few dormant notochord cells) can start to grow and produce a chordoma (Choi et al. 2008). These notochord-derived tumors have a striking similarity to cartilage-derived tumors such as chondromas (Romeo and Hogendorn 2006). Comparing gene expression profiles from notochord and cartilage tumors, Vujovic et al. (2006) concluded that chordomas resemble cartilaginous neoplasms. On the other hand, vimentin, S-100 protein and cytokeratins, characteristic molecules expressed in extracellular matrices (Chapters 4 and 8) and in chordomas, usually are associated with epithelial rather than mesenchymal cells, indicative of the epithelial nature of the notochord.

Brachyury continues to be expressed in cells of the nucleus pulposus in adult mammals (Risbud and Shapiro 2011). As noted above, *Brachyury* causes chondrogenic differentiation of cells *in vitro* and *in vivo* (Hoffmann et al. 2002) and is a canonical notochord gene even in the most basal chordates (Satoh et al. 2012). *Brachyury* is expressed in chordomas and has a causal role in chordoma formation (Romeo and Hogendoorn 2006; Vujovic et al. 2006; Nibu et al. 2013). Duplication of *Brachyury* in mice results in the formation of familial chordomas, which develop from remnants of the notochord (Yang et al. 2009).

C. The Origins of Intervertebral and Intravertebral Cartilages

The discussions in Chapters 7 and 9 demonstrating the notochord and somitic mesodermal origin of vertebral bodies speak volumes to the formation of homologous tissues (vertebral bodies) from different cellular populations and to the shared properties of notochord and sclerotome that allow them to form these cartilages. Witten and Hall (2021) conclude that vertebral bodies are homologous but it is of course also a matter of level of homology and what defines a vertebral body. This matter was discussed by Fleming et al. (2015) under the heading "Broad Sense Homology." We concluded that a vertebral body is defined by more than just the mode of centrum matrix formation, e.g., the notochord, the intervertebral joint and the intervertebral ligaments all are parts of a vertebral body (see Chapter 9).

The developmental sequence from notochord → chordoid → chondroid → cartilage in development and regeneration of the gecko vertebral column discussed above provides clear evidence of a transformational series between notochord and cartilage and establishes chordoid and chondroid as intermediate and transitory steps in the transformation, as imagined by Schaffer (1930). Is this series an exception or is it a well-worked out example of more extensive transformational relationships between cartilage and notochord? We take up this question as we draw this book to a close with a discussion of the origins of *intervertebral* and *intravertebral cartilage*. From our analysis of the literature, we concluded that the *closest developmental relationships between notochord and cartilage in normal development and across the vertebrates* are:

  i. The formation of *intervertebral* cartilage that *replaces* the notochord *between* the vertebrae, and
  ii. the formation of *intravertebral* cartilage *within* the vertebrae by the *transformation* of notochord to cartilage.

a. Intervertebral Cartilage

Studies published in the 1890s by several researchers show that controversy surrounded the mode of formation of the intervertebral cartilages that take the place of the notochord between the vertebrae. This is perhaps the closest association between notochord and cartilage in any individual and stands in contrast to the separate modes of development of vertebral bodies in widely separated taxa that belong to the actinopterygian and to the sarcopterygian (including tetrapods) lineage of osteichthyans. From his studies in many bony fish species, Gegenbaur (1862) concluded:

> In *Syngnathus* [a pipefish from the teleost family Syngnathidae] I clearly observe cartilage inside the notochord sheath and with large chorda cells located anterior and posterior to this cartilage. The transformation of notochord into cartilage, or perhaps one should say the production of intercellular matrix by notochord cells, is common. This I have observed in several genera of bony fish from different orders, and in various modifications; not further detailed in this chapter.
>
> *(Gegenbaur 1862, p. 60)*

By the 1890s, interpretations of further studies across the vertebrates had solidified into two divergent schools of thought:

1. Intervertebral cartilage *forms directly from notochord cells — the notochord transforms into cartilage* — the view held by Gegenbaur (1864), Field (1895), von Ebner (1896), Klaatsch (1893a, b, 1895) and Schauinsland (1906).
2. Intervertebral cartilage forms *from migrating cartilaginous cells* that rupture the outer elastin layer of the notochord to form cartilage within the notochord sheath — *the notochord is replaced by cartilage* — the view held by Lwoff (1893), Zykoff (1893) and Gadow and Abbott (1895).

The German histologist Victor von Ebner, as early as 1896, concluded that intervertebral cartilage was notochord in origin:

> I have no doubt that notochord cells transform into cartilage cells. This has been shown by Baer and by his disciple Victor Schmidt confirming previous reports from Heinrich Müller about notochord to cartilage transformation in the caudal end of the notochord in urodeles. I personally have studied the notochord that extends into the skull in salamanders and in triton larvae. The transition from cells of the notochord epithelium into cartilage can be clearly observed inside a completely intact notochord sheath.
>
> *(von Ebner 1896, p. 154).*

Development of cartilage inside the notochord surrounded by an intact notochord sheath should not be confused with the transformation of the notochord sheath into cartilage that typifies vertebral body development in several elasmobranchs. The best example for the latter process is again outlined in Goodrich's treatise (Goodrich 1930) in his discussion of the catshark *Scyliorhinus canicula*, although before Goodrich, evidence had been accumulated by Hasse (1892), Klaatsch (1893a, b), Gadow and Abbott (1895) and Schauinsland (1906). Cartilage precursor cells/ fibroblasts, back then interpreted as cells from the bases of the vertebral and neural arches, become mobile and *dedifferentiate* into migrating chondrocytes/fibroblasts with abundant cell processes. These cells migrate through the outer elastic membrane into the notochord sheath where they redifferentiate into chondrocytes, resulting in a cartilage in which the matrix is provided by the notochord, as shown in Figures 5.5 and 5.6. The proponents of the arcualia hypothesis see these cells as derived from the bases of the vertebral arches. The proponents of the autocentrum model regard these cells as derived from the connective tissue sleeve that surrounds the notochord sheath (see also Sections 6.3 and 9.2).

Because cell invasion and cartilage development in the notochord sheath defines the early anlage of vertebral centra, Arratia et al. (2001) designated them as chordacentra (Chapter 7). Invasion of the notochord sheath by fibroblasts has also been observed in basal osteichthyans (Arratia et al. 2001, and see Figure 5.6).

A different process occurs in developing embryos of the common chicken *Gallus domesticus*. The chicken notochord produces the ECM *before* sclerotomal cells begin to migrate to surround the notochord. Therefore, the migrating cells move into an extracellular environment produced by the notochord rich in collagen fibers and sulfated and unsulfated glycosaminoglycans. Some of this pre-existing notochord matrix may indeed be integrated into the matrix which later surrounds the sclerotomal cells (a hybrid matrix, if you will).

Development of "short-tails" is a common problem in farmed Atlantic salmon (*Salmo salar*). These are fish in which the vertebral column is compressed along the A-P axis to such an extent that body length is reduced and body depth so enlarged that body form hardly resembles that of a salmon. The defect at the cellular level is metaplastic transformation of the outer part of the intervertebral tissues to cartilage, thus constricting the further expansion of notochord in the intervertebral space (Figure 8.3). The result is the development of ectopic cartilage and fibrocartilage in the intervertebral joints, similar to what occurs in chicken early in ontogeny and similar to the cartilaginous outer portion of the mammalian intervertebral disk, the annulus fibrosus. Curiously, also the vertebral body endplates flatten (Witten et al. 2005; Fraser et al. 2019). Such a metaplastic transformation of mature cells in adult salmon can be compared with the intervertebral cartilages in birds and other sauropsids that derive from sclerotomal cells that enter the intervertebral space and subsequently constrict or replace the notochord.

b. Intravertebral Cartilage

In urodeles (Wake and Lawson 1973) and lizards (Werner 1967), intravertebral cartilage develops *inside the closed notochord and is derived from the notochord*. Wake and Lawson (1973) and Lawson (1966) described the formation of notochord (intravertebral) cartilage in the centra of the northern two-lined salamander *Eurycea bislineata* and in the Frigate Island caecilian *Hypogeophis rostratus*. In both species, the cartilage developed from cells of the notochord epithelial sheath and not from invading mesenchymal cells derived from the sclerotome. Indeed, absence of a distinct sclerotome and the presence of plentiful notochord cartilage characterizes vertebral development in salamanders (Figures 7.5 and 8.2).

Whether vertebral centra initially arise as cartilage, bone, or mineralized notochord sheath has been argued to be a function of the speed of development in

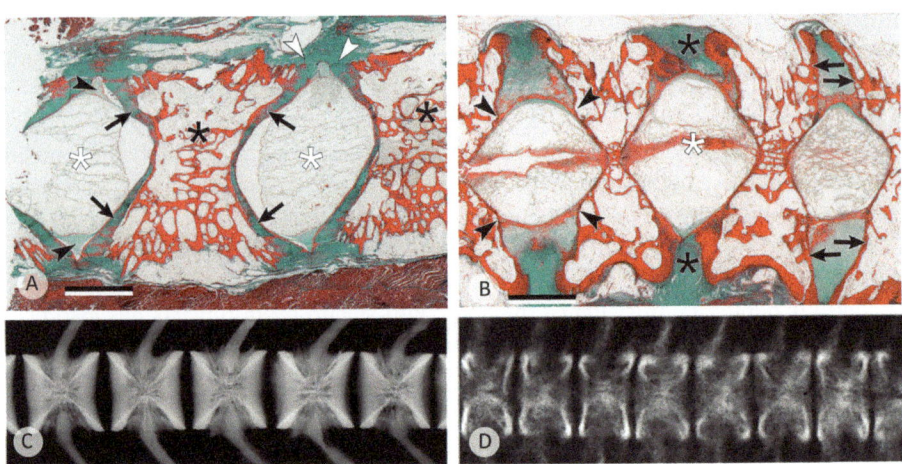

FIGURE 8.3 Parasagittal histological sections and radiographs of normal (A, C) and compressed (B, D) vertebral centra in Atlantic salmon, *Salmo salar*. Compression of vertebral centra (radiograph C, non-compressed; radiograph D compressed) in salmonid species is discussed as a consequence of mechanical overloading or of mineral deficiency. The compression phenotype (D) develops by transdifferentiation of osteoblasts into cartilaginous cells in the growth zone of the vertebral centra (white arrowheads in A, black asterisk in B). As a result, further extension of the notochord (white asterisks in A and B) in the location of the intervertebral disk is restricted (black arrowheads in B) and the notochord becomes surrounded by a ring of fibrocartilage (black asterisk in B). The flattened vertebral body endplates (normal, black arrows in A, flattened, black arrows in B) and the massive ring of fibrocartilage is a pathology that resembles a mammalian intervertebral disk. Black asterisks in A, intravertebral bone trabeculae. Scale bars in A and B = 2 mm. (A and B From Witten, et al., 2005. *Dis Aquat Org* 64: 237–246; C and D from Witten, et al., 2009. *Aquaculture* 295: 6–14.)

different species. Wake and Wake (2000) compared vertebral body development between gymnophionans and mammals and concluded that the observed major differences can be attributed to vastly different developmental rates. Furthermore, large parts of the vertebral centrum develop in response to mechanical load. As Joshua Laerm puts it: "It is possible to suggest that a centrum is a centrum is a centrum, reflecting the generalised adaptive response of sclerotome tissue to functional demands for vertebral consolidation" (Laerm 1979b, p. 482).

Two further examples, both from teleost fish, illustrate the ability of the notochord to form cartilage. During regeneration after amputation of the tail of the glass knifefish, *Eigenmannia virescens*, cartilage develops at the end of the severed notochord in the absence of notochord cells. The chondrocytes do not hypertrophy, nor does the matrix mineralize. Cartilage on the ventral surface is removed by multinucleated cells and replaced by perichondral bone, resulting in a regenerated tail with 2–3 mm of unmineralized cartilage at the distal tip (Kirschbaum and Meunier 1988).

In a recent study of the repair of injured notochords in zebrafish, Lopez-Baez et al. (2018) demonstrated expression of the gene *WT1* (*Wilms Tumor 1b*) in a population of cells derived from the notochord sheath. These cells formed a "blastema" at the wound surface from which cartilage developed, the cartilage identified by the expression of chondrogenic and mesenchyme genes in a transcriptome analysis. Smaller than normal chordacentra developed at the wound sites. Current studies on injured notochords are linked to the keyword "notochords regeneration" (Seleit et al. 2021). It must, however, be emphasized that the notochord can be repaired (Lopez-Baez et al. 2018) but has no regenerative capacity, another character shared with cartilage (Ashhurst 2004). Apart from experimental notochord injury the notochord can also become damaged connected to severe vertebral column malformations. The cartilage that derives from injured chordocytes represents a scar tissue that may further differentiate into a keratinized and mineralized fibrocartilage-like tissue (Loizides et al. 2014; De Clercq et al. 2018; Krague et al. 2021) (Figure 8.4).

D. Terminal Cartilaginous Notochord Extension

Notochord-generated cartilage is a common phenomenon, as part of normal development and in response to increased mechanical load. Such cartilage is typically located in the intervertebral space (intervertebral cartilage) or inside the vertebral centra (intravertebral cartilage); Section 8.6C. A third location for cartilage that may be notochord-generated is the caudal tip of the notochord. In dipnoans and chondrichthyans, series of cartilaginous elements can be found, whereas in teleosts, the terminal cartilage remains undivided (Arratia and Schultze 1992).

During development of extant basal chondrichthyans and osteichthyans but also in adult basal chondrichthyans and osteichthyans, the notochord, together with the spinal cord, extends to the very tip of the tail fin (Agassiz 1877; Cope 1890; Whitehouse 1910). If caudal vertebral centra are present, the notochord can extend beyond the last vertebral centrum into the fin fold to in between the dermal fin rays. Based on the presence and/or type of material around the notochord, various terms have been applied to such caudal extensions. Indeed, the literature challenges the reader by addressing the terminal cartilaginous extension of the notochord either as postcaudal cartilage, opisthural cartilage, urostyle or *cartilago urostyla* (Figure 8.5).

- Caudal extension of the notochord may become enveloped by a styliform osseous coat termed a *steganochord* (from the Greek *steganos*, covered).
- If the caudal notochord extension remains unprotected, it has been called a *gymnochord* (from the Greek *gymnos*, naked) (Huxley 1859).
- Recent literature uses the terms *notochord appendage* (Carpenter 1975) or *opisthural lobe* (Desvignes et al. 2018) for the terminal notochord extension.
- If the notochord is extended by cartilage, this cartilage is designated as *postcaudal* cartilage (Arratia et al. 2001) or *opisthural* cartilage (Wiley et al. 2015). Norden (1961) defines this cartilage as a *urostyle* (*cartilago urostyla*), pointing out that in a restricted sense, this includes only the cartilaginous termination of the vertebral column, which in salmonid fishes curves dorsally behind the last three upturned vertebrae. Gloria Arratia and Hans-Peter Schultze define the urostyle differently. They follow Huxley (1859) who created the name for a mineralized notochord that is located posterior to the last identifiable vertebral centrum (Arratia and Schultze 1992).

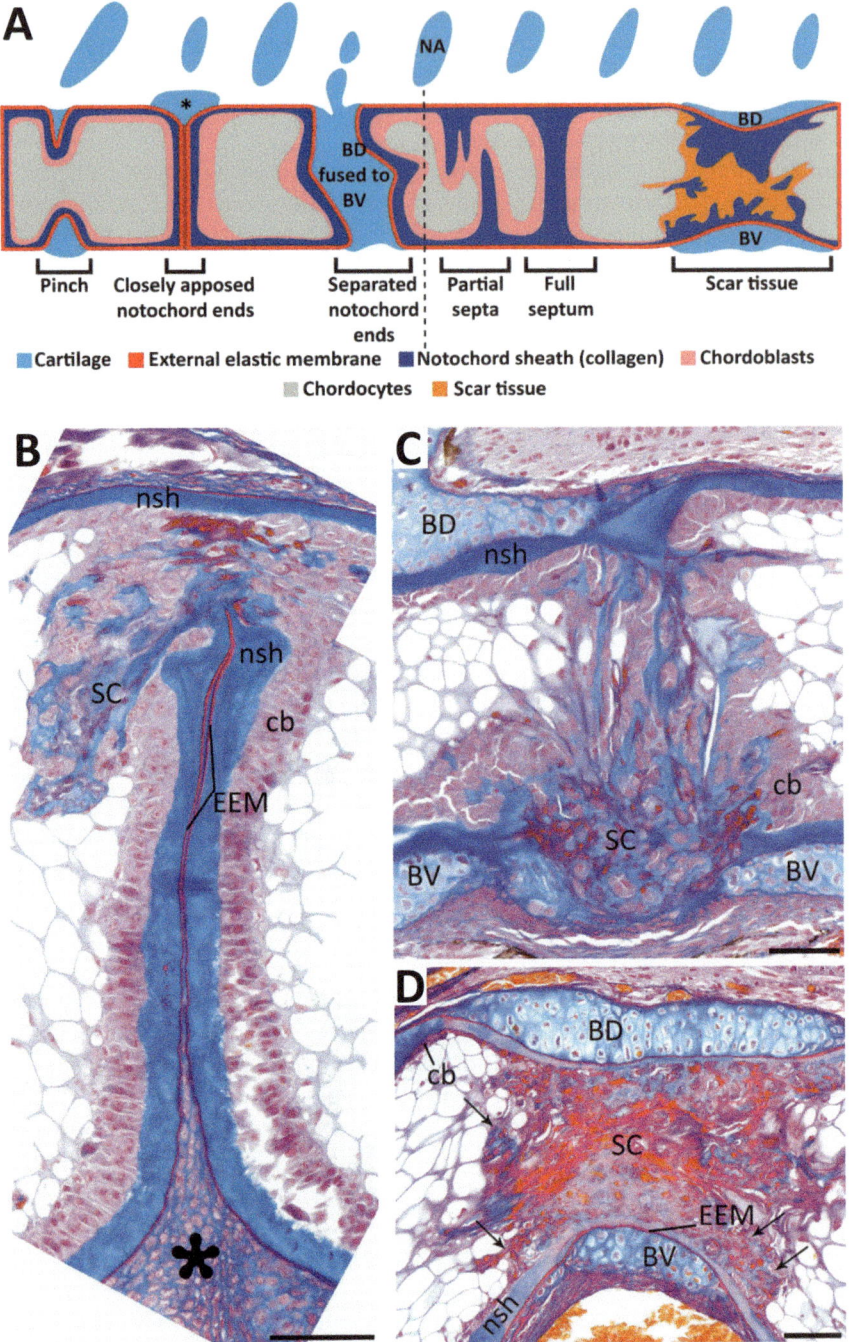

FIGURE 8.4 Types of notochord malformations and notochord scar tissues in first feeding early life stages of Pacific salmon, *Oncorhynchus tshawytscha*.

(*Continued*)

FIGURE 8.4 (*CONTINUED*)

(A) A schematic representation of notochord malformation types. Malformations anterior (left) of the vertical dashed line consist of infoldings of the notochord sheath which leads to pinching of the notochord (Pinch). The resulting depression becomes filled with cartilage from the bases of the neural (BD, basidorsal) and hemal (BV, basiventral) arches. In cases of complete pinching the notochord becomes separated by cartilage. Notochord malformation is shown posterior to (right of) the dashed line are characterized by the transformation of chordoblasts and chordocytes into scar tissue (orange).

(B) A deep pinch along the notochord and hyperplastic chordoblasts (SC). Cartilage (asterisk) fills the space of the inward fold of the external elastic membrane (EEM). Highly polarized (cylindrical) chordoblasts (cb) line the notochord sheath (nsh). Scale bar = 100 µm.

(C) Hyperplastic chordoblasts (cb) produce collagenous notochord sheath material forming scar tissue (SC). The notochord sheath (nsh) and the outer elastic membrane are interrupted by extruding scar tissue. Abbreviations: BD and BV, see A. Scale bar = 100 µm.

(D) Advanced formation of scar tissue (SC) which has replaced the chordocytes (cb) but is contained within the notochord sheath (nsh). Abbreviations: BD and BV, see A; cb, chordoblasts; EEM, External elastic membrane; Scale bar = 100 µm. Sagittal sections, Azan staining. (Modified from De Clercq, et al., 2018. *J Fish Dis* 41 (3): 511–527.)

The term urostyle also can include a series of completely fused terminal vertebral centra in teleosts (Wiley et al. 2015). In anurans, the urostyle is defined as a composite structure, comprising an ossified (initially cartilaginous) hypochord ventrally and a coccyx dorsally (Senevirathne et al. 2020). Below, we use *opisthural lobe* for the terminal notochord and *opisthural cartilage* for a cartilaginous notochord extension.

The origin of the opisthural cartilage is an intriguing subject. The cartilage precisely aligns with the extended terminal notochord; the positions of an opisthural lobe and an opisthural cartilage correspond. Arratia et al. (2001) do not recognize a connection between the urostyle in lungfish and in teleosts. Bartsch (1989) suggests that an opisthural cartilage may form from the invasion of cartilage precursor cells into the notochord sheath. By using the term urostyle, homology between lungfish and teleosts opisthural cartilage is inferred. Bartsch (1989) further points to the fact that regeneration of the caudal tip of the notochord occurs regularly in lungfish. It is thus difficult to distinguish an ontogenetic from a regenerated opisthural cartilage.

Kryvi et al. (2020) suggest that in Atlantic salmon, the anlage of opisthural cartilage buds off from the notochord sheath. Clearly, the authors show that the caudal end of the notochord is not completely closed and connects with the opisthural cartilage via an elastin-rich transition zone. The same situation is found in zebrafish (Figure 8.5). A transition zone between notochord and opisthural cartilage also was shown for the African spotted lungfish *Protopterus dolloi* by Bartsch (1989) and by Arratia et al. (2001), with chondrocytes clearly located inside the caudal-most portion of the notochord. Curiously, figures published by Kryvi et al. (2020) show opisthural cartilage tissue in Atlantic salmon also located below the notochord, in the same position as the anuran urostyle cartilage that derives from the hypochord (Senevirathne et al. 2020).

The structure of the teleost notochord that constitutes the opisthural lobe differs considerably from the structure of the anterior notochord. In zebrafish, the opisthural lobe is separated from the anterior notochord by a mineralized septum (Bensimon-Brito et al. 2012b). In rainbow trout and in Atlantic salmon, this part of the notochord is filled with chordocytes that have lost their vacuoles and shows patchy mineralization (Arratia and Schultze 1992; Kryvi et al. 2020). Clearly, the internal cell composition shifted toward a fibrous connective tissue, which may explain the presence of a notochord to cartilage transition zone. Figure 8.5D shows the opisthural cartilage and the transition between notochord and cartilage in zebrafish. The ultimate origin of the curious opisthural cartilage deserves, however, further investigations.

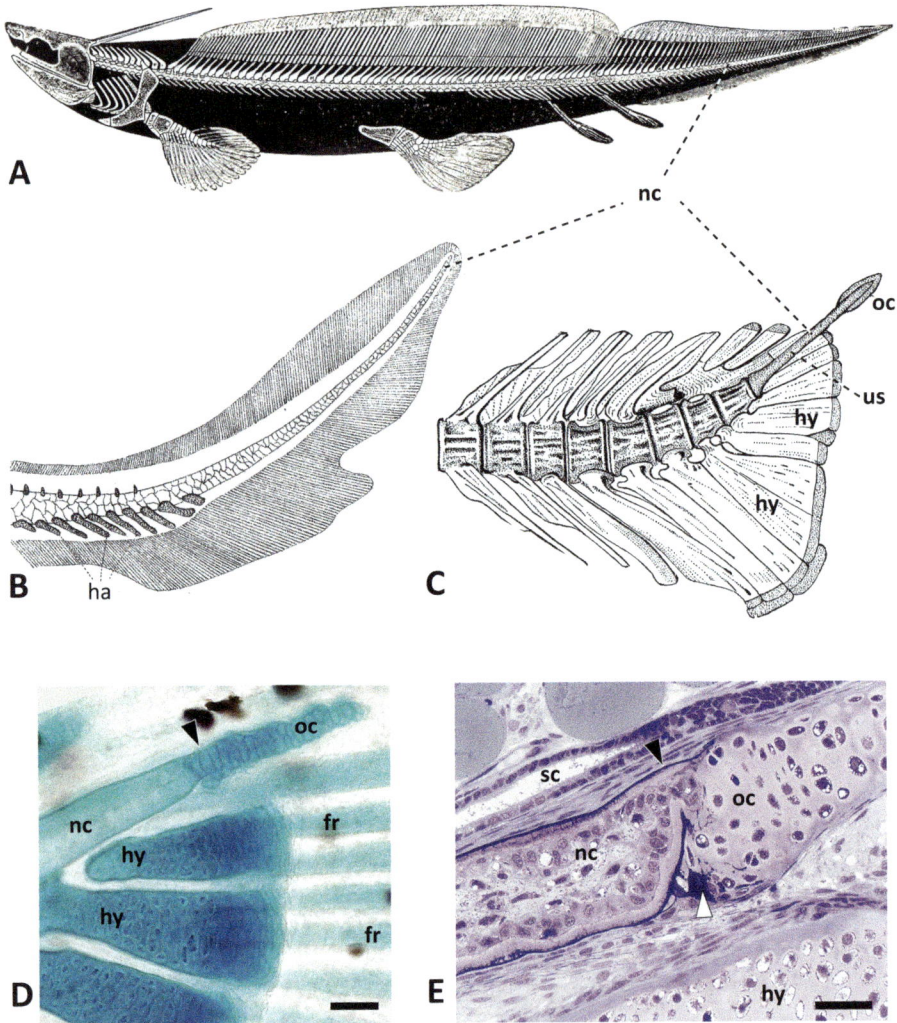

FIGURE 8.5 Terminal extension of the notochord in various aquatic gnathostomes.

(*Continued*)

FIGURE 8.5 (*CONTINUED*)

(A) The notochord (nc) extends to the very tip of the tail in the extinct shark *Pleuracanthus decheni*, from the Permian of the Czech Republic 299–272 MYA.

(B) The notochord (nc) extends to the tip of the tail in a young gar, *Lepisosteus sp.*, 21 mm length.

(C) The notochord (nc) extends to the tip of the tail in a sea trout, *Salmo trutta*.

In the gar and in the sea trout, the notochord extends further than the last hemal (ha) and neural arches, vertebral centra and hypurals (hy). Notochord and spinal cord (sc in E) extend into fin fold that is supported by dermal fin rays. In C only endoskeletal elements are shown, not the dermal fin rays (fr in D) that extend from the hypurals (hy). In the sea trout and other basal teleosts (e.g., Clupeids, Salmonids, Cyprinids), the caudal tip of the notochord is cartilaginous, forming what is known as an opisthural cartilage. See Arratia and Schultze (1992) for a detailed description of all skeletal elements.

(D) Shows the terminal notochord (nc) in a zebrafish of 8.3 mm total length. The notochord is extended by the opisthural cartilage (oc). Whole mount staining with Alcian blue visualizes the cartilage of the hypurals (hy), the notochord (nc) and proteoglycans in dermal fin rays (fr) prior to mineralization. The base of the opisthural cartilage connects to the notochord (see E) and is enclosed by the notochord sheath (black arrowhead), similar to what has been observed in lungfish (see the text). Scale Bar = 25 µm.

(E) A microscopic image that shows the connection between the terminal notochord (nc) and the opisthural cartilage (oc) in a juvenile zebrafish of 16 mm total length. Note that the notochord and opisthural cartilage are continuous. Elastin, the material that constitutes the outer wall of the notochord sheath (black arrowhead) is also abundant in the matrix of the cartilage that abuts the notochord (white arrowhead). Epon section, toluidine blue staining, Scale Bar = 25 µm. hy, hypural. ([(A) From Goodrich (1909) Vertebrata Craniata (First Fascicle: Cyclostomes and Fishes). In: *A Treatise on Zoology*, Part IX. Lancaster, R. (ed.). Adam and Charles Black, London. (B and C) After Cope, 1890. *Am Nat* 24 (281): 401–423. (D and E) Provided by PEW.])

8.7 NOTOCHORD AS A TYPE OF CARTILAGE

The concept of notochord (in this case the notochord sheath) contributing ECM to another tissue (vertebral cartilage) is a conceptually different issue from whether notochord should be regarded as a cartilaginous tissue. As the notochord may be phylogenetically older than chordate cartilage and certainly is older than vertebrate cartilage, it has been proposed that the ability to synthesize and deposit type II collagen originated in dorsal, mesodermally derived notochord and was transferred to (co-opted by?) mesodermal vertebral cartilage (Section 8.6 and see Hall 2009; Hall and Kerney 2012; Hall and Gillis 2013). Such a proposal — an *evolutionary transformation series* — establishes a phylogenetic relationship between notochord and vertebral cartilage as two embryonic tissues that synthesize and deposit fibrils of type II collagen, without necessarily requiring that we regard the two tissues as "the same."

There is no doubt that cartilage is an active skeletogenic tissue. Despite morphological features that distinguish notochord cells from typical hyaline cartilage, genes that are expressed by cartilage are expressed by the notochord (Schmitz 1998a, b). Furthermore, in cartilaginous fish (chondrichthyans) and in basal bony fish, such as

Polypterus and lungfish, the notochord sheath is cellular, as shown in Figure 5.6. The sheath is acellular in other basal bony fish such as sturgeons and also acellular in extant agnathans and in elasmobranchs (Schmitz 1998a, b). In sharks, the notochord was recognized as producing the first cartilage matrix of the early vertebral centrum; the matrix of the notochord sheath. The cartilage cells inside the notochord sheath, however, derive from chondrocyte precursor cells that invade the notochord sheath. These cells are of sclerotomal origin and belong to the autocentrum (Section 8.6C).

As already discussed and summarized in Table 5.1, cartilage and notochord share many specialized gene products (including collagen type II, vimentin, versican, aggrecan and chondroitin sulfate) and transcription factors (Pax, Sox). Do these similarities reflect independent evolution of notochord and cartilage or is there a close relationship between notochord and cartilage, reflecting either a shared origin or co-option of features of one from the other?

With respect to our recognition of tissues/tissue types and/or cells/cell types, two possibilities exist:

   i. If evidence of similarity of cell structure/ultrastructure and tissue function
      is sufficient to regard notochord and cartilage as related, then we could con-
      sider *cartilage and notochord as two classes of skeletal (or connective)*
      tissues. Cartilage and bone are recognized as skeletal tissues with far less
      similarity than exists between notochord and cartilage.
  ii. If evidence of similarity of gene products and proteins is sufficient to regard
      notochord and cartilage as related, and given that the notochord originated
      before vertebrate cartilage, we could *consider notochord as a type of carti-*
      lage in the same way and using the same evidence by which cardiac, smooth
      and striated muscles are recognized as muscle.

Using four classes of evidence, Stemple (2005a) concluded, reminiscent of Schaffer (1930), that: "As a tissue, it [the notochord] is most closely related to cartilage and is likely to represent a primitive form of cartilage" (p. 2503). In a quick guide published in *Current Biology* in the same year Stemple stated the relationship to cartilage even more strongly: "By the time the notochord is mature it expresses a constellation of genes that is essentially identical to that of developing cartilage" (Stemple 2005b, p. R874). His explanation was shared function: "This [shared constellation of genes] is consistent with the mechanical role of the notochord" (ibid, p. R874).

Such shared function (and/or shared genes), however, is not sufficient evidence to claim a homology of notochord and cartilage or to place the notochord as a skeletal tissue within a category of cartilage; recall the similar disclaimer in Chapter 2 when comparing shared genes in the annelid axochord and vertebrate notochord. While the phylogenetic relationship between proto- and deuterostomes rules out homology between the annelid axochord and the notochord (Hejnol and Lowe 2014), shared functions or genes can be strong arguments for homology. Based on operational criteria, we regard organs or structures as homologous with increasing certainty (i) if they share the same position and relationships to other structures; (ii) if they share specific qualities (e.g., microstructure, biochemical composition, gene expression patterns);

(iii) if either ontogenetic or phylogenetic intermediate stages exist (Remane 1952; Futuyma 1998; Ridley 2004; Zachos and Hoßfeld 2006; Schultze and Arratia 2013). Homologous structures need not meet all three criteria; for instance, homologous structures can develop through different pathways — because development evolves — but still be recognized as homologous structures (Hall 2003).

Stemple's (2005a) four classes of evidence for notochord — cartilage homology or for notochord as a precursor of cartilage (all of which are included in the discussion above) are:

i. shared genes (collagens II and IX, aggrecan, *Sox9*, chondromodulin-1), discussed above;
ii. similar structural components, among those elastin, discussed in Chapter 2;
iii. expression of type X collagen associated with final cell fate as cartilage is resorbed and replaced by bone and as notochord is resorbed in birds and mammals or transformed into the intervertebral nucleus pulposus in mammals (a process discussed above when discussing chordomas, which are notochord tumors). Absence of type X expression in those regions of the notochord between vertebrae that are transformed to the nucleus pulposus is cited by Stemple as supporting evidence. While such absence has been documented by Aszódi et al. (1998) and by Smits and Lefebvre (2003), and is discussed in Section 8.4C), the absence of type X expression is not a reliable criterion upon which to identify cartilage;
iv. absence of blood vessels, nerves and lymphatic vessels in notochord and cartilage in contrast to all other connective tissues.

Stemple contrasted the ECM of cartilage with the thickened basement membrane and sheath of the notochord as important differences between cartilage and notochord, although he softened this distinction by concluding that retention of hydrated material in notochord vacuoles served the function of retention of water by cartilage ECM (again a function and not a phylogenetic explanation). *Indeed, the notochord sheath is an extracellular matrix and not just an expanded basement membrane.* The notochord sheath in some vertebrates becomes very thick (see also Chapter 7) and can be cellular (see above) and so certainly is much more than a basement membrane, expanded or not. Given this and taking notochord as a member of the cartilage family, it is interesting to consider a thick acellular notochord sheath as a type of acellular cartilage similar to acellular bone (Ekanayake and Hall 1991; Franz-Odendaal et al. 2006; Hall 2015).

8.8 SUMMARY

The close relationships between notochord and cartilage, the formation of notochord tumors (chordomas) with features of notochord and cartilage, the transformation of notochord cells into chondroblasts by de- and redifferentiation or by metaplasia, and the transformation series notochord → chordoid → chondroid → cartilage, all demonstrate shared developmental features of this class of cells and tissues.

Notochord is an unusual tissue with large vacuolated cells connected by desmosomes and surrounded by an expanded basement membrane as a notochord sheath. Notochord and cartilage cells both synthesize and deposit type II collagen, the collagen type that has always been referred in the literature of developmental biology as "cartilage-type collagen." Given that no invertebrate cartilage contains type II collagen and that the notochord is a more ancient vertebrate tissue than is cartilage, it may be that type II collagen originated in the chordate or vertebrate notochord and was co-opted by cartilage-forming cells. This would make type II collagen *notochord-type collagen* and not *cartilage-type collagen* in a phylogenetic context and the deposition of type II collagen into the notochord sheath a precursor of the deposition of type II collagen into the ECM of vertebrate cartilage.

9 The Notochord and Hypotheses about the Evolution of the Vertebral Column

CONTENTS

9.1   THE NOTOCHORD AS VIEWED IN LIFE SCIENCE DISCIPLINES

Several biological disciplines are concerned with the notochord. Each discipline has its own view, which one could describe in short, and thus of course inexactly (as when three blind men are asked to identify an elephant)[1] as follows:

- For human and veterinary medicine, the notochord is part of the intervertebral disk.
- For developmental biology, the notochord is foremost a signaling center.
- For taxonomy, the notochord is a structure that is present or replaced by vertebral bodies. Paleontology has a similar view of the notochord as does taxonomy but with the complication that the notochord is often poorly preserved or not preserved at all in fossils.
- For biomechanics, the notochord is a hydroskeleton.
- For evolutionary biology, the notochord is a key component of the chordate *Bauplan*, providing evidence for common decent.

---

[1] Three blind men come across an elephant. The first man touches its leg, and concludes it's a tree, the second man touches its trunk and concludes it's a snake, the third man touches its tail and concludes it's a broom. This parable of the blind men and an elephant originated in the ancient Indian subcontinent. The Rigveda, dated to have been composed between 1500 and 1200 BCE, states "Reality is one, though wise men speak of it variously. ... After arguing, they decided to *find one and determine what it was like by direct experience*".

DOI: 10.1201/9781315155975-13

Originating as an early midline structure and as the first axial skeleton, the craniate notochord becomes part of the vertebral column. The function of the notochord for vertebral column development is pivotal as a signaling center, and to provide a structural and functional component at all stages of vertebral column development.

The increasing use of zebrafish and medaka as model organisms in developmental biology and genetics brought substantial progress toward the understanding of vertebral column development. The zebrafish notochord actively determines the location of the intervertebral joints and regulates the formation of vertebral body anlagen independent from the somites (Fleming et al. 2015; Pogoda et al. 2018; Lleras-Forero et al. 2018). We now understand the cellular and molecular mechanisms that facilitate what Thomas Huxley and Albert Kölliker observed over 160 years ago: teleost vertebral bodies arise as segmented mineralizations of the notochord sheath (Huxley 1859; Kölliker 1859). Indeed, in teleosts and some other taxa, the notochord sets the pattern and mineralizes its own sheath.

9.2   THE NOTOCHORD'S STRUGGLE WITH THE ARCUALIA HYPOTHESIS

The active notochord, as outlined above, contrasts with a classical view about the evolution of vertebral bodies, represented by the *arcualia hypothesis*, which assigns a passive role to the notochord, which is viewed as a placeholder for the developing vertebral bodies and not as an active, skeletogenic, organ.[2]

According to the arcualia hypothesis, vertebral centra evolved from sclerotome-derived paired cartilaginous elements that abut dorsally and ventrally against the notochord sheath. These cartilages represent the bases of the neural and hemal arches that eventually give rise to the vertebral centra.

Gadow and Abbott (1895) introduced *arcualia* as a general term for these cartilages. They further subdivided arcualia into dorsalia and ventralia for the dorsal and ventral cartilaginous elements, respectively. There is more than one pair of dorsalia and ventralia in each segment. The assumed plesiomorphic (shared ancestral) condition is the presence of two dorsal and two ventral pairs of these elements per segment. Gadow and Abbott (1895) identified a small anteriorly located pair of interdorsalia and a larger posteriorly located pair of basidorsalia. Likewise, at the ventral side, interventralia and basiventralia were recognized.

Gadow and Abbott concluded that no centra are formed in the notochord sheath or outside the sheath, in the skeletogenous layer, without the prior development of cartilage. This conclusion appears to fit with the mode of vertebral centra development in amniotes and chondrichthyans. Moreover, the situation in sturgeons that have arcualia but no vertebral centra was considered as a pre-vertebral centra condition. Despite their knowledge about the formation of teleost chordacentra, Gadow and Abbott expanded the arcualia hypothesis to include teleosts. As a consequence, many textbooks endorse the arcualia hypothesis as a general scheme for gnathostome vertebral centra development.

Developmental studies that focused on vertebral column development in amniotes, recognized the notochord as an active early signaling center. Beyond this early

---

[2] Also see Section 7.2 for more information about the arcualia hypothesis.

phase, however, the amniote notochord appears to be simply surrounded by cartilage. Moreover, the cartilage anlage of the amniote centrum forms as a unit with the anlagen of the neural arches. Such a pattern of development is not in conflict with the predictions of the arcualia hypothesis. As a consequence, and up to the present day, the arcualia hypothesis has formed the basis for the interpretation of data about the development and evolution of the vertebral column in osteichthyans and chondrichthyans (Ota et al. 2011; Renn et al. 2013; Peskin et al. 2020). Zangerl's remark that paired mineralizations within the notochord sheath in several extinct chondrichthyans have been wrongly interpreted as ribs (Zangerl 1981) may exemplify how data have been interpreted in the context of the arcualia hypothesis. In extant jawless vertebrates, cartilaginous elements in the caudal region of the notochord are interpreted either as evidence of a basal vertebral column or as being highly derived and degenerate vertebrae. Hall (2015) cautions that in lampreys, the sclerotome is sparse and hard to visualize, that arcualia do not develop until metamorphosis and that they do not enclose the dorsal wall of the spinal cord as vertebral arches as they do in tetrapods.

The fact that in teleosts, which constitute about half of all extant species of vertebrates, the onset of vertebral centra development is in the notochord sheath and regulated by the notochord clearly is in conflict with the arcualia hypothesis. The conflict could be resolved if one considers this mode of vertebral centrum development as a derived feature of teleosts (Pescin et al. 2020). However, as discussed in Section 6.3 and above, mineralized notochord sheaths as vertebral body anlage were already present in extinct chondrichthyans, sarcopterygians (*Neoceratodus*, Laerm 1979b) and in palaeoniscoids, a group of chondrosteans.

In the Senegal bichir *Polypterus senegalus*, an extant member of the ancient group of chondrosteans, vertebral centrum formation begins with mineralization of the notochord sheath. Since notochord sheath mineralization is faint and transient, the process is difficult to recognize in fossils or even during development of extant vertebrates (Witten and Hall 2021). The arcualia hypothesis did not take development into account; it was formulated in the context of taxonomical and paleontological research. Back in those days, these disciplines focused on more solid skeletal elements and thus viewed the notochord simply as a structure that is replaced by vertebral bodies. As described in Section 7.3, the arcualia hypothesis was eventually rejected by most scientists for osteichthyans, very strongly rejected for tetrapods by Williams (1959), and finally rejected by most workers after the middle of the 19th century (Danto et al. 2017). Williams (1959) cites Remane (1936) (translated by PEW):

> Gadow's arcualia theory suffers from two errors. It accepts only the arch elements as basic components and assumes that vertebral bodies only derive from arcualia. The possibility of autocentric vertebral body formation is not even discussed. This is a matter of concern as we cannot prove the presence of interdorsals and interventrals anywhere in tetrapods, let alone the claim that these elements are involved in vertebral body formation. Based on the findings of ontogenetic studies we must assume autocentric vertebral body formation. Thus the arch theory of vertebral body formation becomes completely obsolete.

Remane's and Williams' rebuttal of the arcualia hypothesis set the stage for the prominence given to the *autocentrum* in which sclerotome-derived cells form a layer around the notochord, a layer that is initially unsegmented (Scaal 2016). The next section addresses the *hypothesis that vertebral centra derive from autocentra*. This hypothesis assigns a much larger function to the notochord within the process of vertebral body development than did (does) the arcualia hypothesis.

9.3   THE NOTOCHORD AND AUTOCENTRA

The arcualia hypothesis suggested that from the four pairs of arcualia (two dorsal and two ventral), different sets of pairs give rise to vertebral centra in different classes of vertebrates. Accordingly, vertebral centra would be non-homologous and convergent, across gnathostomes. In contrast, Remane (1936), Mookerjee (1936) and Lauder (1980) argued that centra in all vertebrates are of autocentral origin; arcualia give rise to neural arches, hemal arches and ribs but not to vertebral centra. Autocentra derive from a fibrous, initially unsegmented, tube of varying thickness that surrounds the outer elastic membrane of the notochord in all gnathostomes (Gardiner 1983; Figure 9.1 (and see also Figure 1.2B). Gardiner further commented:

> A glance at any paper on amniote development shows that early on, a perichordal ring of fibrous tissue encloses the notochord and that skeletogenic material forms within this perichordal tube and proceeds to constrict the notochord, just as in selachians. This perichordal tube in early development appears identical in form and relationships in all vertebrate groups and it is difficult to see why it should not be considered homologous.
>
> *(Gardiner 1983, p. 12)*

Joshua Laerm discussed whether it would be possible, as an alternative to the confusion about the origin of vertebral bodies, to suggest that

> a centrum is a centrum is a centrum reflecting a generalized adaptive response of sclerotome tissues to functional demands for vertebral consolidation.
>
> *(Laerm 1979b, p. 482)*

The cells that form the autocentrum around the notochord originate from the ventral sclerotome and migrate to the perinotochordal space upon signaling from the notochord. In amniotes, an unsegmented, multilayered mantle of *Pax1*-expressing cells encases the notochord in its entire circumference. Because the migrating sclerotome cells do not cross the midline of the embryo, this apparently continuous layer of skeletogenic tissue has 'hidden' left and right halves.

In birds, both future vertebral bodies *and* intervertebral disks derive completely from the autocentrum. In other vertebrates, the autocentrum forms the bulk of the vertebral centrum, contributing to the intervertebral tissues to various degrees.

Segmentation of the autocentrum in amniotes first becomes visible as metameric cell condensations that represent the anlagen of the future intervertebral disks. Subsequently, loose mesenchyme between the condensations gives rise to the anlagen of the vertebral bodies (Scaal 2016). Subdivision of the autocentrum into vertebral centra and disks is regulated by Shh produced by the notochord. It has been suggested that notochord-derived Shh is also responsible for guiding the sclerotomal cells to the notochord. *Noggin*, which induces *Pax1* expression, is involved in this process. *Shh* induces *Pax1/9* and *Meox1/2* (*Mesenchyme Homeobox 1*) in the sclerotome, which in

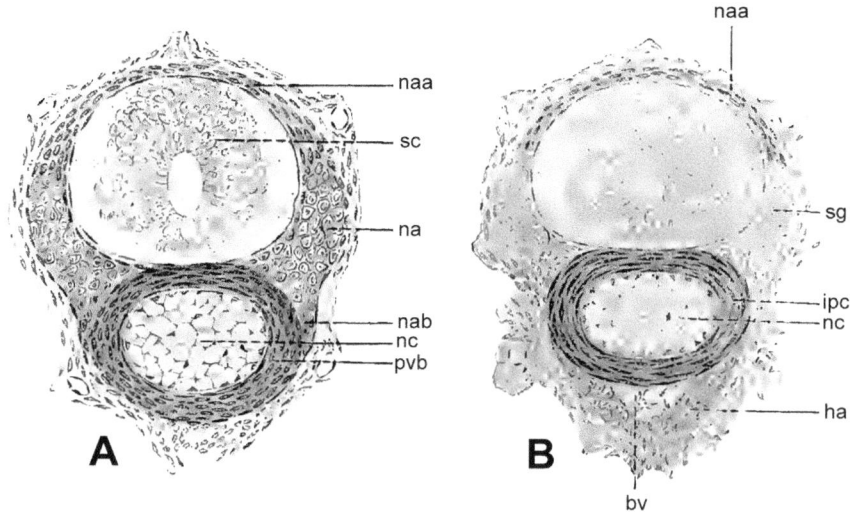

FIGURE 9.1 Vertebral body anlagen arise from perichordal cells. Two cross sections of the anlage of the vertebral column in the tail of a young tuatara embryo, *Sphenodon sp.* Section (A) cuts through the middle of an early vertebral body, section (B) through the anlage of the intervertebral space. pvb labels the primary vertebral centrum that derives from the cells around the notochord (autocentrum). The cells have a cartilaginous character. Schauinsland (1906) designated these as perichordal cells. Section (B) shows intervertebral perichordal cells (ipc), which have the characteristics of fibroblasts. Abbreviations in A and B: bv, blood vessel; ha, hemal arch; ipc, intervertebral perichordal cells; na, neural arch; naa, neural arch anlage; nab, neural arch basis; nc, notochord; pvb, primary vertebral body; sc, spinal cord; sg, spinal ganglion. (Adapted from Schauinsland, H., 1906. Die Entwicklung der Wirbelsäule nebst Rippen und Brustbein. In: Hertwig O (ed.) *Handbuch der vergleichenden und experimentellen Entwicklungslehre der Wirbeltiere* 3(2). Gustav Fischer Verlag, Jena. Figure labels translated by PEW.)

turn activate expression of *Nkx3.2* (*NK3 Homeobox 2*). Under the influence of ongoing *Bmp* signaling, *Nkx3.2* allows for *Sox9* expression, which in turn activates *Nkx3.2* in a positive feedback loop that promotes chondrogenesis in amniotes (Scaal and Wiegreffe 2006; Choi and Harfe 2011; Choi et al. 2012; Scaal 2016).

Hall (1977) summarized the experimental evidence for the notochord's function to attract sclerotomal cells, to induce cartilage formation, and to subdivide the autocentrum into vertebral centrum anlage. Studies go back to the 1950. Holtzer (1952a, b), by examining development in several urodele amphibians (*Ambystoma punctatum*, *Ambystoma tigrinum*, *Triturus torosus*), showed that removal of the notochord from embryos resulted in formation of a massive undivided cartilaginous rod. Holtzer and Detwiler (1953) further substantiated these observations. Watterson et al. (1954) showed that transplantation of the chicken notochord into a position lateral to the somites induced the formation of vertebral centrum-like cartilage around the notochord. Kosher and Lash (1975) showed that the embryonic chicken notochord stimulates or "induces" the chondrogenic differentiation of somites *in vivo* and *in vitro*. They concluded that proteoglycans, produced by the notochord, control somitic chondrogenesis. Kosher and Lash (1975) observed impairment of the notochord's ability to support chondrogenesis after treating notochords with enzymes (chondroitinases and hyaluronidases) that specifically degrade proteoglycans. Subsequently, Fleming et al. (2004) prevented the formation of vertebral centra in zebrafish by ablating notochord cells. Finally, Lleras-Forero et al. (2018) underscored the independence of the autocentrum from the somites by showing that even triple mutant zebrafish, with disturbed somite formation plus ablated somitic clock genes, still maintain separated vertebral centra. Separation is maintained in later stages when the chordacentrum is at best a vestige and the bone of the vertebral centrum is derived from the autocentrum (Figures 9.2 and 9.3).

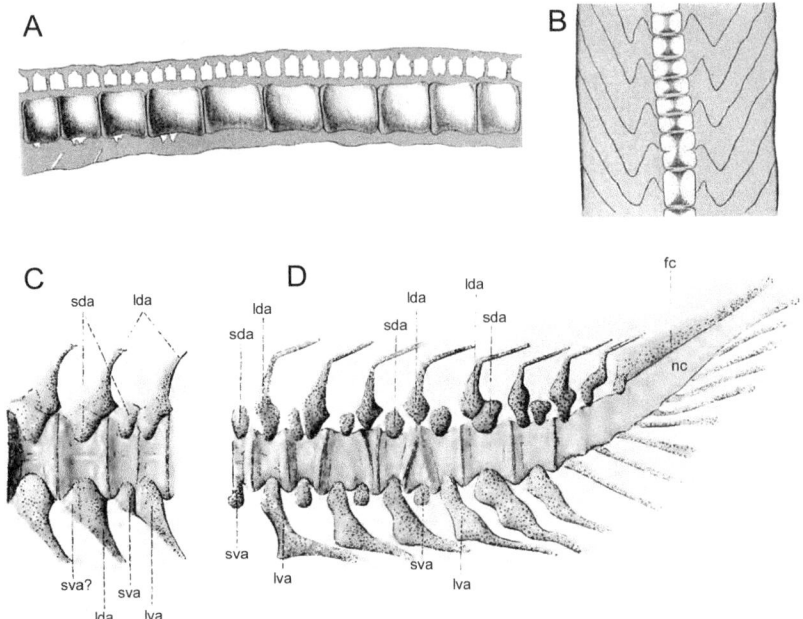

FIGURE 9.2 Examples of the arrangement of vertebral centra in relation to arches and myosepta that deviate form a one-to-one relationship in chondrichthyans and osteichthyans. (A) Smooth hammerhead shark, *Sphyra zygaena*. (B) Spiny dogfish *Squalus acanthias*. (C and D) A young bowfin, *Amia sp.* (7.5 cm total length).

(A) In the smooth hammerhead shark, several neural arches (white) are associated with each vertebral centrum, but the pattern is not regular.

(B) The variable arrangement of vertebral centra (white) and myosepta (chevrons) representing former somite boundaries in the spiny dogfish.

(C) The transition between the abdominal and caudal vertebral column in *Amia* showing the change from monospondyly (one centrum per segment, left) to diplospondyly (two centra per segment, right).

(D) The variable arrangement of arches and centra in the caudal-most part of the vertebral column of *Amia*. Similar variations occur in the caudal-most region of the teleost vertebral column (Bensimon-Brito et al. 2012b; Martini et al. 2021). Abbreviations in C and D: fc, fused cartilage element; lda, large dorsal arch; lva, large ventral arch; nc, notochord; sda, small dorsal arch; sva, small ventral arch. ((A and B) From Šećerov, 1911. *Arb Zool Inst Wien* Bd. 19: 1–28. (C and D) Adapted from Schauinsland, H., 1906. In: Hertwig O (ed.) *Handbuch der vergleichenden und experimentellen Entwicklungslehre* der Wirbeltiere 3(2). Gustav Fischer, Jena. Figure labels translated by PEW.)

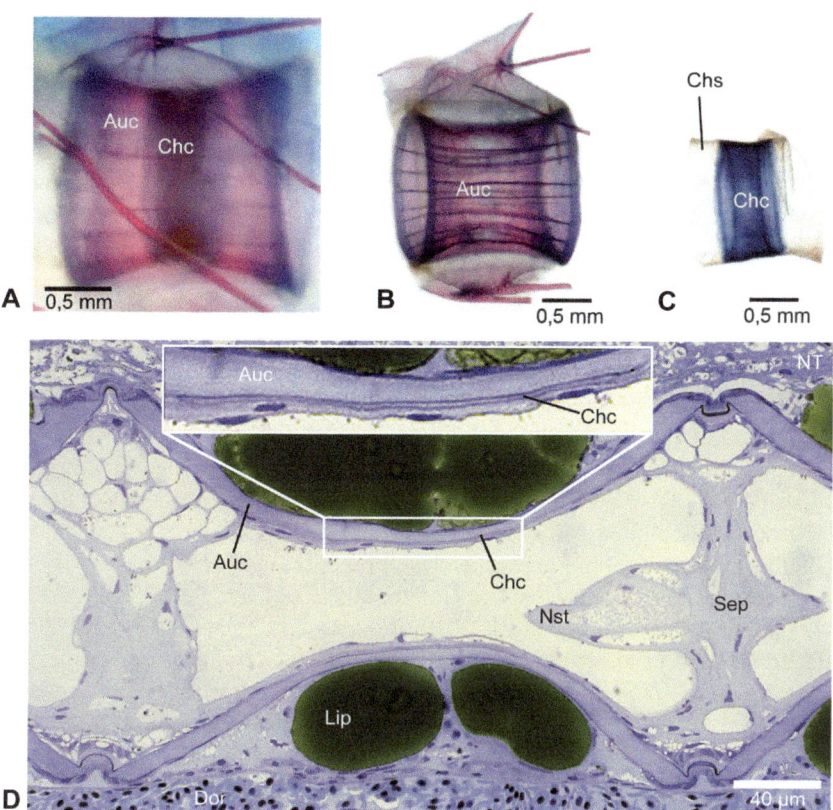

FIGURE 9.3 (A-C) Centra formation through chorda- and autocentra in Sloane's viperfish, *Chauliodus sloani* (230 mm length), a deep sea species from the order Stomiiformes
(A) A vertebral centrum formed by the autocentrum (Auc) and the chordacentrum (Chc).
(B, C) Separated autocentrum (Auc) and chordacentrum (Chc) with notochord sheath (Chs). The ability of the chordacentrum to separate from the autocentrum likely is related to the low degree of mineralization of vertebral centra. In species with a greater degree of centrum mineralization, chorda and autocentra fuse.
(D) A fully mineralized vertebral body of the Japanese medaka, *Oryzias latipes* from the order Cyprinodontiformes. The chordacentrum (Chc) is fully mineralized and fused with the bone of the autocentrum (Auc). Here, vestiges of the outer elastin layer of the notochord sheath (Chs) are still visible inside the bone as a dark blue line. In many other species, it is no longer possible to recognize traces of the autocentrum. Other abbreviations: Dor, dorsal aorta; Lip, lipids; NT, neural tube. (A–C Whole mount double staining for bone and cartilage after Dingerkus and Uhler (1977). D, Epon embedding, sagittal section, Toluidine blue staining, after Huysseune et al. 2022b. A–C From Schnell, et al., 2010. *J Morphol* 271: 1006–1022, permission and images provided by Nalani Schnel. D provided by PEW.)

However, unlike zebrafish, in which mutations of somite or clock genes primarily affect the formation of the arches, vertebral centra also are affected by these mutations in chicken and mice (Andrade et al. 2007).

Arguably, as also discussed by Ward and Stern (2018), the influence of the notochord on vertebral centrum development in teleosts and amphibians is stronger than it is in chickens and mice. There appears to be a trend within the amniotes for the co-option of initially notochord-related induction pathways to the somites. Although the function of the notochord outlined above was based on zebrafish, chickens and mice, detailed descriptions of the formation of the unsegmented perinotochordal tube (autocentrum) and its subdivision to give rise to vertebral centra independent from the arches exist for all groups of vertebrates.[3]

9.4   CENTRA FROM NOTOCHORD, ARCHES FROM SOMITES

Modules are defined as discrete anatomical units or groups of cells that can change without affecting other modules. Modules for hind- and forelimb, toes and fingers are examples and show that modules can be hierarchical. Similarly, developmental processes can be modular; one process may function and change without affecting another process (Ward and Brainerd 2007). Cell division and cell differentiation are two examples. Studies on zebrafish and medaka revealed the developmental and genetic mechanisms by which the notochord determines vertebral body identity, independently of the somites and independently of the somitic clock.

Gnathostome centra and arches as developmental modules have long been recognized (Strudel 1953; Detwiler and Holtzer 1956; DeClercq et al. 2017, 2018; Hall 1977; Lauder 1980; Hautier et al. 2010; de Azevedo et al. 2012; Hall 2015; Yu et al. 2017; Ward et al. 2018; Witten and Hall 2021). Numerous examples of this modularity exist. *Diplospondyly* (two centra per segment), which is widespread in the caudal region of selachians and non-teleost actinopterygians such as halecomorphs (the bowfin *Amia calva* as an extant example) and chondrosteans, is an obvious example (Hay 1895; Lauder 1980; Figure 9.3). Gardiner (1983) argues that diplospondyly is the basic condition for amniotes and that acceptance of this analysis would render the resegmentation theory redundant. The complete range of spondyly encompasses aspondyly, monospondyly, diplospondyly and polyspondyly (Zhang 2009) and selachian vertebral centra that extend across several somite boundaries (Šećerov 1911).

The vertebral column of essentially all gnathostomes is functionally and morphologically regionalized and this regionalization requires the uncoupling of arch and centra development. Key evidence for arch and centra modularity, reviewed by Hall (2015), was obtained from ablation experiments, from mutants, and from vertebral column malformations (Strudel 1955; Takeuchi 1966; Van Eeden 1996; Fleming et al. 2004; Witten et al. 2006; Sporendonk et al. 2008). Hautier et al. (2010), who tested the patterns of events in the sequence of development of arches and centra in the brown-throated sloth *Bradypus variegatus*, identified significant modularity in this mammalian species.

---

[3] Support for these conclusions is based on studies of selachians (Grassé 1958), dipnoans (Andrew and Westoll 1970), chondrosteans (Patterson 1968), holosteans (Schultze and Arratia 1986), teleosts (Schaeffer 1967; Laerm 1976, 1982; Bensimon Brito et al. 2012b; Wang et al. 2013), amphibians (Lawson 1966; Danto et al. 2017), birds (Piiper 1928; Williams 1942; Kosher and Lash 1975; Shapiro (1992), mammals (Verbout 1976; Carlier 1980; Götz et al. 1995; Scaal 2016) and discussion of several gnathostome groups (Gardiner 1983).

Today, modularity between arches and centra is generally accepted, which makes it a concept that is clearly in conflict with the still widely endorsed arcualia hypothesis, although this conflict is often not addressed in the literature. The widespread observation that vertebral body anlagen arise independently from the arches refutes the arcualia hypothesis that views cartilaginous arch elements as the units from which vertebral bodies are derived (Ota et al. 2011, 2014).

Modularity of centra and arches triggers the question of whether the notochord or the somites were the first segmented structure in vertebrates, or indeed in chordates. The absence of somites in basal chordates such as tunicates, led Claudio Stern to suggest that the notochord may have been the first segmented vertebrate structure (Stern 1990). Ward et al. (2018) and Peskin et al. (2020) argue that both somites and notochord are important for the segmented development of vertebral bodies with somites taking the lead in amniotes and the notochord taking the lead in teleosts. Perhaps we will not be able to answer the question "did somitic or intrinsic notochord segmentation evolve first?" Still, the rediscovery of the active segmented notochord and the new knowledge about the molecular mechanisms by which the notochord establishes vertebral centra anlagen in teleosts may encourage scientists to revisit notochord segmentation, including studies on non-teleost vertebrates. This in turn may influence the interpretation of the fossil record. Not only primary aquatic osteichthyans, but amphibians and stemward amniotes (including early mammals) have amphicoelous (hourglass-shaped) vertebral centra and a continuous notochord (Wintrich et al. 2020) (Figure 7.2).

9.5   THE FUSION OF CHORDACENTRA AND AUTOCENTRA

The literature concerning autocentra and chordacentra can be nebulous. This is because the commonly used term 'notochord sheath' can refer to the prospective chordacentrum, the notochord sheath proper (Jonasson et al. 2012; Trapani et al. 2017), or the autocentrum that forms around the notochord sheath (Christ and Ordahl 1995; Cole 2011). Some literature addresses the autocentrum as the notochord sheath without mentioning whether there is also a notochord sheath proper.

It can indeed be very difficult to decide if a fibrous sheath that encases the notochord is the product of the notochord epithelium or the product of mesenchymal cells that surround the notochord. Mineralization of the former would constitute a chordacentrum, mineralization of the latter would constitute an autocentrum. The truth is, chordacentra and autocentra have essentially identical functions. They are rings of mineralized connective tissue around the notochord that are separated by non-mineralized sections. Moreover, as we have learned from classical ablation experiments and the recent studies on zebrafish notochord development, both mineralization and septation of autocentra and chordacentra is controlled by signals from the notochord. Chordacentra and autocentra eventually fuse, often early during development. Below we elaborate the arguments listed above as we argue that the evolution and development of the vertebral column is better understood if we regard the two structures as equal: rings of mineralized connective tissue that form around the notochord (Figures 6.3 and 9.4).

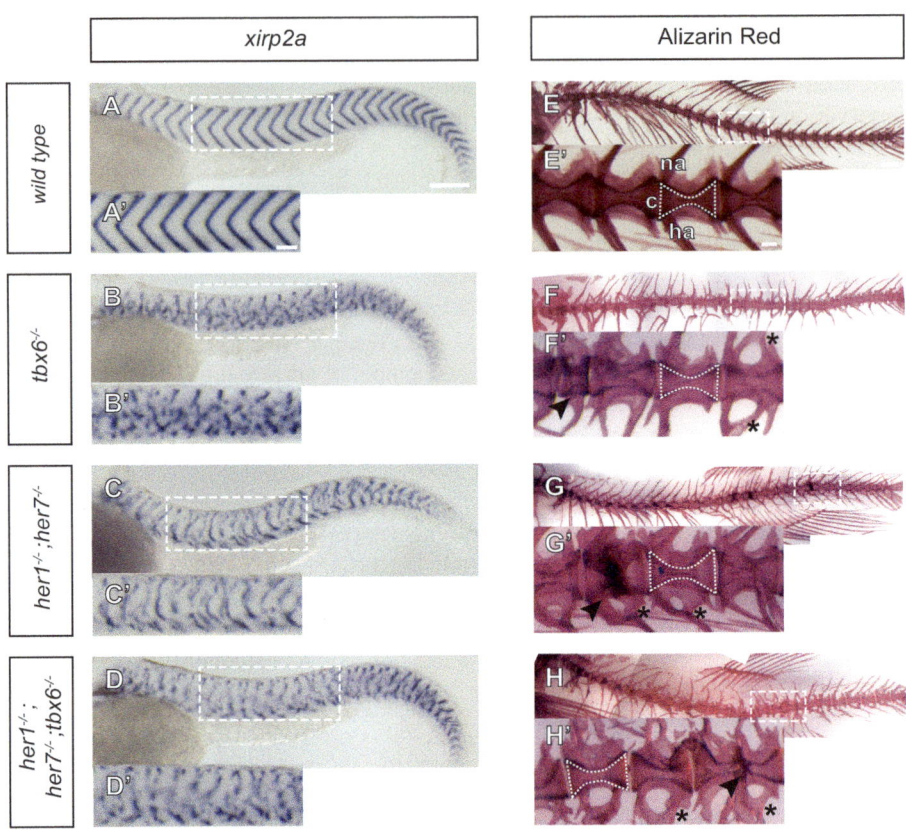

FIGURE 9.4 Myotome boundaries are disrupted in *fused somite* mutant zebrafish and in animals with additional segmentation clock mutations, but the vertebral centra (chordacentra plus autocentra) are still patterned and separated. (A–D) (wild-type) and (A′–D′) (mutant) *in situ* hybridization for myotome boundaries using the marker *Xirp2a*. Myotome boundaries are severely disrupted in all mutants shown at 40 days post-fertilization. (E–H) (wild-type) and (E′–H′) (mutant) vertebrae with Alizarin red staining for bone in animals between two and six months old. (E) Wild-type zebrafish with well aligned myosepta. (F, F′) Centra are well-formed and separated in *tbx6* $^{-/-}$ mutant (*fused somite mutant* zebrafish) while the patterning of neural and hemal arches clearly is disturbed. Centra are also well-formed and separated in *her1*$^{-/-}$, *her7*$^{-/-}$ mutants (G, G′) and in *her1*$^{-/-}$, *her7*$^{-/-}$, *tbx6*$^{-/-}$ mutants (H, H′). Occasional vertebral compression (arrowhead in F′), ectopic mineralization (arrowhead in G′) and fusion of adjacent vertebrae (arrowhead in H′) occur. Asterisks highlight fused neural and hemal arches. Abbreviations: na, neural arch; hr, hemal arch; c, centrum. Scale bars in A and A′ are 150 and 100 mm respectively and apply to A–D and A′–D′ respectively. Scale bar in E is 1 mm and applies to E–H, scale bar in E′ is 200 mm and applies to E′–H′. (From Leras Forero, et al., 2018. *eLife* 7: e33843. Used with permission from Stefan Schulte-Merker.)

Traditionally, chordacentra and autocentra are considered as different starting points for the formation of vertebral centra. Chordacentra can be mineralized segments of the notochord sheath, which is a common type of actinopterygian vertebral centra anlage but occurs also outside the actinopterygians. In elasmobranchs, chordacentra are formed by the segmented invasion of cartilage precursor cells into the notochord sheath (Figures 5.5 and 5.6). Like the notochord sheath, autocentra consist of initially non-segmented skeletogenic tissue that forms close around the notochord sheath. The tissue is of sclerotomal origin and develops into either cartilage or bone. The cartilage that contributes to vertebral centra can be transformed into bone via endochondral bone formation, a typical process in amniotes. Bone can also form directly as intramembranous ossification without a cartilaginous precursor (several amphibians including apodans, teleosts, *Lepisosteus* and *Polypterus*) (Gardiner 1983). Formation of vertebral body anlagen from chordacentra *or* from autocentra is commonly viewed as being based on distinct pathways which may suggest convergent evolution of vertebral centra in different gnathostome lineages. Another argument for the convergent evolution of vertebral centra in different groups of vertebrates is the patchy presence and absence of vertebral centra in extinct primitive and extant basal osteichthyans. Ossified vertebral centra in essentially all members of the group are only common in teleosts and tetrapods (Arratia et al. 2001).

The development of vertebral bodies from chordacentra is best studied in teleosts and elasmobranchs (Fleming et al. 2015). Despite their importance as centrum anlagen, teleost chordacentra are faint and transient structures. The main material for the vertebral centra is eventually contributed by the autocentra, as is generally the case in vertebrates (Williams 1959) (Figure 9.2). After the start of autocentrum mineralization, chordacentra fuse with autocentra and usually become unidentifiable (Bensimon-Brito et al. 2012a, b; Fleming et al. 2015).

The situation in elasmobranchs is remarkably similar. The cartilaginous chordacentra fuse with the surrounding cartilage of the autocentra. Only remains of the fenestrated outer elastin layer indicate the previous border between the cartilage of the chordacentra and the cartilage of the autocentra (Figure 5.6C). Some advanced teleosts groups have lost chordacentra altogether, while chordacentra are usually not recognized in amphibians and amniotes. Yet, vertebral centrum development in amphibians can involve the mineralization of the notochord sheath (Welsch and Storch 1971) and the fusion of notochord-derived cartilage with cartilage from the autocentrum. The latter occurs when *notochord-derived cartilage inside the notochord* undergoes endochondral bone formation as does the surrounding cartilage of the autocentrum (Wake 1970, and see Section 8.6A). Whether mineralization of the notochord sheath in human embryos of 32 mm length that occurs together with the mineralization of the autocentrum (Williams 1908) qualifies as chordacentrum formation remains to be elucidated. Studies on chickens and mice recognize only the development of an autocentrum.

The hypothesis according to which vertebral centra in gnathostomes derive from autocentra is nowadays widely accepted (see Sections 9.1 and 9.2). If, or to what extent, chordacentra were present earlier in development is difficult to evaluate. Evidently, the faint and transient nature of chordacentra also contributed to the lack of recognition of chordacentra in parts of the scientific literature for nearly 140 years (see Chapter 7). In those vertebrate clades in which we do not recognize chordacentra, either the structures were never present or rudiments are present but have not yet been discovered. A third option would be that chordacentra have been lost as is clearly the case in birds and other amniotes in conjunction with reduced notochord signaling activity (Ward et al. 2018).

The presence of chordacentra in extant chondrosteans and of ring vertebrae in early osteichthyans could suggest that chordacentra once had a more prominent role in vertebral body development. Fossils with mineralized chordacentra have been identified (Gardiner 1983). Still, it is very difficult to distinguish mineralized notochord sheath segments from mineralized autocentra in fossils (John Maisey, personal communication). In extant vertebrates, ring vertebrae can be autocentra, with or without a chordacentrum fused to the autocentrum. Eventually the distinction between chordacentra and autocentra becomes less important if one considers the function of the notochord for chordacentrum and autocentrum patterning, addressed in Section 9.3.

When developmental biology started to use zebrafish and medaka as model organisms in place of chicken and mice, the molecular science community rediscovered teleost chordacentra. What followed was a chain of groundbreaking studies that revealed the mechanisms by which the notochord controls both, the segmented mineralization of its own sheath and mineralization of the autocentrum. The role of the notochord turned out to be extensive and somite-independent. Eventually, classical embryological experiments that had shown the importance of the notochord for vertebral centra formation and molecular studies came together (see Chapter 7).

Excitingly, all classical and modern studies with embryos old enough to develop autocentra, or with species that only form autocentra, show that the patterning function of the notochord is not restricted to chordacentra. The notochord also patterns the autocentrum and carries out this role independently of the somites. *Fused somite* mutant zebrafish have separated autocentra. The same mutant with additional ablated somitic clock genes still maintains separate autocentra (Lleras-Forero et al. 2018). This is intriguing because arches and autocentra are both of sclerotomal origin but they are under different control: Arch patterning is controlled by the somites; autocentra patterning is controlled by the notochord (Figure 9.4). Cell signaling apparently does not care about the origin of the mesenchyme to be patterned. Possibly, signaling was co-opted from one set of mesenchymal cells to another early in vertebrate evolution.

Concerning location, function, and control of development, one can view the autocentrum as an outer jacket of the notochord sheath. This extends the notochord sheath and turns it into a structure of mixed origin, somitic- and notochord-derived mesenchyme. This is not as odd a suggestion as it may seem; organs and skeletal elements of mixed origin are not unusual for vertebrates, examples being the skull (neural crest and mesoderm), the middle ear (neural crest and endoderm), the pharyngeal arches (neural crest and mesoderm) and the oral epithelium (ectoderm and endoderm) (Couly et al. 1993; Thompson and Tucker 2013; Hall 2014; Sefton et al. 2015; Oralova et al. 2020; Huysseune et al. 2022a). Not least, as discussed in Chapter 3, there is evidence that the anterior-most section of the notochord is endoderm-derived while posterior parts derive from mesoderm and the caudal-most part from the chordoneural hinge. Gardiner (1983) put the discussion about the origin of the layers of the notochord sheath into perspective:

> Since both mesoderm and notochord have a similar embryonic origin (from chorda-mesodermal plate) and the notochord behaves like a mesoderm derivative having the same type II collagen as cartilage, it seems pointless to argue whether or not the mesenchyme of the sheath is chordal or mesodermal in origin, because it will possess the same

potentiality for skeletogenic development whichever its source. It suffices to point out that there is an elastic membrane surrounding the notochord in all gnathostomes and that on the outside of this membrane there is a fibrous layer of varying thickness (pp. 11–12).

If we consider chordacentra and autocentra as a unit, it is possible to view sarcopterygian and actinopterygian vertebral centra as homologous. The autocentrum — with or without a transient chordacentrum — unites ring vertebrae of basal gnathostomes with ring vertebrae of sarcopterygians and actinopterygians. Ring vertebrae grow into amphicoelic, hourglass-shaped vertebral centra, the common vertebral centrum type for actinopterygians and the basal type for all tetrapod groups.

Mineralized vertebral centra connect the fossil record with extant vertebrates. The shared function of the notochord for initiation of vertebral column development goes, of course, beyond the induction of segmented mineralization. The notochord (i) induces neurulation, (ii) patterns the presomitic mesoderm, and (iii) induces sclerotome differentiation. In amniotes, however, the intrinsic segmentation of the sclerotome becomes dominant over segmental information from the notochord (Ward et al. 2018). Still, amniote vertebral centra cannot subdivide without the notochord (Stern 1990, and see Section 9.3). Clearly, the amniote notochord also retains patterning information. This becomes more obvious if one shifts the focus from vertebral centra formation to the initiation of intervertebral joints.

9.6   JOINT FORMATION, A KEY FOR VERTEBRAL COLUMN HOMOLOGY

Studies of the development and the evolution of the vertebral column focus on vertebral centra. Centra are preserved in the fossil record and neontology has numerous techniques to visualize mineralized tissues in high detail (Bruneel and Witten 2015). The vertebral column is, however, a metameric structure composed of alternating intervertebral joints and vertebral bodies (Symmons 1979). For the function of the vertebral column, the joints are perhaps more important than the vertebral bodies. The notochord alone can be a fully functional flexible axial skeleton and flexibility must be maintained when vertebral centra arise (Figure 7.2). Setting up the positions of the non-mineralized intervertebral joints is a function of the notochord and a highly conserved process throughout the vertebrates. This has been clearly shown in mammals (Leeson and Leeson 1958) where the notochord persists and has a lifelong function to maintain intervertebral joints (Figure 7.8). Prior to vertebral centrum formation, the notochord extends in size in the location of the prospective intervertebral joint. Moreover, the notochord epithelium increases the formation of the notochord sheath matrix (Figures 5.2, 6.3, and 6.5).

Studies on zebrafish and salmon have provided new insights into the cellular and molecular mechanisms underlying intervertebral joint formation (Grotmol et al. 2003; Wopat et al. 2018; Lleras-Forero et al. 2018). Ridewood (1921) showed the process for elasmobranchs. Even in birds, where the notochord disappears in the course of development, the notochord extends intra- or intervertebrally. Intervertebral extension shapes cervical intervertebral joints in the common gull (*Larus canus*) similar to other gnathostomes, as beautifully shown by Piiper (1928) (Figure 9.5).

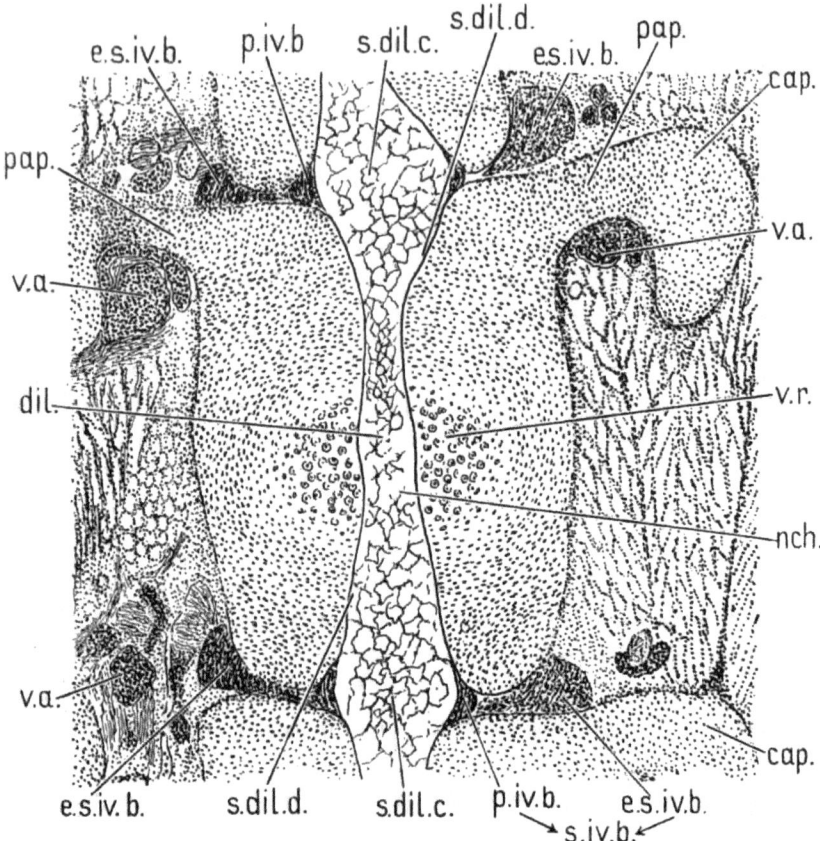

FIGURE 9.5 A frontal section of a vertebra from the cervical region of a ten-day-old common gull (*Larus canus*) embryo. The notochord extends intravertebrally (dil) and intervertebrally (s.dil.c.). The section passes through the middle portion of the notochord, anterior above. Abbreviations: cap., capitular portion of the rib; dil., dilatation; e.s.iv.b., external portion of secondary intervertebral body; nch., notochord; p.iv.b., primary intervertebral body; pap., parapophysis; s.dil.c., secondary dilatation of constricted portion of notochord; s.iv.b., secondary intervertebral body; v.a., vertebral artery; v.r., vertebral ring. (From Piiper, 1928. *Philos Trans R Soc London B* 216 (431–439): 285–351.)

While information for the subdivision of the autocentrum was transferred from the notochord to the somites in amniotes, the function of the notochord for generating intervertebral joints is obviously conserved.

Witten and Hall (2021) discuss the homology of vertebral bodies in different classes of vertebrates linked to processes in early vertebral column development, a typical approach from the perspective of the discipline evolutionary developmental biology. Apart from the early patterning function of the notochord for setting up intervertebral spaces, other early events in development could also be considered as the start of vertebral column formation. Considered as characters, these developmental events differ from characters that are commonly used in discussions about the homology of vertebral bodies across vertebrates. Morphological characters, for example, are typically viewed as two or three dimensional structures. But, as pointed out by Wallace Arthur, all characters (morphological or molecular) have an additional dimension which is time and time translates into development (Arthur 1997).

As stated above, classical embryology considers the condensation of somite-derived chondrogenic cells around the notochord as the start of vertebral body development. Paleontology largely relies on the recognition of mineralized vertebral bodies, be it mineralized cartilage or bone. Comparative anatomy and taxonomy largely follow this approach. Still, each of the following conserved events could equally be defined as key characters of vertebral body development (Witten and Hall 2021):

1. Notochord signaling to subdivide the presomitic mesoderm into somites.
2. Molecular subdivision of the notochord into intravertebral notochord and intervertebral spaces.
3. Expansion of the notochord to establish intervertebral joints.
4. Notochord signaling (*brachyury*) to induce the differentiation of skeletal precursor cells.
5. Mineralization of the notochord sheath.

Conserved early events of vertebral column development and the use of a hierarchical homology concept (Hall 2003; West-Eberhard 2003) can shed new light on whether a vertebral column evolved once or multiple times (also discussed by Fleming et al. 2015). If we recognize evidence from classical notochord ablation experiments along with evidence from new molecular studies that the notochord is a segmented structure (Stern 1990), and if we recognize notochord segmentation as the basis for vertebral column development, then we must consider *vertebral column development as homologous across the vertebrates*.

9.7 SUMMARY

In this last chapter in our journey through notochord and vertebral development and evolution, we addressed the perennial question of the role and fate of the notochord across the vertebrates. Each group of specialists has approached the notochord from the perspective of their own discipline. We tease apart those perspectives to conclude that the function of the notochord for vertebral column development is threefold and essential: (i) as a signaling center, and to provide (ii) structural and (iii) functional components at all stages of vertebral column development.

Since 1895, one hypothesis — the arcualia hypothesis — has dominated the interpretation of the relationship between the notochord and vertebral centra development. According to this hypothesis, vertebral centra evolved from sclerotome-derived paired cartilaginous elements (arcualia) that abut dorsally and ventrally against the notochord sheath to form the bases of the neural and hemal arches and eventually the vertebral centra.

In contrast, based on developmental studies, the now widely accepted autocentra hypothesis claims that a perichordal ring of fibrous tissue forms a second layer around the notochord and that skeletogenic material of the vertebral bodies forms within this perichordal tube. In about half of all extant vertebrate species (teleosts), vertebral centra (chordacentra) development is initiated in the notochord sheath and then proceeds to the second layer, the autocentrum. The formation of chordacentra and autocentra is regulated by the notochord, two processes clearly in conflict with the arcualia hypothesis. Chordacentra have not been identified in tetrapods. Rather, sclerotome-derived cells form an unsegmented layer around the notochord that subsequently separates into vertebral centra (autocentra). There is molecular evidence that defining the intervertebral spaces is the early step of vertebral body formation. Extension of the notochord sheath and proliferation of the notochord epithelium are visible characters in this process. After considering the available evidence, we conclude that, although chordacentra arise from the notochord, and although autocentra are sclerotomal, they are functionally equivalent and represent one developmental module. The notochord controls mineralization and patterning of both structures. We conclude that chordacentra and autocentra, and the process of intervertebral joint initiation, are homologous across the vertebrates. Therefore, vertebral columns are homologous across the vertebrates.

References

Adachi, N., and Kuratani, S. 2012. Development of head and trunk mesoderm in the dogfish, *Scyliorhinus torazame*: I. Embryology and morphology of the head cavities and related structures. *Evol Dev* 14: 234–256.

Adams, D. S., Keller, R., and Koehl, M. A. 1990. The mechanics of notochord elongation, straightening and stiffening in the embryo of *Xenopus laevis*. *Development* 110: 115–130.

Agassiz, A. 1877. On the young stages of some osseous fishes. *Proc Am Acad Arts Sci* 13: 117–127.

Aires, R., de Lemos, L., Nóvoa, A., Jurberg, A. D., Masccrez, B., Duboule, D., and Mallo, M. 2019. Tail bud progenitor activity relies on a network comprising *Gdf11*, *Lin28*, and *Hox13* genes. *Dev Cell* 48: 383–395. doi: 10.1016/j.dvcel.2018.12.004.

Akiyama, H., Chaboissier, M.-C., Martin, J. F., Schedl, A., and de Crombrugghe, B. 2002. The transcription factor Sox9 has essential roles in successive steps of the chondrocyte differentiation pathway and is required for expression of *Sox5* and *Sox6*. *Genes Dev* 16: 2813–2828.

Alenghat, F. J., and Ingber, D. E. 2002. Mechanotransduction: all signals point to cytoskeleton, matrix, and integrins. *Sci STKE* 119: pe6. doi: 10.1126/stke.2002.119.pe6.

Amacher, S. L., and Kimmel, C. B. 1998. Promoting notochord fate and repressing muscle development in zebrafish axial mesoderm. *Development* 125: 1397–1406.

Anderson, C., and Stern, C. D. 2016. Organizers in development. *Curr Top Dev Biol* 117: 435–454.

Anderson, C., Bartlett, S. J., Gansner, J. M., Wilson, D., He, L., Gitlin, J. D., Kelsh, R. N., and Dowden, J. 2007. Chemical genetics suggests a critical role for lysyl oxidase in zebrafish notochord morphogenesis. *Mol BioSyst* 3: 51–59.

Andrade, R. P., Palmeirim, I., and Bajanca, F. 2007. Molecular clocks underlying vertebrate embryo segmentation: a 10-year-old hairy-go-round. *Birth Defects Res* 81: 65–83.

Andrews, S. M., and Westoll, T. S. 1970. XII. — The postcranial skeleton of rhipidistian fishes excluding *Eusthenopteron*. *Trans R Soc Edinburgh* 68: 391–489.

Ang, S. L., and Rossant, J. 1994. HNF-3 beta is essential for node and notochord formation in mouse development. *Cell* 78: 561–574.

Annona, G., Holland, N. D., and D'Aniello, S. 2015. Evolution of the notochord. *EvoDevo* 6: 30.

Apschner, A., Huitema, L. F. A., Ponsioen, B., Peterson-Maduro, J., and Schulte-Merker, S. 2014. Zebrafish enpp1 mutants exhibit pathological mineralization, mimicking features of generalized arterial calcification of infancy (GACI) and pseudoxanthoma elasticum (PXE). *Disease Models Mech* 7: 811–822.

Arendt, D., and Nübler-Jung, K. 1996. Common ground plans in early brain development in mice and flies. *BioEssays* 18: 255–259.

Arratia, G., and Schultze, H. P. 1992. Reevaluation of the caudal skeleton of certain actinopterygian fishes: III. Salmonidae. Homologization of caudal skeletal structures. *J Morph* 214: 187–249.

Arratia, G., Schultze, H. P., and Casciotta, J. 2001. Vertebral column and associated elements in dipnoans and comparison with other fishes: development and homology. *J Morph* 250: 101–172.

Arthur, W. 1997. *The Origin of Animal Body Plans. A study in Evolutionary Developmental Biology*. p. 339. Cambridge University Press, Cambridge, UK.

Ashhurst, D. E. 2004. The cartilaginous skeleton of an elasmobranch fish does not heal. *Matrix Biol* 23: 15–22.

Aszódi, A., Chan, D., Hunziker, E., Bateman, J. F., and Fassler, R. 1998. Collagen II is essential for the removal of the notochord and the formation of intervertebral discs. *J Cell Biol* 143: 1399–1412.

Babić, M. S. 1990. Relationship between notochord and the bursa pharyngea in early human development. *Cell Differ Dev* 32: 125–130.

Babić, M. S. 1991. Development of the notochord in normal and malformed human embryos and fetuses. *Int J Dev Biol* 35: 345–352.

Balfour, F. M. 1876. On the development of the spinal nerves in elasmobranch fishes. *Proc R Soc London* 24: 135–136 (abstract).

Balfour, F. M. 1877. On the development of the spinal nerves in elasmobranch fishes. *Philos Trans R Soc London B* 166: 175–195 (reprinted in *Studies, Physiol. Lab. Camb* 1877, 3: 54–81).

Balfour, F. M. and Parker, W. 1882. On the structure and development of *Lepidosteus*. *Philos Trans R Soc London* 173: 359–442.

Balling, R., Neubüser, A., and Christ, B. 1996. Pax genes and sclerotome development. *Sem. Cell Dev Biol* 1: 129–136.

Balmer, S., Nowotschin, S., and Hadjantonakis, A. K. 2016. Notochord morphogenesis in mice: current understanding and open questions. *Dev Dyn* 245: 547–557.

Balon, E. K. 1990. Epigenesis of an epigeneticist: the development of some alternative concepts on the early ontogeny and evolution of fishes. *Guelph Ichthyol Rev* 1: 1–48.

Balon, E. K. 1991. The mystery of a persistent life form. *Env Biol Fishes* 32: 9–13.

Balon, E. K. 2003. Alternative ontogenies and evolution: a farewell to gradualism. In *Environment, Development, and Evolution. Toward a Synthesis* (B. K. Hall, R. D. Pearson, and G. B. Müller, eds), pp. 37–66. The MIT Press, Cambridge.

Bancroft, M., and Bellairs, R. 1976. The development of the notochord in the chick embryo, studied by scanning and transmission electron microscopy. *J Embryol Exp Morph* 35: 383–401.

Barnes, M. J., Constable, B. J. and Kodicek, E. 1969. Studies in vivo on the biosynthesis of collagen and elastin in ascorbic acid-deficient guinea pigs. *Biochem J* 113 (2): 387–397.

Barrionuevo, F., Taketo, M. M., Scherer, G., and Kispert, A. 2006. Sox9 is required for notochord maintenance in mice. *Dev Biol* 295: 128–140.

Barteczko, K., and Jacob, M. 2002. The morphology of the rostral notochord in embryos of *Ichthyophis kohtaoensis* (Amphibia, Gymnophiona) is comparable to that of higher vertebrates. *Anat Embryol* 205: 99–112.

Bartsch, P. 1989. Some remarks on the axial skeleton in *Protopterus* (Pisces, Dipnoi). *Gegenbaurs Morphol Jahrb* 135: 419–426.

Bartsch, P., and Gemballa, S. 1992. On the anatomy and development of the vertebral column and the pterygophores in *Polypterus senegalus* CUVIER, 1892 ("Pisces", Polypteriformes). *Zool Jb Anat* 122: 497–529.

Bassham, S., and Postlethwait, J. 2000. *Brachyury* (T) expression in embryos of a larvacean urochordate, *Oikopleura dioica*, and the ancestral role of T. *Dev Biol* 220: 322–332.

Bastiani, M., and Parton, R. G. 2010. Caveolae at a glance. *J Cell Sci* 123: 3831–3836.

Bateson, W. 1884a. The early stages in the development of *Balanoglossus (sp. incert.)*. *Q J Microsc Sci* 24: 206–236.

Bateson, W. 1884b. On the development of *Balanoglossus*. *Ann Mag Nat Hist* 13: 65.

Bateson, W. 1886. The ancestry of the Chordata. *Q J Micros Sci* 26: 535–571.

Bazin, S., and Strudel, G. 1972. Mise en évidence de collagêne dans le matériel extracellulaire des organes axiaux de jeunes embryons de poulet. *CR Acad Sci* 275: 1167–1170.

Bazin, S., and Strudel, G. 1973. Biosynthesis of collagen in the axial organs of young chick embryos. In *Biology of Fibroblasts* (E. Kulonen and J. Pikkarairen, eds), pp. 411–416. Academic Press, New York, NY.

Beck, C. W., and Slack, J. M. W. 1998. Analysis of the developing *Xenopus* tail bud reveals separate phases of gene expression during determination and outgrowth. *Mech Dev* 72: 41–52.

Beck, C. W., and Slack, J. M. W. 1999. A developmental pathway controlling outgrowth of the *Xenopus* tail bud. *Development* 126: 1611–1620.

Beddington, R. S. P., and Robertson, E. J. 1998. Anterior patterning in mouse. *Trends Gen* 14: 277–284.

Benjamin, M. 1989. Hyaline-cell cartilage (chondroid) in the heads of teleosts. *Anat Embryol* 179: 285–303.

Benjamin, M. 1990. The cranial cartilages of teleosts and their classification. *J Anat* 169: 153–172.

Bensimon-Brito, A., Cancela, M. L., Huysseune, A., and Witten, P. E. 2012b. Vestiges, rudiments and fusion events: the zebrafish caudal fin endoskeleton in an evo-devo perspective. *Evol Dev* 14: 116–127.

Bensimon-Brito, A., Cardeira, J., Cancela, M. L., Huysseune, A., and Witten, P. E. 2012a. Distinct patterns of notochord mineralization in zebrafish coincide with the localization of Osteocalcin isoform 1 during early vertebral centra formation. *BMC Dev Biol* 12: 28.

Bensimon-Brito, A., Cardeira, J., Dionísio, G., Huysseune, A., Cancela, M. L., and Witten, P. E. 2016. Revisiting in vivo staining with Alizarin Red S – a valuable approach to analyse zebrafish skeletal mineralization during development and regeneration. *BMC Dev Biol* 16: 2, doi: 10.1186/s12861-016-0102-4.

Beresford, W. A. 1981. *Chondroid Bone, Secondary Cartilage and Metaplasia*. Urban and Schwarzenberg, Munich.

Beresford, W. A. 1983. Ectopic cartilage, Neoplasia and Metaplasia. In *Cartilage, Volume 3. Biomedical Aspects* (B. K. Hall, ed.), pp. 1–48. Academic Press, New York, NY.

Berio, F., Broyon, M., Enault, S., Pirot, N., López-Romero, F. A., and Debiais-Thibaud, M. 2021. Diversity and evolution of mineralized skeletal tissues in Ccondrichthyans. *Frontiers in Ecology and Evolution,* 9: 223. doi: 10.3389/fevo.2021.660767.

Bertrand, S., Petillon, Y. L., Somorjai, I. M. L., and Escriva, H. 2017. Developmental cell-cell communication pathways in the cephalochordate amphioxus: actors and functions. *Int J Dev Biol* 61: 697–722.

Bevilacqua, C., Sanchez Iranzo, H., Richter, D., Diz-Muñoz, A., and Prevedel, R. 2019. Imaging mechanical properties of sub-micron ECM in live zebrafish using Brillouin microscopy. *Biomed Opt Express* 10: 1420–1431.

Bijtel, J. H. 1931. Über die Entwicklung des Schwanzes bei Amphibien. *W Roux' Arch Entwickl Mech Org* 125: 448–486.

Bird, N. C., and Mabee, P. M. 2003. Developmental morphology of the axial skeleton of the zebrafish *Danio rerio*. *Dev Dyn* 228: 337–357.

Blagden, C. S., Currie, P. D., Ingham, P. W., and Hughes, S. M. 1997. Notochord induction of zebrafish slow muscle mediated by Sonic hedgehog. *Genes Dev* 11: 2163–2175.

Blieck, A., Turner, S., Burrow, C. J., Schultze, H.-P., Rexroad, C. B., Bultynck, P., and Nowlan, G. S. 2010. Fossils, histology, and phylogeny: why conodonts are not vertebrates. *Episodes* 33: 234–241.

Bǒcina, I., and Saraga-Babić, M. 2006. Immunohistochemical study of cytoskeletal and extracellular matrix components in the notochord and notochordal sheath of amphioxus. *Int J Biol Sci* 2: 73–78.

Boeke, J. 1902. Über die ersten Entwicklungsstadien der Chorda dorsalis. Ein Beitrag zur Centrosomlehre. Petrus Camper. *Nederl Bijdr Anat* 1: 568–586.

Boeke, J. 1908. Das Geldrollenstadium der Vertebraten-Chorda und des Squelette der Mundcirren von *Branchiostoma lanceolatum*, und seine cytomechanische Bedeutung. *Anat Anz* 33: 541–556: 574–580.

Boos, N., Nerlich, A. G., Wiest, I., von der Mark, K., and Aebi, M. 1997. Immunolocalization of type X collagen in human lumbar intervertebral discs during ageing and degeneration. *Histochemistry* 108: 471–480.

Bowler, P. J. 1996. *Life's Splendid Drama. Evolutionary Biology and the Reconstruction of Life's Ancestry 1860–1940*. The University of Chicago Press, Chicago.

Briggs, D. E. G. 1992. Conodonts: a major extinct group added to the vertebrates. *Science* 256: 1285–1286.

Briggs, D. E. G. 1995. Experimental taphonomy. *Palaios* 10: 539–550.

Bronner-Fraser, M., and Fraser, S. 1997. Differentiation of the vertebrate neural tube. *Curr Opin Cell Biol* 9: 885–891.

Bruggeman, B. J., Maier, J. A., Mohiuddin, Y. S., Powers, R., Lo, Y., Guimaraes-Camboa, N., Evans, S. M., and Harfe, B. D. 2012. Avian intervertebral disc arises from rostral sclerotome and lacks a nucleus pulposus: implications for evolution of the vertebrate disc. *Dev Dyn* 241: 675–683.

Bruneel, B., and Witten, P. E. 2015. Power and challenges of using zebrafish as a model for skeletal tissue imaging. *Conn Tissue Res* 56: 161–173.

Bruns, R. R., and Gross, J. 1970. Studies on tadpole tail 1. Structure and organization of notochord and its covering layers in *Rana catesbeiana*. *Am J Anat* 128: 193–223.

Bumcrot, D. A., and McMahon, A. P. 1995. Somite differentiation. Sonic signals somites. *Curr Biol* 5: 612–614.

Bundy, J., Rogers, R., Hoffman, S., and Conway, S. J. 1998. Segmental expression of aggrecan in the non-segmented perinotochordal sheath underlies normal segmentation of the vertebral column. *Mech Dev* 79: 213–217.

Burger, A., Vasilyev, A., Tomar, R., Selig, M. K., Nielsen, G. P., Peterson, R. T., Drummond, I. A. and Haber, D. A. 2014. A zebrafish model of chordoma initiated by notochord-driven expression of HRASV12. *Dis Models Mech* 7: 907–913.

Capellini, T. D., Dunn, M. P., Passamaneck, Y. J., Selleri, L., and Di Gregorio, A. 2008. Conservation of notochord gene expression across chordates: insights from the Leprecan gene family. *Genesis* 46: 683–696.

Carlier, E. W. 1890. Fate of the notochord and development of the intervertebral disc in the sheep, with observations on the structure of the adult disc in these animals. *J Anat Physiol* 24(4): 573–584.

Carlson, E. C. 1973. Periodic fibrillar material in membrane-bound bodies in notochordal epithelium of the early chick embryo. *J Ultrastruct Res* 42: 287–297.

Carlson, E. C., Upson, R. H., and Evans, D. K. 1974. The production of extracellular connective tissue fibrils by chick notochordal epithelium *in vitro*. *Anat Rec* 179: 361–374.

Carpenter, C. C. 1975. Functional Aspects of the notochordal appendage of the young-of-the-year gar (*Lepisosteus*). *Proc Okla Acad Sci* 55: 57–64.

Carroll, R. L. 1988. *Vertebrate Paleontology and Evolution*. p. 698. W. H. Freeman and Company, New York.

Carroll, R. L. 1989. Developmental aspects of lepospondyl vertebrae in Paleozoic tetrapods. *Hist Biol* 3: 1–25.

Cartier, O. 1875. Beiträge zur Entwicklungsgeschichte der Wirbelsäule. *Z Zool* 25: S65–70.

Cerdà, J., Gründ, C., Franke, W. W., and Brand, A. M. 2002. Molecular characterization of Calymmin, a novel notochord sheath-associated extracellular matrix protein in the zebrafish Embryo. *Dev Dyn* 224: 200–209.

Cetinkaya, E. A. 2018. Thornwaldt cyst. *J Craniofac Surg* 29(6): e560–e562.

Cheah, K. S. E., Lau, E. T., Au, P. K. C., and Tam, P. P. L. 1991. Expression of the mouse alpha-1(II) collagen gene is not restricted to cartilage during development. *Development* 111: 945–953.

Chiang, C., Litingtung, Y., Lee, E., Young, K. E., Corden, J. L., Westphal, H., and Beachy, P. A. 1996. Cyclopia and defective axial patterning in mice lacking Sonic hedgehog gene function. *Nature* 383: 407–413.

Chimal-Monroy, J., Rodriguez-Leon, J., Montero, J. A., Gañan, Y., Macias, D., Merino, R., and Hurlé, J. M. 2003. Analysis of the molecular cascade for mesodermal limb chondrogenesis: Sox genes and BMP signaling. *Dev Biol* 257: 292–301.

Choi K. S., and Harfe, B. D. 2011. Hedgehog signalling is required for formation of the notochord sheath and patterning of nuclei pulposi within the intervertebral discs. *Proc Natl Acad Sci USA* 108: 9484–9489.

Choi, K. S., Cohn, M. J., and Harfe, B. D. 2008. Identification of nucleus pulposus precursor cells and notochordal remnants in the mouse: implications for disk degeneration and chordoma formation. *Dev Dyn* 237: 3953–3958.

Choi, K. S., Lee, C., and Harfe, B. D. 2012. Sonic hedgehog in the notochord is sufficient for patterning of the intervertebral discs. *Mech Dev* 129: 255–262.

Christ, B., and Ordahl, C. R. 1995. Early stages of chick somite development. *Anat Embryol* 191: 381–396.

Christ, B., Huang, R., and Scaal, M. 2004. Formation and differentiation of the avian sclerotome. *Anat Embryol* 208: 333–350.

Christiansen, H. E., Lang, M. R., Pace, J. M., and Parichy, D. M. 2009. Critical early roles for col27a1a and col27a1b in zebrafish notochord morphogenesis, vertebral mineralization and post-embryonic axial growth. *PLoS One* 4: e8481.

Christopherson, L. R., Rabin, B. M., Hallam, D. K., and Russell, E. J. 1999. Persistence of the notochordal canal: MR and plain film appearance. *AJNR Am J Neuroradiol* 20: 33–36.

Claus, C. 1894. Über die Herkunft der die Chordascheide der Haie begrenzenden äusseren Elastica. *Anz Kaiserl Akad Wiss Math-Nat Classe* 31: 118–124.

Cleaver, P., and Krieg, P. A. 2001. Notochord patterning of the endoderm. *Dev Biol* 234: 1–12.

Cohen, A. M., and Hay, E. D. 1971. Secretion of collagen by embryonic neuroepithelium at the time of spinal cord–somite interaction. *Dev Biol* 26: 578–605.

Cole, A. G. 2011. A review of diversity in the evolution and development of cartilage: the search for the origin of chondrocytes. *Eur Cells Mater* 21: 122–129.

Cole, A. G., and Hall, B. K. 2004a. Cartilage is a metazoan tissue; integrating data from non-vertebrate sources. *Acta Zool* 85: 65–80.

Cole, A. G., and Hall, B. K. 2004b. The nature and significance of invertebrate cartilages revisited: distribution and histology of cartilage and cartilage-like tissues within the Metazoa. *Zoology* 107: 261–274.

Cole, A. G., and Hall, B. K. 2009. Cartilage differentiation in cephalopod molluscs. *Zoology* 112: 2–15.

Conway Morris, S. 1994. Why molecular biology needs palaeontology. *Development* 1994: 1–13.

Conway Morris, S., and Caron, J. B. 2012. *Pikaia gracilens* Walcott, A stem-group chordate from the Middle Cambrian of British Columbia. *Biol Rev Camb Philos Soc* 87: 480–512.

Cooper, G. W. 1965. Induction of somite chondrogenesis by cartilage and notochord: a correlation between inductive activity and specific stages of cytodifferentiation. *Dev Biol* 12: 185–212.

Cope, E. D. 1889. Synopsis of the families of Vertebrata. *Am Nat* 23(274): 849–877.

Cope, E. D. 1890. The homologies of the fins of fishes. *Am Nat* 24(281): 401–423.

Corallo, D., Schiavinato, A., Trapani, V., Moro, E., Argenton, F., and Bonaldo, P. 2013. Emilin3 is required for notochord sheath integrity and interacts with Scube2 to regulate notochord-derived Hedgehog signals. *Development* 140: 4594–4601.

Corallo, D., Trapani, V., and Bonaldo, P. 2015. The notochord: structure and functions. *Cell Mol Life Sci* 72: 2989–3008.

Corsin, J. 1974. Matériel extracellulaire et chondrogenèse chez les amphibiens. *Arch Anat Microsc Morphol Exp* 63: 231–238.

Cortes, D. H. and Elliott, D. M. 2014. The intervertebral disc: overview of disc mechanics. In *The Intervertebral Disc* (I. M. Shapiro and M. V. Risbud, eds), pp. 17–31. Springer-Verlag, Wien.

Cotti. S., Huysseune, A., Koppe, W., Rücklin, M., Marone, F., Wölfel, E. M., Fiedler, I. A. K., Busse, B., Forlino, A., and Witten, P. E. 2020. More bone with less minerals? The effects of dietary phosphorus on the post-cranial skeleton in zebrafish. *Int J Mol Sci* 21: 5429. doi: 10.3390/ijms21155429.

Couly, G. F., Coltey, P. M., and Le Dourarin, N. M. 1993. The triple origin of skull in higher vertebrates: a study in quail-chick chimeras. *Development* 117: 409–429.

Coutinho, P., Parsons, M. J., Thomas, K. A., Hirst, E. M. A., Saúde L, Campos, I., Williams, P. H., and Stemple, D. L. 2004. Differential requirements for COPI transport during vertebrate early development. *Dev Cell* 7: 547–558.

Cox, M. K., and Serra, R. 2014. Development of the intervertebral disc. In *The Intervertebral Disc* (I. M. Shapiro and M. V. Risbud, eds), pp. 33–51. Springer-Verlag, Wien.

Criswell. K. E., Coates, M. I., Gillis, J. A. 2017. Embryonic origin of the gnathostome vertebral skeleton. *Proc R Soc London B* 284: 20172121. doi: 10.1098/rspb.2017.2121.

Cumplido, N., Allende, M. L., and Arratia, G. 2020. From Devo to Evo: patterning, fusion and evolution of the zebrafish terminal vertebra. *Front Zool* 17: 18, https://doi.org/10.1186/s12983-020-00364-y.

Dale, R. M., and Topczewski, J. 2011. Identification of an evolutionarily conserved regulatory element of the zebrafish col2a1a gene. *Dev Biol* 357: 518–531.

Daniel, J. F. 1934. *The Elasmobranch Fishes.* p. 332. University of California Press, Berkeley, CA.

Danos, M. C., and Yost, H. J. 1995. Linkage of cardiac left-right asymmetry and dorsal-anterior development in *Xenopus*. *Development* 121: 1467–1474.

Danto, M., Witzmann, F., Kamenz, S. K., and Fröbisch, N. B. 2019. How informative is vertebral development for the origin of lissamphibians? *J Zool* 307: 292–305.

Danto, M., Witzmann, F., Pierce, S. E., and Fröbisch, N. B. 2017. Intercentrum versus pleurocentrum growth in early tetrapods: a paleohistological approach. *J Morph* 278: 1262–1283.

Darwin, C. 1859. *The Origin of Species.* Reprint of the sixth edition from 1872. The Harvard Classics, p. 552, P. F. Collier & Son 1909, New York.

Darwin, C. 1871. *The Descent of Man and Selection in Relation to Sex. Volume I.* p. 409. Appleton and Company, New York.

Davis, R. L., and Kirschner, M. W. 2000. The fate of cells in the tailbud of *Xenopus laevis*. *Development* 127: 255–267.

Dawes, B. 1930. The Development of the vertebral column in mammals, as illustrated by its development in Mus musculus. *Philos Trans R Soc London Ser B* 218: 115–170.

de Azevedo, T. P., Witten, P. E., Huysseune, A., Bensimon-Brito, A., Winkler, C., To. T. T., and Palmeirim, I. 2012. Interrelationship and modularity of notochord and somites: A comparative view on zebrafish and chicken vertebral body development. *J Appl Ichthyol* 28(3): 316–319.

de Beer, G. R. 1928. *Vertebrate Zoology: An Introduction to the Comparative Anatomy, Embryology, and Evolution of Chordate Animals.* Macmillan, New York.

Debiais-Thibaud, M. 2019. The evolution of endoskeletal mineralisation in Chondrichthyan fish. Development, cells and molecules. In *Evolution and Development of Fishes* (Z. Johanson, C. Underwood, and M. Richter, eds), pp. 110–125. Cambridge University Press, Cambridge.

Debiais-Thibaud, M., Simion, P., Venteo, S., Munoz, D., Marcellini, S., Mazan, S., and Haitina, T. 2019. Skeletal mineralization in association with type X collagen expression is an ancestral feature for jawed vertebrates. *Mol Biol Evol* 36(10): 2265–2276.

de Bree, K., de Bakker, B. S., and Oostra, R.-J. 2018. The development of the human notochord. *PLoS One* 13(10): e0205752. doi: 10.1371/journal.pone.0205752.

De Clercq, A. 2018. The effect of incubation temperature on early malformation, regionalisation and meristic characters of the vertebral column in farmed Chinook salmon (*Oncorhynchus tshawytscha*). PhD thesis, Massey University, Manawatū, Palmerston North, New Zealand. p. 271.

De Clercq, A., Perrott, M. R., Davie, P. S., Preece, M. A., Huysseune, A., and Witten, P. E. 2018. The external phenotype-skeleton link in post-hatch farmed Chinook salmon (*Oncorhynchus tshawytscha*). *J Fish Dis* 41(3): 511–527.

De Clercq, A., Perrott, M. R., Davie, P. S., Preece, M. A., Wybourne, B., Ruff, N., Huysseune, A., and Witten, P. E. 2017. Vertebral column regionalisation in Chinook salmon, *Oncorhynchus tshawytscha*. *J Anat* 231: 500–514.

De Robertis, E. M., Fainsod, A., Gont, L. K., and Steinbeisser, H. 1994. The evolution of vertebrate gastrulation. *Development* 1 Suppl: 17–124.

Denison, R. H. 1978. Placodermi. In *Handbook of Paleoichthyology, Volume 2* (H.-P. Schultze, ed.), pp. 1–138. Gustav Fischer Verlag, Stuttgart.

Detwiler, S. R. 1937. Observations upon the migration of neural crest cells, and upon the development of the spinal ganglia and vertebral arches in Amblystoma. *Am J Anat* 61: 63–94.

Desvignes, T., Carey, A., and Postlethwait, J. H. 2018. Evolution of caudal fin ray development and caudal fin hypural diastema complex in spotted gar, teleosts, and other neopterygian fishes. *Dev Dyn* 2478(6): 832–853.

Detwiler, S. R., and van Dyke, R. H. 1934. The development and functions of deafferented fore limbs in *Amblystoma*. *J Exp Zool* 68: 321–346.

Detwiler, S. R., and Holtzer, H. 1956. The developmental dependence of the vertebral column upon the spinal cord in the urodeles. *J Exp Zool* 132: 299–310.

Deutsch, U., Dressler, G. R., and Gruss, P. 1988. Pax 1, a member of a paired box homologous murine gene family, is expressed in segmented structures during development. *Cell* 53: 617–625.

Dewit, J., Witten, P. E., and Huysseune, A. 2011. The mechanism of cartilage subdivision in the reorganization of the zebrafish pectoral fin endoskeleton. *J Exp Zool (B) Mol Dec Evol* 316: 584–597

Diedhiou, S., and Bartsch, P. 2009. Staging of the Early Development of *Polypterus* (Cladistia: Actinopterygii). In *Development of Non-Teleost Fishes* (Y. W. Kunz, C. A. Luer, and B. G. Kapoor, eds), p. 289. Science Publishers, New Hampshire.

Dietz, V. H., Ziegelmeier, G., Bittner, K., Bruckner, P., and Balling, R. 1999. Spatio-temporal distribution of chondromodulin-I mRNA in the chicken embryo: expression during cartilage development and formation of the heart and eye. *Dev Dyn* 216: 233–243.

Dingerkus, G., and Uhler, L. D. 1977. Enzyme clearing of alcian blue stained whole small vertebrates for demonstration of cartilage. *Stain Technol* 52: 229–232.

Dohrn, A. 1875. *Der Ursprung der Wiebelthiere und das Princip des Functionswechsels*. Verlag von Wilhelm Engelmann, Leipzig.

Domowicz, M., Li, H., Hennig, A., Henry, J., Vertel, B. M., and Schwartz, N. B. 1995. The biochemically and immunologically distinct CSPG of notochord is a product of the aggrecan gene. *Dev Biol* 171: 655–664.

Donoghue, P. C. J., and Rücklin, M. 2016. The ins and outs of the evolutionary origin of teeth. *Evol Dev* 18: 19–30.

Drost, H.-G., Janitza, P, Grosse, I., and Quint, M. 2017. Cross-kingdom comparison of the developmental hourglass *Curr Opin Gen Dev* 45: 69–75.

Du, S. J., and Dienhart, M. 2001. Zebrafish Tiggy-Winkle Hedgehog promoter directs notochord and floor plate green fluorescence protein expression in transgenic zebrafish embryos. *Dev Dyn* 222: 655–666.

Dutel, H., Galland, M., Tafforeau, P., Long, J. A., Fagan, J. M., Janvier, P., Herrel, A., Santin, M. D., Clément, G., and Herbin, M. 2019. Neurocranial development of the coelacanth and the evolution of the sarcopterygian head. *Nature* 569: 556–559.

Dutel, H., Herrel, A., Clément, G., and Herbin, M. 2013. A reevaluation of the anatomy of the jaw-closing system in the extant coelacanth *Latimeria chalumnae*. *Naturwissenschaften* 100: 1007–1022.

Eakin, R. M., and Westfall, J. 1962. Fine structure of the notochord of Amphioxus. *J Cell Biol* 12: 646–651.

Eastman, J. T., Witmer, L. M., Ridgely, R. C., and Kuhn, K. L. 2014. Divergence in skeletal mass and bone morphology in Antarctic notothenioid fishes. *J Morph* 275: 841–861.

Economides, K. D., Zeltser, L., and Capecchi, M. R. 2003. Hoxb13 mutations cause overgrowth of caudal spinal cord and tail vertebrae. *Dev Biol* 256: 317–330.

Edeling, M. A., Smith, C., and Owen, D. 2006. Life of a clathrin coat: insights from clathrin and AP structures. *Nat Rev Mol Cell Biol* 7: 32–44.

Ekanayake, S., and Hall, B. K. 1991. Development of the notochord in the Japanese medaka, *Oryzias latipes* (Teleostei, Cyprinidontidae) with special reference to desmosomal connections and functional integration with adjacent tissues. *Can J Zool* 69: 1171–1177.

Elinson, R. P., and Kezmoh, L. 2010. Molecular Haeckel. *Dev Dyn* 239: 1905–1918.

Ellis, K., Bagwell, J., and Bagnat, M. 2013. Notochord vacuoles are lysosome-related organelles that function in axis and spine morphogenesis. *J Cell Biol* 200: 667–679.

Fan, C.-M., and Tessier-Lavigne, M. 1994. Patterning of mammalian somites by surface ectoderm and notochord: evidence for sclerotome induction by a hedgehog homolog. *Cell* 79: 1175–1186.

Fernández-Lloris, R., Viñals, F., López-Rovira, T., Harley, V., Bartrons, R., Rosa, J. L., and Ventura, F. 2003. Induction of the Sry-related factor SOX6 contributes to bone morphogenetic protein-2-induced chondroblastic differentiation of C3H10T1/2 cells. *Mol Endocrinol* 17: 1332–1343.

Fiaz, A. W., Léon-Kloosterziel, K. M., Gort, G., Schulte-Merker, S., van Leeuwen, J. L., and Kranenbarg, S. 2012. Swim-training changes the spatio-temporal dynamics of skeletogenesis in zebrafish larvae (*Danio rerio*). *PLoS ONE* 7(4) e34072.1–13.

Field, H. H. 1895. Bemerkungen über die Entwickelung der Wirbelsäule bei den Amphibien. *Morph Jahrb* 22: 340–356.

Finarelli, J. A., and Coates, M. I. 2014. *Chondrenchelys problematica* (Traquair, 1888) redescribed: a lower carboniferous, eel-like holocephalan from Scotland. *Earth Environ Sci Trans R Soc Edinburgh* 105: 35–59.

Fleming, A., Keynes, R. J., and Tannahill, D. 2001. The role of the notochord in vertebral column formation. *J Anat* 199: 177–180.

Fleming, A., Keynes, R., and Tannahill, D. 2004. A central role for the notochord in vertebral patterning. *Development* 131: 873–888.

Fleming, A., Kishida, M. G., Kimmel, C. B., and Keynes, R. J. 2015. Building the backbone: the development and evolution of vertebral patterning. *Development* 142: 1733–1744.

Forey, P. L. 1991. *Latimeria chalumnae* and its pedigree. *Envt Biol Fishes* 32: 75–97.

Forlino, A., and Marini, J. C. 2016. Osteogenesis imperfecta. *Lancet* 387(10028): 1657–1671.

Fouquet, B., Weinstein, B. M., Serluca, F. C., and Fishman, M. C. 1997. Vessel patterning in the embryo of the zebrafish: guidance by notochord. *Dev Biol* 183: 37–48.

Fox, H. 1973. Degeneration of the tail notochord of *Rana temporaria* at metamorphic climax. Examination by electron microscopy. *Z Zellforsch* 138: 371–386.

Franco-Browder, D., de Rydt, J., and Dorfman, A. 1963. The identification of a sulfated mucopolysaccharide in chick embryos stages 11–23. *Proc Natl Acad Sci USA* 49: 643–647.

François, Y. 1958. Quelques points de la structure et du développement de la vertébre de *Salmo*. *Bull Soc Zool Fr* 80: 247–248.

François, Y. 1966. Structure et développement de la vertèbre de *Salmo* et des téléosténs. *Arch Zool Exp Gén* 107: 287–232.

Franz-Odendaal, T. A., Hall, B. K., and Witten, P. E. 2006. Buried alive: how osteoblasts become osteocytes. *Dev Dyn* 235: 176–190.

Fraser, R. C. 1960. Somite genesis in the chick. III. The role of induction. *J Exp Zool* 126: 349–400.

Fraser, T. W. K., Witten, P. E., Albrektsen, S., Breck, O., Fontanillas, R., Nankervis. L., Thomsen, T. H., Koppe, W., Sambraus, F., and Fjelldal, P. G. 2019. Phosphorus nutrition in farmed Atlantic salmon (*Salmo salar*): life stage and temperature effects on bone pathologies. *Aquaculture* 511(2019): 734246. doi: 10.1016/j.aquaculture.2019.734246.

Frederickson, R. G., and Low, F. N. 1971. The fine structure of perichordal microfibrils in control and enzyme-treated chick embryos. *Am J Anat* 130: 347–376.

Froese, R., and Pauly, D. (Eds) 2019. *FishBase*. World Wide Web Electronic Publication. www.fishbase.org, (12/2019).

Fujimoto, S., Yamanaka, K., Tanegashima, C., Nishimura, O., Kuraku, S., Kuratani, S., and Irie, N. 2021. Measuring potential effects of the developmental burden associated with the vertebrate notochord. *J Exp Zool B Mol Dev Evol* doi: 10.1002/jez.b.23032 (Epub ahead of print).

Furumoto, T.-A., Miura, N., Akasaka, T., Mizutani-Koseki, Y., Sudo, H., Fukuda, K., Maekawa, M., Yuasa, S., Fu, Y., Moriya, H., Taniguchi, M., Imai, K., Dahl, E., Balling, R., Palvlova, M., Gossler, A., and Koseki, H. 1999. Notochord-dependent expression of MFH1 and Pax1 cooperates to maintain the proliferation of sclerotome cells during the vertebral column development. *Dev Biol* 210: 15–29.

Futuyma, D. J. 1998. *Evolutionary Biology*. Third Edition. Sinauer Associates, Sunderland, Massachusetts.

Gadow, H. 1896. On the evolution of the vertebral column of Amphibia and Amniota. *Philos Trans R Soc London B* 187: 1–57.

Gadow, H., and Abbott, E. C. 1895. On the evolution of the vertebral column of fishes. *Philos Trans R Soc London B* 186: 163–221.

Gajović, S., Kostović-Knczěvić, L., and Švajger, A. 1989. Origin of the notochord in the rat embryo tail. *Anat Embryol* 179: 305–310.

Gansner, J. M., Mendelsohn, B. A., Hultman, K. A., Johnson, S. L., and Gitlin, J. D. 2007. Essential role of lysyl oxidases in notochord development. *Dev Biol* 307: 202–213.

Garcia, J., Bagwell, J., Njaine, B., Norman, J., Levic, D. S., Wopat, S., Miller, S. E., Liu, X., Locasale, J. W., Stainier, D. Y. R., and Bagnat, M. 2017. Sheath cell invasion and trans-differentiation repair mechanical damage caused by loss of caveolae in the zebrafish notochord. *Curr Biol* 27: 1–8.

Gardiner, B. G. 1982. Tetrapod classification. *Zool J Linn Soc London* 74: 207–232.

Gardiner, B. G. 1983. Gnathostome vertebrae and the classification of the Amphibia. *Zool J Linn Soc London* 79: 1–59.

Gardiner, B. G. 1984. Devonian Palaeoniscid fishes: new specimens of Mimia and Moythomasia from the upper Devonian of Western Australia. *Bull Br Mus Nat Hist* 37: 173–428.

Gardiner, B. G., and Schaeffer B. 1989. Interrelationships of lower actinopterygian fishes. *Zool J Linn Soc London* 97: 135–187.

Garstang, W. 1928. The morphology of the Tunicata and its bearing on the phylogeny of the Chordata. *Q J Microsc Sci* 72: 51–187.

Gegenbaur, K. 1862. *Untersuchungen zur vergleichenden Anatomie der Wirbelsäule bei Amphibien und Reptilien*. W. Engelmann, Leipzig.

Gegenbaur, K. 1864. Ueber die Bildung Knochenstruktur. *Jena Zeitsch Med Nat* 1: 343–369.

Germain, D., Schnell, N. K., and Meunier, F. J. 2019. Histological data on bone and teeth in two dragonfishes Stomiidae; Stomiiformes): *Borostomias panamensis* Regan & Trewavas 1929 and *Stomias boa* Reinhardt, 1842. *Cybium* 43: 103–107.

Ghanem, E. 1996. Immunohistochemical localization of type I and II collagens in the involuting chick notochords in vivo and in vitro. *Cell Biol Intern* 20: 681–685.

Gistelinck, C., Witten, P. E., Huysseune, A., Symoens, S., Malfait, F., Larionova, D., Simoens, P., Dierick, M., Van Hoorebeke, L., De Paepe, A., Kwon, R. Y., Weis, M-A., Eyre, D. R., Willaert, A., and Coucke, P. J. 2016. Loss of type I collagen telopeptide lysyl hydroxylation causes musculoskeletal abnormalities in a zebrafish model of Bruck Syndrome. *J Bone Min Res* 31: 1930–1942.

Glickman, N. S., Kimmel, C. B., Jones, M. A., and Adams, R. J. 2003. Shaping the zebrafish notochord. *Development* 130: 873–887.

Godsave, S. F., Anderton, B. H., and And Wylie, C. C. 1986. The appearance and distribution of intermediate filament proteins during differentiation of the central nervous system, skin and notochord of *Xenopus laevis*. *J Embryol Exp Morph* 97: 201–223.

Goette, A. 1879. Beiträge zur vergleichenden Anatomie des Skeletsystemes der Wirbelthiere. II. Die Wirbelsäule und ihre Anhänge. *Arch Mikrosc Anat* XVI: 117–152.

Goldberg, R. L., and Toole, B. P. 1984. Pericellular coat of chick embryo chondrocytes: structural role of hyaluronate. *J Cell Biol* 99: 2114–2122.

Goldstein, A. M., and Fishman, M. C. 1998. Notochord regulates cardiac lineage in zebrafish embryos. *Dev Biol* 201: 247–252.

Gomez-Navarro, N., and Miller, E. A. 2016. COP-coated vesicles. *Curr Biol* 26: 54–57.

Goodrich, E. S. 1909. Vertebrata Craniata (First Fascicle: Cyclostomes and Fishes). In *A Treatise on Zoology*, Part IX (R. Lancaster, ed.), p. 518. Adam and Charles Black, London.

Goodrich, E. S. 1930. *Studies on the Structure and Development of Vertebrates*. p. 837. MacMillan and Co., London.

Goodrich, E. S. 1958. *Studies on the Structure and Development of Vertebrates. Volumes 1 and 2*. Dover Publications Inc., New York.

Götz, W., Osmers, R., and Herken, R. 1995. Localisation of extracellular matrix components in the embryonic human notochord and axial mesenchyme. *J Anat* 186: 111–121.

Grassi, G. B. 1883. Developpement de la colonne vertébrale chez les poissons osseuses. *Arch Ital Biol* 4: 251–268.

Gray, R. S., Wilm, T. P., Smith, J., Bagnat, M., Dale, R. M., Topczewski, J., Johnson, S. L., and Solnica-Krezel, L. 2014. Loss of col8a1a function during zebrafish embryogenesis results in congenital vertebral malformations. *Dev Biol* 386: 72–85.

Greco, T. L., Takada, S., Newhouse, M. M., McMahon, J. A., McMahon, A. P., and Camper, S. A. 1996. Analysis of the vestigial tail mutation demonstrates that Wnt-3a gene dosage regulates mouse axial development. *Genes Dev* 10: 313–324. doi: 10.1101/gad.10.3.313.

Greenwood, P. H. 1988. *A Living Fossil Fish: The Coelocanth* Latimeria Chalumnae. British Museum (Natural History), London.

Griffith, C. M., Wiley, M. J., and Sanders, E. J. 1992. The vertebrate tail bud: three germ layers from one tissue. *Anat Embryol* 185: 101–113.

Griffith, R. W., Mathews, M. B., Umminger, B. L., Grant, B. F., Pang, P. K. T., Thomson, K. S., and Pickford, G. E. 1975. Composition of fluid from the notochordal canal of the coelacanth, *Latimeria chalumnae*. *J Exp Zool* 192: 165–172.

Grodzinsky, A. 1983. Electromechanical and physiochemical properties of connective tissues. *CRC Critical Rev. Biomed Engin* 9: 133–199.

Grotmol, S., Kryvi, H., Keynes, R., Grossøy, C., Nordvik, K., and Totland, G. K. 2006. Stepwise enforcement of the notochord and its intersection with the myoseptum: an evolutionary path leading to development of the vertebra? *J Anat* 209: 339–357.

Grotmol, S., Kryvi, H., Nordvik, K., and Totland, G. K. 2003. Notochord segmentation may lay down the pathway for the development of the vertebral bodies in the Atlantic salmon. *Anat Embryol* 207: 263–272.

Grotmol, S., Kryvi, H., Nordvik, K., and Totland, G. K. 2005. A segmental pattern of alkaline phosphatase (ALP) activity within the notochord coincides with the initial formation of the vertebral bodies. *J Anat* 206: 427–436.

Haeckel, E. 1866. *Generelle Morphologie der Organismen: Allgemeine Grundzüge der organischen Formen-Wissenschaft, mechanisch begründet durch die von Charles Darwin reformite Descendenz-Theorie. Volume 2*. Georg Reimer, Berlin.

Haeckel, E. 1868. *Natürliche Schöpfungsgeschichte. Gemeinverständliche wissenschaftliche Vorträge über die Entwickelungslehre im Allgemeinen und diejenige von Darwin, Goethe und Lamarck im Besonderen*. Georg Reimer, Berlin.

Haga, Y., Dominique III, V. J. D., and Du, S. J. 2009. Analyzing notochord segmentation and intervertebral disc formation using the twhh:gfp transgenic zebrafish model. *Transgenic Res* 18: 669–683.

Hall, B. K. 1967. The formation of adventitious cartilage by membrane bones under the influence of mechanical stimulation applied in vitro. *Life Sci* 6: 663–667.

Hall, B. K. 1968. In vitro studies on the mechanical evocation of adventitious cartilage in the chick. *J Exp Zool* 168: 283–306.

Hall, B. K. 1971. Histogenesis and morphogenesis of bone. *Clin Orthop Rel Res* 74: 249–268.

Hall, B. K. 1973. Correlations between the concentrations of acid mucopolysaccharides and collagen in the tibia of the embryonic chick. *Can J Zool* 51: 771–776.

Hall, B. K. 1977. Chondrogenesis of the somitic mesoderm. *Adv Anat Embryol Cell Biol* 53: 1–50.

Hall, B. K. 1982. Bone in the cartilaginous fishes. *Nature* 298: 324.

Hall, B. K. 1997. Phylotypic stage or phantom: is there a highly conserved embryonic stage in vertebrates? *Tree* 12: 461–463.

Hall, B. K. 1998a. *Evolutionary Developmental Biology*. 2nd Edition. p. 491. Chapman and Hall, London/Kluwer, Academic Publishers, Netherlands.

Hall, B. K. 1998b. Germ layers and the germ-layer theory revisited. *Evol Biol* 30: 121–186.

Hall, B. K. 2000. A role for epithelial-mesenchymal interactions in tail growth/morphogenesis and chondrogenesis in embryonic mice. *Cell Tissues Organs* 166: 6–14.

Hall, B. K. 2002. Palaeontology and evolutionary developmental biology: a science of the 19th and 21st centuries. *Palaeontology* 45: 647–669.

Hall, B. K. 2003. Descent with modifications: the unity underlying homology and homoplasy as seen through an analysis of development and evolution. *Biol Rev Camb Philos Soc* 78: 409–433.

Hall, B. K. 2005. Consideration of the neural crest and its skeletal derivatives in the context of novelty/innovations. *J Exp Zool* 304B: 548–557.

Hall, B. K. 2009. *The Neural Crest and Neural Crest Cells in Vertebrate Development and Evolution*. Springer, Boston, MA. https://doi.org/10.1007/978-0-387-09846-3_3.

Hall, B. K. 2013. Homology, homoplasy, novelty and behavior. *Dev Psychobiol* 55: 4–12.

Hall, B. K. 2014. Endoskeleton/exo (dermal) skeleton — Mesoderm/neural Crest: two pair of problems and a shifting paradigm. *J Appl Ichthyol* 30: 608–615.

Hall, B. K. 2015. *Bones and Cartilage: Developmental and Evolutionary Skeletal Biology*. 2nd Edition. p. 920. Academic Press, London.

Hall, B. K. 2018. Germ layers, the neural crest and emergent organization in development and evolution. *Genesis* 56: e23103. doi: 10.1002/dvg.23103.

Hall, B. K., and Gillis, J. A. 2013. Incremental evolution of the neural crest, neural crest cells and neural crest-derived skeletal tissues. *J Anat* 222: 19–31.

Hall, B. K., and Kerney, R. 2012. Levels of biological organization and the origin of novelty. *J Exp Zool* 318B: 428–437.

Hall, B. K., and Witten, P. E. 2007. The origin and plasticity of skeletal tissues in vertebrate evolution and development. In *Major Transitions in Vertebrate Evolution* (J. S. Anderson and H.-D. Sues, eds), Festschrift for Dr Robert L. Carroll. pp. 13–57. Indiana University Press, Bloomington, IN.

Hall, B. K., and Witten, P. E. 2019. Plasticity of skeletal cells and tissues and the evolutionary development of fishes. In *Evolution and Development of Fishes* (Z. Johanson, C. Underwood, and M. Richter, eds), pp. 126–143. Cambridge University Press, Cambridge.

Hamburger, V., and Hamilton, H. L. 1951. A series of normal stages in the development of the chick embryo. *J Morph* 88: 49–92.

Handrigan, G. R. 2003. Concordia discors: duality in the origin of the vertebrate tail. *J Anat* 202: 255–267.

Hashimshony, T., Feder, M., Levin, M., Hall, B. K., and Yanai, I. 2015. Spatiotemporal transcriptomics reveals the evolutionary history of the endoderm germ layer. *Nature* 519(7542): 219–222.

Hasse, C. 1892. Die Entwicklung der Wirbelsäule von Triton taeniatus. Erste Abhandlung über die Entwicklung der Wirbelsäule. *Zwissen Zool* 53 (Suppl): 1–20.

Hasse, C. 1893. Die Entwicklung der Wirbelsäule der Dipnoi. Vierte Abhandlung über die Entwicklung der Wirbelsäule. *Zwissen Zool* 55: 533–542.

Hatschek, B. 1893. *The Amphioxus and Its Development.* p. 181. Swan Sonnenschein & Co., London.

Hautier, L., Weisbecker, V., Sánchez-Villagra, M. R., Goswami, A., and Asher, R. J. 2010. Skeletal development in sloths and the evolution of mammalian vertebral patterning. *Proc Natl Acad Sci USA* 107 (44): 18903–18908.

Hay, E. D., and Meier, S. 1974. Glycosaminoglycan synthesis by embryonic inductors - neural tube, notochord, and lens. *J Cell Biol* 62: 889–898.

Hay, P. O. 1895. On the structure and development of the vertebral column of *Amia*. Field Columbian Mus Publ 5, *Zool Ser* 1(1): 1–54.

Hayashi, M., Hayashi, K., Iyama, K.-I., Trelstad, R. L., Linsenmayer, T. F., and Mayne, R. 1992. Notochord of chick embryos secretes short-form type IX collagen prior to the onset of vertebral chondrogenesis. *Dev Dyn* 194: 169–176.

Heathfield, S. K., Le Maitre, C. L., and Hoyland, J. A. 2008. Caveolin-1 expression and stress-induced premature senescence in human intervertebral disc degeneration. *Arthritis Res Ther* 10: R87.

Hejnol, A., and Lowe, C. J. 2014. Animal evolution: stiff or squishy notochord origins? *Curr Biol* 24: R1131–R1133.

Hejnol, A., and Lowe, C. J. 2015. Embracing the comparative approach: how robust phylogenies and broader developmental sampling impacts the understanding of nervous system evolution. *Philos Trans R Soc London B* 370: 20150045.

Hensen, V. 1876. Beobachtungen über die Befruchtung und Entwicklung des Kaninchens und Meerschweinchens. *Z Anat Entwickl* 1: 213–273.

Hérault, Y., Beckers, J., Gérard, M., and Duboule, D. 1999. Hox gene expression in limbs: colinearity by opposite regulatory controls. *Dev Biol* 208: 157–165.

Hertwig, O. 1915. *Lehrbuch der Entwicklungsgeschichte des Menschen und der Wirbeltiere.* 10th Edition, p. 782. Verlag von Gustav Fischer, Jena.

Hilton, E. J., Britz, R., Johnson, G. D., and Forey, P. L. 2007. Clarification of the occipito-vertebral region of *Arapaima gigas* (Osteoglossomorpha: Osteoglossidae) through developmental osteology. *Copeia* 1: 218–224.

Hirsinger, E., Duprez, D., Jouve, C., Malapert, P., and Pourquié, O. 1997. Noggin acts downstream of Wnt and Sonic Hedgehog to antagonize BMP4 in avian somite patterning. *Development* 124: 4605–4614.

Hoffmann, A., Czichos, S., Kaps, C., Bachner, D., Mayer, H., Zilberman, Y., Turgeman, G., Pelled, G., Gross, G., and Gazit, D. 2002. The T-box transcription factor Brachyury mediates cartilage development in mesenchymal stem cell line C3H10T1/2. *J Cell Sci* 115: 769–781.

Holder, L. A. 1960. The comparative morphology of the axial skeleton in the Australian Gekkonidae. *J Linn Soc London Zool* 44: 300–335.

Holland, N. D. 2005. Chordates. *Curr Biol* 15: R911–R914.

Holland, N. D., and Holland, L. Z. 2017. The ups and downs of amphioxus biology: a history. *Int J Dev Biol* 61: 575–583.

Holland, N. D., Holland, L. Z., and Holland, P. W. H. 2015. Scenarios for the making of vertebrates. *Nature* 520: 450–455.

Holley, S. A. 2007. The genetics and embryology of zebrafish metamerism. *Dev Dyn* 236: 1422–1449.

Hollins, A. J., Campbell, L., Gumbleton, M., and Evans, D. J. R. 2002. Caveolin expression during chondrogenesis in the Avian Limb. *Dev Dyn* 225: 205–211.

Holmdahl, D. E. 1928. Die Enstehung und weitere Entwicklung der Neuralleiste (Ganglienleiste) bei Vogeln und Saugetieren. *Z Mikrosk-Anat Forsch* 14: 99–298.

Holtzer, H. 1952a. An experimental analysis of the development of the spinal column. Part I. Response of pre-cartilage cells to size variations of the spinal cord. *J Exp Zool* 121: 121–147.

Holtzer, H. 1952b. An experimental analysis of the development of the spinal column. Part II. The dispensability of the notochord. *J Exp Zool* 121: 573–591.

Holtzer, H., and Detwiler, S.R. 1953. An experimental analysis of the development of the spinal column. III. Induction of skeletogenous cells. *J. Exp. Zool.* 123: 335–369.

Honer, W., and Komnick, H. 1990. The cell junctions of the notochord of *Xenopus laevis* tadpoles. *Tissue Cell* 22: 149–155.

Hörstadius, S. 1944. Ueber die Folgen von Chordaexstirpation an spaeten Gastrulae und Neurulae von *Amblystoma punctatum*. *Acta Zool* 25: 75–87.

Huang, R., Zhi, Q., Brand-Saberi, B., and Christ, B. 2000. New experimental evidence for somite resegmentation. *Anat Embryol* 202: 195–200.

Huber, G. C. 1918. On the anlage and morphogenesis of the chorda dorsalis in Mammalia, in particular in the guinea pig (*Cavia cobaya*). *Anat Rec* 14: 217–264.

Hughes, A. F., and Freeman, R. B. 1974. Comparative remarks on the development of the tail cord among higher vertebrates. *J Embryol Exp Morph* 32: 355–363.

Hunter, C. J., Matyas, J. R., and Duncan, N. A. 2003. The notochordal cell in the nucleus pulposus: a review in the context of tissue engineering. *Tissue Eng* 9(4): 667–677.

Hurley, I. A., Scemama, J.-L., and Prince, V. E. 2007. Consequences of Hoxb1 duplication in teleost fish. *Evol Dev* 9: 540–554.

Huxley, T. H. 1859. Observations on the development of some parts of the skeleton of fishes. *Q J Microsc Sci* 7: 33–46.

Huysseune, A. 1989. Morphogenetic aspects of the pharyngeal jaws and neurocranial apophysis in postembryonic *Astatotilapia elegans* (Trewavas 1933) (Teleostei, Cichlidae). *Med Kon Acad Wet Lett* 51: 11–35.

Huysseune, A. 1990. Development of the anterior part of the mandible and the mandibular dentition in two species of cichlidae (Teleostei). *Cybium* 14: 327–344.

Huysseune, A. 2000. Skeletal systems. In *Microscopic Functional Anatomy* (G. K. Ostrander, ed.), pp. 307–317. Academic Press, San Diego.

Huysseune, A., Cerny, R., and Witten, P. E. 2022a. The conundrum of pharyngeal teeth origin: The role of germ layers, pouches, and gill slits. *Biol Rev* 97: 414–447.

Huysseune, A., Soenens, M., Sire, J-Y., and Witten, P. E. 2022b. High resolution histology for craniofacial studies on zebrafish and other teleost models. Methods Molec Biol 2403: 249–262.

Inohaya, K., Takano, Y., and Kudo, A. 2007. The teleost intervertebral region acts as a growth center of the centrum: in vivo visualization of osteoblasts and their progenitors in transgenic fish. *Dev Dyn* 236: 3031–3046.

Irie, N., and Kuratani, S. 2014. The developmental hourglass model: a predictor of the basic body plan? *Development* 141: 4649–4655.

Ishikawa, T., Okada, T., Ishikawa-Fujiwara, T., Todo, T., Kamei, Y., Shigenobu, S., Tanaka, M., Saito. T. L., Yoshimura, J., Morishita, S., Toyoda. A., Sakaki, Y., Taniguchi, Y., Takeda, S., and Mori, K. 2013. ATF6α/β-mediated adjustment of ER chaperone levels is essential for development of the notochord in medaka fish. *Mol Biol Cell* 24: 1387–1395.

Ishikawa, T., Toyama, T., Nakamura, Y., Tamada, K., Shimizu, H., Ninagawa, S., Okada, T., Kamei, Y., Ishikawa-Fujiwara, T., Todo, T., Aoyama, E., Takigawa, M., Harada, A., and Mori, K. 2017. UPR Transducer BBF2H7 allows export of type II collagen in a cargo- and developmental stage specific manner. *J Cell Biol* 216: 1761–1774.

Iwamatsu, T. 2004. Stages of normal development in the medaka *Oryzias latipes*. *Mech Dev* 121: 605–618.

James, H. F. 2009. Repeated evolution of fused thoracic vertebrae in songbirds. *The Auk* 126: 862–872.

Jandzik, D., Garnett, A. T., Square, T. A., Cattell, M. V., Yu, J.-K., and Medeiros, D. M. 2015. Evolution of the new vertebrate head by co-option of an ancient chordate skeletal tissue. *Nature* 518: 534–538.

Janvier, P. 2015. Facts and fancies about early fossil chordates and vertebrates *Nature* 520: 483–489.

Jiang, D., and Smith, W. C. 2007. Ascidian notochord morphogenesis. *Dev Dyn* 236: 1748–1757.

Johanson, Z., Ericsson, R., Long, J., Evans, F., and Joss, J. 2009. Development of the axial skeleton and median fin in the Australian Lungfish, *Neoceratodus forsteri*. *Open Zool J* 2: 91–101.

Johanson, Z., Kearsley, A., den Blaauwen, J., Newman, M., and Smith, M. M. 2012. Ontogenetic development of an exceptionally preserved Devonian cartilaginous skeleton. *J Exp Zool B* 318B: 50–58.

Johnson, R. L., Riddle, R. D., Laufer, E., and Tabin, C. 1994. Sonic hedgehog: a key mediator of anterior-posterior patterning of the limb and dorso-ventral patterning of axial embryonic structures. *Biochem Soc Trans* 22: 569–574.

Jonasson, K. A., Russell, A. P., and Vickaryous, M. K. 2012. Histology and histochemistry of the gekkotan notochord and their bearing on the development of notochordal cartilage. *J Morph* 273: 596–603.

Jurand, A. 1962. The development of the notochord in chick embryos. *J Embryol Exp Morph* 10: 602–621.

Jurand, A. 1974. Some aspects of the development of the notochord in mouse embryos. *J Embryol Exp Morph* 32: 1–33.

Kague, E., Turci, F., Newman, E., Yang, Y., Robson Brown, K., Aglan, M. S., Otaify, G. A., Temtamy, S. A., Ruiz-Perez, V. L., Cross, S., Royall, C. P., Witten, P. E., and Hammond, C. L. 2021. 3D assessment of intervertebral disc degeneration in zebrafish identifies changes in bone density that prime disc disease. *Bone Research* 9: 39. https://doi.org/10.1038/s41413-021-00156-y

Kaneko, T., Freeha, K., Wu, X., Mogi, M., Uji, S., Yokoi, H., and Suzuki, T. 2016. Role of notochord cells and sclerotome-derived cells in vertebral column development in fugu, Takifugu rubripes: histological and gene expression analyses. *Cell Tissue Res* 366: 37–49.

Kanki, J. P., and Ho, R. K. 1997. The development of the posterior body in zebrafish. *Development* 124: 881–893.

Katz, M. J. 1983. Comparative anatomy of the tunicate tadpole, *Ciona intestinalis*. *Biol Bull* 164: 1–27.

Keller, R. E. 1976. Vital dye mapping of the gastrula and neurula of *Xenopus laevis*. II. Prospective areas and morphogenetic movements of the deep layer. *Dev Biol* 51: 118–137.

Keller, R., Cooper, M. S., Danilchik, M., Tibbetts, P., and Wilson, P. A. 1989. Cell intercalation during notochord development in *Xenopus laevis*. *J Exp Zool* 251: 134–154.

Kemp, A. 1981. Rearing of embryos and larvae of the Australian lungfish, *Neoceratodus forsteri*, under laboratory conditions. *Copeia* 4: 776–784.

Kerr, J. G. 1909. Normal plates of the development of *Lepidosiren paradoxa* and *Protopterus annectens*. In *Normentafeln zur Entwicklungsgeschichte der Wirbeltiere*. Part 10 (F. Keibel, ed.), pp. 1–31. Verlag Gustav Fischer, Jena.

Kim, S. K., Hebrok, M., and Melton, D. A. 1997. Notochord to endoderm signaling is required for pancreas development. *Development* 124: 4243–4252.

Kimmel, C. B., and Warga, R. M. 1986. Tissue specific cell lineages originate in the gastrula of the zebrafish. *Science* 231: 365–368.

Kimmel, C. B., Miller, C. T., and Moens, C. B. 2001. Specification and morphogenesis of the zebrafish larval head skeleton. *Dev Biol* 233: 239–257.

Kimmel, C. B., Warga, R. M., and Schilling, T. F. 1990. Origin and organization of the zebrafish fate map. *Development* 108: 581–594.

Kimura, S., and Kamimura, T. 1982. The characterisation of lamprey notochord collagen with special reference to its skin collagen. *Comp Biochem Phys* 73B: 335–339.

Kirschbaum, F., and Meunier, F. J. 1988. South American Gymnotiform fishes as model animals for regeneration experiments. In *Control of Cell Proliferation and Differentiation during Regeneration* (H. J. Anton, ed.), pp. 112–123. S. Karger, Basel.

Kitchin, I. C. 1949. The effects of notochordectomy in *Amblystoma mexicanum*. *J Exp Zool* 112: 393–415.

Klaatsch, H. 1893a. Beiträge zur vergleichenden Anatomie der Wirbelsäule. I. Über den Urzustand der Fischwirbelsäule. *Morph Jahrb* 19: 649–680.

Klaatsch, H. 1893b. Beitrage zur vergleichenden Anatomie der Wirbelsäule. II. Uber die Bildung knorpeliger Wirbelkörper bei Fischen. *Morph Jahrb* 20: 143–186.

Klaatsch, H. 1895. Beitrage zur vergleichenden Anatomie der Wirbelsäule. III. Zur Phylogenese der Chordascheiden und zur Geschichte der Umwandlung der Chordastruktur. *Morph Jahrb* 22: 514–560.

Kobus, K., Ammar, D., Nazari, E. M., and Müller, Y. M. R. 2013. Homocysteine causes disruptions in spinal cord morphology and changes the expression of Pax 1/9 and Sox 9 gene products in the axial mesenchyme. *Birth Defects Res* 97: 386–397.

Kocher, W. 1957. Vakuolisierung der Chorda dorsalis und Wirkung exttrachordaler Defekte auf die Differenzierung von Chorda und Neuralstrukturen bei *Triton alpestris*. *Roux Arch Entwickl* 149: 443–503.

Koehl, M. A. R., Quillin, K. J., and Pell, C. A. 2000. Mechanical design of fiber-wound hydraulic skeletons: the stiffening and straightening of embryonic notochords. *Am Zool* 40: 28–41.

Kölliker, A. 1858. Entwicklung der Ligamenta intervertebralia. *Verh Phys-Med Ges Würzburg* 9(X): XLVIII–XLIX.

Kölliker, A. 1859. On the structure of the chorda dorsalis of the plagiostomus and some other fishes, and on the relation of its proper sheath to the development of the vertebrae. *Proc R Soc London* 10: 214–222.

Kölliker, A. 1861. *Entwicklungsgeschichte der Menschen und der höheren Thiere*. p. 468. Verlag von Wilhelm Engelmann, Leipzig.

Kölliker, A. 1863. *Weitere Beobachtungen über die Wirbel der Selachier, insbesondere über die Wirbel der Lamnoidei, nebst allgemeinen Bemerkungen über die Bildung der Wirbel der Plagiostomen*. Verlag HL, Bröuuer, Frankfurt a. Main. p. 51 + 5 figure tables.

Kölliker, A. 1864. *Über die Darwin'sche Schöpfungstheorie*. pp. 1–15. Verlag von Wilhelm Engelmann, Leipzig.

Kölliker, A. 1884. Die Embryonalen Keimblatter und die Gewebe. *Zeit Wiss Zool* 40: 179–213.

Kondylis, V., Pizette, S., and Rabouille, C. 2009. The early secretory pathway in development: a tale of proteins and mRNAs. *Sem Cell Dev Biol* 20: 817–827.

Koob, T. J., and Long, J. H. Jr. 2000. The vertebrate body axis: evolution and mechanical function. *Am Zool* 40: 1–18.

Kosher, R. A., and Lash, J. W. 1975. Notochordal stimulation of *in vitro* somite chondrogenesis before and after enzymatic removal of perinotochordal materials. *Dev Biol* 42(2): 362–378.

Kosher, R. A., and Solursh, M. 1989. Widespread distribution of type II collagen during embryonic chick development. *Dev Biol* 131: 558–566.

Kovalevsky, A. O. 1866a. Anatomie des *Balanoglossus*. *Mém Acad Imp Sci St Petersb* 7: 1–10.

Kovalevsky, A. O. 1866b. Entwicklungsgeschichte der einfachen Ascidien. *Mém Acad Imp Sci St Petersb* 10: 1–19.

Kovalevsky, A. O. 1867. Entwickelungsgeschichte des *Amphioxus lanceolatus*. *Mém Acad Sci St. Petersbourg* 11: 1–17.

Kovalevsky, A. O. 1871a. Weitere Studien über die Entwicklung der einfachen Ascidien. *Arch Mikrosk Anat* 7: 101–130.

Kovalevsky, A. O. 1871b. Embryologische Studien an Würmen und Arthropoden. *Mém Acad Imp Sci St Pétersb (Sér) VII* 161: 1–70.

Kovalevsky, A. O. 1877. Weitere Studien über die Entwickelungsgeschichte des *Amphioxus lanceolatus,* nebst einen Beitrage zur Homologie des Nervensystems der Würmen und Wierbelthiere. *Arch Mikrosk Anat* 13: 181–204.

Krauss, F. 1909. Über die Genese des Chordaknorpels der Urodelen und die Natur des Chordage-webes. *Archiv für Mikroskopische Anatomie und Entwicklungsgeschichte* 73: 69–116.

Krauss, S., Concordet, J.-P., and Ingham, P. W. 1993. A functionally conserved homolog of the *Drosophila* segment polarity gene hh is expressed in tissues with polarizing activity in zebrafish embryos. *Cell* 75: 1431–1444.

Krol, A. J., Roellig, D., Dequéant, M.-L., Tassy, O., Glynn, E., Hattem, G., Mushegian, A., Oates, A. C., and Pourquié, O. 2011. Evolutionary plasticity of segmentation clock net-works. *Development* 138: 2783–2792.

Krstić, R. V. 1985. *General Histology of the Mammal.* p. 404. Springer Verlag, Berlin.

Kryvi. H., Nordvik, K., Fjelldal, P. G., Eilertsen, M., Helvik, J. V., Storen, E. N., and Long, J. H. 2020. Heads and tails: the notochord develops differently in the cranium and caudal fin of Atlantic Salmon (*Salmo salar*, L.). *Anat Rec* 1–21. doi: 10.1002/ar.24562.

Kryvi, H., Rusten, I., Fjelldal, P. G., Nordvik, K., Totland, G. K., Karlsen, T., Wiig H, and Long, J. H. 2017. The notochord in Atlantic salmon (*Salmo salar* L.) undergoes profound morphological and mechanical changes during development. *J Anat* 231: 639–654.

Kvellestad, A., Høie, S., Thorud, K., Tørud, B. and Lyngøy, A. 2000. Platyspondyly and short-ness of vertebral column in farmed Atlantic salmon *Salmo salar* in Norway – descrip-tion and interpretation of pathological changes. *Dis Aquat Organ* 39: 97–108.

Kvist, T. N., and Finnegan, C. V. 1970. The distribution of glycosaminoglycans in the axial region of the developing chick embryo. I. Histochemical analysis. *J Exp Zool* 175: 221–240.

Laerm, J. 1976. The development, function, and design of amphicoelous vertebrae in teleost fishes. *Zool J Linn Soc London* 58: 237–254.

Laerm, J. 1979a. On the origin of Rhipidistian vertebrae. *J Paleont* 53: 175–186.

Laerm, J. 1979b. On the origin of the chondrostean vertebral centrum. *Can J Zool* 57: 475–485.

Laerm, J. 1982. The origin and homology of the neopterygian vertebral centrum. *J Paleont* 56: 191–202.

LaFlamme, S. E., Jamrich, M., Richter, K., Sargent, T. D., and Dawid, I. B. 1988. Xenopus endo B is a keratin preferentially expressed in the embryonic notochord. *Genes Dev* 2: 853–862.

Lamarck, J. B. 1809. *Philosophie Zoologique.* Dentu Libraire, Paris.

Lang, M. R., Lapierre, L. A., Frotscher, M., Goldenring, J. R., and Knapik, E. W. 2006. Secretory COPII coat component Sec23a is essential for craniofacial chondrocyte mat-uration. *Nat Gen* 38: 1198–1203.

Lankester, E. R. 1877. Notes on the embryology and classification of the animal kingdom: comprising a revision of speculations relative to the origin and significance of the germ-layers. *Q J Microsc Sci* 68: 399–454.

Lash, J. W. 1963. Tissue interaction and specific metabolic responses: chondrogenesis induc-tion and differentiation. In *Cytodifferentiation and Macromolecular Synthesis* (M. Locke, ed.), pp. 235–260. Academic Press, New York, NY.

Lash, J. W., Glick, M. C., and Madden, J. W. 1964. Cartilage induction in vitro and sulfate-activating enzymes. *Natl Cancer Inst Monogr* 13: 39–49.

Lash, J. W., Holtzer, S., and Holtzer, H. 1957. An experimental analysis of the development of the spinal column. VI. Aspects of cartilage induction. *Exp Cell Res* 13: 292–303.

Lauder, G. V. 1980. On the relationship of the myotome to the axial skeleton in vertebrate evolution. *Paleobiology* 6: 51–56.

Lauri, A., Brunet, T., Handberg-Thorsager, M., Fischer, A. H., Simakov, O., Steinmetz, P. R., Tomer, R., Keller, P. J., and Arendt, D. 2014. Development of the annelid axochord; insights into notochord evolution. *Science* 345: 1365–1368.

Lawson, K. A., Meneses, J. J., and Pedersen, R. A. 1991. Clonal analysis of epiblast fate during germ layer formation in the mouse embryo. *Development* 113: 891–911.

Lawson, L., and Harfe, B. D. 2015. Notochord to nucleus pulposus transition. *Curr Osteoporosis Rep* 13: 336–341.

Lawson, R. 1966. The development of the centrum of *Hypogeophis rostratus* (Amphibia, Apoda) with special reference to the notochordal (intravertebral) cartilage. *J Morph* 118: 137–148.

Leeson, T. S., and Leeson, C. R. 1958. Observations and the histochemistry and fine structure of the notochord in rabbit embryos. *J Anat* 92: 278–285.

Lefebvre, V. 2002. Toward understanding the functions of the two highly related Sox5 and Sox6 genes. *J Bone Miner Metab* 20: 121–130.

Lefebvre, V., and de Crombrugghe, B. 1998. Toward understanding SOX9 function in chondrocyte differentiation. *Matrix Biol* 16: 529–540.

Lefebvre, V., Li, P., and de Crombrugghe, B. 1998. A new long form of Sox 5 (L-Sox5), Sox6 and Sox9 are coexpressed in chondrogenesis and cooperatively activate the type II collagen gene. *EMBO J* 17: 5718–5733.

Lehtonen, E., Stefanovic, V., and Saraga-Babic, M. 1995. Changes in the expression of intermediate filaments and desmoplakins during development of human notochord. *Differentiation* 59: 43–49.

Leitges, M., Neidhardt, L., Haenig, B., Herrmann, B. G., and Kispert, A. 2000. The paired homeobox gene Uncx4.1 specifies pedicles, transverse processes and proximal ribs of the vertebral column. *Development* 127: 2259–2267.

Leprévost, A., Azaïs, T., Trichet, M., and Sire, J. Y. 2017. Identification of a new mineralized tissue in the notochord of reared Siberian sturgeon (*Acipenser baerii*). *J Morph* 278: 1586–1597.

Lewis, J., Hanisch, A., and Holder, M. 2009. Notch signaling, the segmentation clock, and the patterning of vertebrate somites. *J Biol* 8: 44. doi: 10.1186/jbiol145.

Lim, Y-W., Lo, H. P., Ferguson, C., Martel, N., Giacomotto, J., Gomez, G. A., Yap, A.S., Hall, T. E. and Parton, R. G. 2017. Caveolae protect notochord cells against catastrophic mechanical failure during development. *Curr Biol* 27: 1–14.

Linsenmayer, T. F., Toole, B. P., and Trelstad, R. L. 1973. Temporal and spatial transitions in collagen types during embryonic chick limb development. *Dev Biol* 35: 232–239.

Linsenmayer, T., Gibney, E., and Schmidt, T. M. 1986. Segmental appearance of type X collagen in the developing avian notochord. *Dev Biol* 113: 467–473.

Lleras-Forero, L., Narayanan, R., Huitema, L. F. A., VanBergen, M., Apschner, A., Peterson Maduro, J., Logister, I., Valentin, G., Morelli, L. G., Oates, A., and Schulte-Merker, S. 2018. Segmentation of the zebrafish axial skeleton relies on notochord sheath cells and not on the segmentation clock. *elife* pii: e33843. doi: 10.7554/eLife.33843.

Locket, N. A. 1980. Review lecture: Some advances in coelacanth biology. *Proc R Soc London Ser B* 208: 265–307.

Lohr, J. L., Danos, M. C., and Yost, H. J. 1997. Left-right asymmetry of a nodal-related gene is regulated by dorsoanterior midline structures during *Xenopus* development. *Development* 124: 1465–1472.

Loizides, M., Georgiou, A. N., Somarakis, S., Witten, P. E., and Koumoundouros, G. 2014. A new type of lordosis and vertebral body compression in Gilthead seabream (*Sparus aurata* Linnaeus 1758): aetiology, anatomy and consequences for survival. *J Fish Dis* 37: 949–957.

Lolas, M., Pablo, D. T., Valenzuela, R. T., and Liu, Z. 2014. Charting Brachyury-mediated developmental pathways during early mouse embryogenesis. *Proc Natl Acad Sci USA* 111: 4478–4483.

Lopez–Baez, J. C., Simpson, D. J., Lleras-Forero, L., Zheng, S., Zhiqiang, Z Brunsdon, H., Salzano, A., Brombin, A., Wyatt, C., Rybski, W., Huitema, L. F. A., Dale, R. M., Kawakami, K., Englert, C., Chandra, T., Schulte-Merker, S., Hastie, N. D., and Patton, E. E. 2018. Wilms Tumor 1b defines a wound-specific sheath cell subpopulation associated with notochord repair. *eLife* 7: e30657.

López-Cuevas, P., Deane, L., Yang, Y., Hammond, C. L., and Kague, E. 2021. Transformed notochordal cells trigger chronic wounds in zebrafish, destabilizing the vertebral column and bone homeostasis. *Dis Models Mech* 14, dmm047001. doi:10.1242/dmm.047001

Lotz, J. C., Colliou, O. K., Chin, J. R., Duncan, N. A., and Liebenberg E. 1998. Compression-induced degeneration of the intervertebral disc: an in vivo mouse model and finite-element study. *Spine* 23(23): 2493–2506.

Lotz, J. C., and Hsieh, A. H. 2014. The effects of mechanical forces on nucleus pulposus and annulus fibrosus cells. In *The Intervertebral Disc. Molecular and Structural Studies of the Disc in Health and Disease* (I. M. Shapiro and M. V. Risbud, eds), pp. 109–123. Springer-Verlag, Wien, Heidelberg, New York, Dordrecht, London.

Lotz J. C., Hsieh, A. H., Walsh, A. L., Palmer, E. I., and Chin, J. R. 2002. Mechanobiology of the intervertebral disc. *Biochem Soc Trans* 30(6): 853–858.

Lowe, C. J. 2008. Molecular genetic insights into deuterostome evolution from the direct developing hemichordate *Saccoglossus kowwalevskii*. *Philos Trans R Soc London* 363: 1569–1578.

Lowe, C. J., Clarke, D. N., Medeiros, D. M., Rokhsar, D. S., and Gerhart, J. 2015. The deuterostome context of chordate origins. *Nature* 520: 456–465.

Lowe, C. J., Wu, M., Salic, A., Evans, L., Lander, E., Strange-Thomann, N., Gruber, C. E., Gerhart, J., and Kirschner, M. 2003. Anterioposterior patterning in hemichordates and the origins of the chordate nervous system. *Cell* 113: 853–865.

Luo, H.-L., Hu, S-X., and Chen, L-Z. 2001. New early Cambrian chordates from Haikou, Kunming. *Acta Geol Sin* 75: 345–348. [English Language Edition]

Lwoff. B. 1893. Über den Zusammenhang von Markrohr und Chorda beim Amphioxus und ähnliche Verhältnisse bei Anneliden. *Z Wiss Zool* 56: 299–309.

Maienschein, J. 1994. 'It's a long way from Amphioxus.' Anton Dohrn and late nineteenth century debates about vertebrate origins. *Hist Philos Life Sci* 16: 465–478.

Maisey, J. G. 1986. Heads and tails: a chordate phylogeny. *Cladistics* 2: 201–256.

Maisey, J. G. 1988. Phylogeny of early vertebrate skeletal induction and ossification patterns. *Evol Biol* 32: 1–36.

Maisey, J. G. 2000. *Discovering Fossil Fishes.* p. 223. Westview Press, Boulder Colorado.

Maisey, J. G. 2008. The postorbital palatoquadrate articulation in elasmobranchs. *J Morph* 269: 1022–1040.

Maisey, J. G. 2012. What is an 'elasmobranch'? The impact of palaeontology in understanding elasmobranch phylogeny and evolution. *J Fish Biol* 80: 918–951.

Malacinski, G. M. and Youn, B. W. 1982. The structure of the anuran amphibian notochord and a re-evaluation of its presumed role in early embryogenesis. *Differentiation* 21: 13–21.

Mansfield, J. H., Haller, E., Holland, N. D., and Brent, A. E. 2015. Development of somites and their derivatives in amphioxus, and implications for the evolution of vertebrate somites. *EvoDevo* 6(1), 1–30. doi: 10.1186/s13227-015-0007-5.

Mansouri, A., Voss, A. K., Thomas, T., Yokota, Y., and Gruss, P. 2000. Uncx4.1 is required for the formation of the pedicles and proximal ribs and acts upstream of Pax9. *Development* 127: 2251–2258.

Marcello, C., Ahlgren, S., and Bronner-Fraser, M. 1999. In vivo regulation of somite differentiation and proliferation by sonic hedgehog. *Dev Biol* 214: 277–287.

Martini, A., Huysseune, A., Witten, P. E., and Boglione, C. 2021. Plasticity of the skeleton and skeletal deformations in zebrafish (*Danio rerio*) linked to rearing density. *J Fish Biol* 98: 971–986.

Mathews, M. B. 1971. Comparative biochemistry of chondroitin sulphate proteins of cartilage and notochord. *Biochem J* 125: 37–46.

Mathews, M. B. (Ed) 1975. Connective tissue macromolecular structure and evolution. In *Molecular Biology Biochemistry and Biophysics. Volume 19.* pp. 1–318.

McBirney, A., Cook, S., and Retallack, G. 2009. *The Philosophy of Zoology before Darwin.* A translated and annotated version of the original French text by Edmond Perrier. p. 224. Springer Dordrecht.

McCann, M. R., and Séguin, C. A. 2016. Notochord cells in intervertebral disc development and degeneration. *J Dev Biol* 4: 3. doi: 10.3390/jdb4010003.

McCoy, V. E., Saupe, E. E., Lamsdell, J. C., Tarhan, L. G., McMahon, S., Lidgard, S., Mayer, P., Whalen, C. D., Soriani, C., Finney, L., Vogt, S., Clark, E. G., Anderson, R. P., Petermann, H., Locatelli, E. R., and Briggs, D. E. G. 2016. The 'Tully monster' is a vertebrate. *Nature* 532: 496–499.

McLean, K. E., and Vickaryous, M. K. 2011. A novel model of epimorphic regeneration: the leopard gecko, *Eublepharis macularius. BMC Dev Biol* 11: 50.

McMahon, J. A., Takada, S., Zimmerman, L. B., Fan, C. M., Harland, R. M., and McMahon, A. P. 1998. Noggin-mediated antagonism of BMP signaling is required for growth and patterning of the neural tube and somite. *Genes Dev* 12: 1438–1452.

McMaster, M. L., Goldstein, A. M., Bromley, C, M., Ishibe, N., and Parry, D. M. 2001. Chordoma: incidence and survival patterns in the United States, 1973–1995. *Cancer Causes Control* 12: 1–11.

Meinertzhagen, I. A. 2010. The organisation of invertebrate brains: cells, synapses and circuits. *Acta Zool* 91: 64–71.

Melby, A. E., Warga, R. M., and Kimmel, C. B. 1996. Specification of cell fates at the dorsal margin of the zebrafish gastrula. *Development* 122: 2225–2237.

Melville, D. B., Montero-Balaguer, M., Levic, D. S., Bradley, K., Smith, J. R., Hatzopoulos, A. K., and Knapik, E. W. 2011. The feelgood mutation in zebrafish dysregulates COPII-dependent secretion of select extracellular matrix proteins in skeletal morphogenesis. *Disease Models Mech* 4: 763–776.

Metchnikoff, V. E. 1866. A Kowalevsky: Le développement de *l'Amphioxus lanceolatus. Arch Sci Phys Nat* 27: 193–195.

Metchnikoff, V. E. 1881. Über die systematische Stellung von Balanoglossus. *Zool Anz* 4: 139–157.

Meunier, F. J. M., Cupello, C., and Clément, G. 2019. The skeleton and the mineralized tissues of the living coelacanths. *Bull Kitak Mus Natl Hist Hum His Ser A* 17: 37–48.

Miller, E. J., and Mathews, M. B. 1974. Characterization of notochord collagen as a cartilage-type collagen. *Biochem Biophys Res Commun* 60: 424–430.

Millot, J., and Anthony, J. 1958. *Anatomie de Latimeria chalumnae. T. 1. Squelette, muscles et formations de soutien.* p. 122. CNRS Edr., Paris. 80 plates.

Mina, M., Kollar, E. J., and Upholt, W. B. 1991. Temporal and spatial expression of genes for cartilage extracellular matrix proteins during avian mandibular arch development. *Differentiation* 48: 17–24.

Minor, R. P., Rosenbloom, J., Lash, J. W. and von der Mark, K. 1975. Chondrogenic differentiation in cultured somites. In *Extracellular Matrix Influences on Gene Expression* (H. C. Slavkin and R. G. Greulich, eds), pp. 169–174. Academic Press, New York.

Minor, R. R. 1973. Somite chondrogenesis. A structural analysis. *J Cell Biol* 56: 27–50.

Mittapalli, V. R., Huang, R., Patel, K., Christ, B., and Scaal, M. 2005. Arthrotome: a specific joint forming compartment in the avian somite. *Dev Dyn* 234: 48–53.

Moffat, L. A. 1973. The development and adult structure of the vertebral column in *Leiopelma* (Amphibia: Anura). *Proc Linn Soc NSW* 98: 142–174.

Mookerjee, H. K. 1930. On the development of the vertebral column of Urodela. *Philos Trans R Soc London B* 218: 415–446.

Mookerjee, H. K. 1936. The development of the vertebral column and its bearing on the study of organic evolution. *Proc Ind Sci Cong* 23: 307–343.

Mookerjee, H. K., Mitra, G. N., and Mazumdar, S. R. 1940. The development of the vertebral column of a viviparous teleost, *Lebistes reticulates. J Morph* 67: 241–269.

Mookerjee, S. 1953. An experimental study of the development of the notochordal sheath. *J Embryol Exp Morph* 1: 411–416.

Mookerjee, S., Deuchar, E. M., and Waddington, C. H. 1953. The morphogenesis of the notochord in amphibia. *J Embryol Exp Morph* 1: 399–409.

Moore, K. L., Persaud, T. V. N., and Torchia, M. G. 2016. *Before we are Born. Essentials of Embryology and Birth Defects.* 9th Edition. Elsevier, Philadelphia.

Morin-Kensicki, E. M., Melancon, E., and Eisen, J. S. 2002. Segmental relationship between somites and vertebral column in zebrafish. *Development* 129: 3851–3860.

Moss, M. L., and Moss-Salentijn, L. 1983. Vertebrate cartilages. In *Cartilage, Volume 12. Structure, Function and Biochemistry* (B. K. Hall, ed.), pp. 1–30. Academic Press, New York, NY.

Mukhopadhyay, B., Shukla, R. M., Mukhopadhyay, M., Mandal, K. C., Haldar, P., and Benare, A. 2012. Spectrum of human tails: a report of six cases. *J Indian Assoc Pediatr Surg* 17: 23–25.

Müller, F., and O'Rahilly, R. 1987. The development of the human brain, the closure of the caudal neuropore, and the beginning of secondary neurulation at stage 12. *Anat Embryol* 176: 413–430.

Müller, H., and Kölliker, A. 1858. Ueber Reste der Chorda dorsalis. *Verh Phys-Med Ges Würzb* 9(VI): XXXIV.

Munsterberg, A. E., and Lassar, A. B. 1995. Combinatorial signals from the neural tube, floor plate and notochord induce myogenic bHLH gene expression in the somite. *Development* 121: 651–660.

Murdock, D. J. E., Rayfield, E. J., and Donoghue, P. C. J. 2014. Functional adaptation underpinned the evolutionary assembly of the earliest vertebrate skeleton. *Evol Dev* 16: 354–361.

Murray, P. D. F. 1936. *Bones. A Study of the Development and Structure of the Vertebrate Skeleton.* p. 203. Cambridge University Press, Cambridge and London. [Reissued in 1985 with an Introduction by B. K. Hall].

Murtaugh, L. C., Chyung, J. H., and Lassar, A. B. 1999. Sonic hedgehog promotes somitic chondrogenesis by altering the cellular response to BMP signaling. *Genes Dev* 13: 225–237.

Musgrove, J. 1891. Persistence of the notochord in the human subject. *J Anat Physiol* 1891: 386–389.

Mwale, F. 2014. Collagen and other proteins of the nucleus pulposus, annulus fibrosus, and cartilage end plates. In *The Intervertebral Disc* (I. M. Shapiro and M. V. Risbud, eds), pp. 79–92. Springer-Verlag, Wien.

Nanglu, K., Caron, J. B., Morris, S. C., and Cameron, C. B. 2016. Cambrian suspension-feeding tubicolous hemichordates. *BMC Biology* 14 (14), Article No. 56. doi:10.1186/s12915-016-0271-4

Nelsen, O.E. 1953. *Comparative Embryology of the Vertebrates.* p. 982. Constable and Company Ltd, London.

Ng, L. J., Tam, P. P. L., and Cheah, K. S. E. 1993. Preferential expression of alternatively spliced mRNAs encoding type II procollagen with a cysteine-rich amino-propeptide in differentiating cartilage and nonchondrogenic tissues during early mouse development. *Dev Biol* 159: 403–417.

Ng, L.-J., Wheatley, S., Muscat, G. E. O., Conway-Campbell, J., Bowles, J., Wright, E., Bell, D. M., Tam, P. M., Cheah, R. S., and Koopman, P. 1997. Sox9 binds DNA, activates transcription, and co-expresses with type II collagen during chondrogenesis in the mouse. *Dev Biol* 183: 108–121.

Nibu, Y., José-Edwards, D. S., and Di Gregorio, A. 2013. From notochord formation to hereditary chordoma: the many roles of Brachyury. *BioMed Res Intern* 826435.

Nickel, W., Brügger, B., and Wieland, F. T. 2002. Vesicular transport: the core machinery of COPI recruitment and budding. *J Cell Sci* 115: 3235–3240.

Nieuwkoop, P. D., and Ubbels, G. A. 1972. The formation of the mesoderm in urodelean amphibians. IV. Qualitative evidence for the purely 'ectodermal' origin of the entire mesoderm and of the pharyngeal endoderm. *Wilhelm Roux Arch EntwMech* 169: 185–199.

Nobel, P. S. 1970. *Introduction to Biophysical Plant Physiology.* p. 488. W. H. Freeman, San Francisco.

Norden, C. R. 1961. Comparative osteology of representative salmonid fishes, with particular reference to the grayling (*Thymallus arcticus*) and its phylogeny. *J Fish Res Board Can* 18(5): 679–791.

Nordvik, K. 2007. From notochord to vertebral column: studies on Atlantic salmon (*Salmo salar* L.) Dissertation for the degree philosophiae doctor (PhD) at the University of Bergen. Bergen, Norway. p. 37.

Nordvik, K., Kryvi, H., Totland, G. K., and Grotmol, S. 2005. The salmon vertebral body develops through mineralization of two preformed tissues that are encompassed by two layers of bone. *J Anat* 206: 103–114.

Nowroozi, B. N., Harper, C. J., De Kegel, B., Adriaens, D., and Brainerd, E. L. 2012. Regional variation in morphology of vertebral centra and intervertebral joints in striped bass, *Morone saxatilis*. *J Morph* 273: 441–445.

Nübler-Jung, K., and Arendt, D. 1994. Is ventral in insects dorsal in vertebrates? A history of embryological arguments favouring axis inversion in chordate ancestors. *Roux Archiv Dev Biol* 203: 357–366.

O'Connell, J., and Low, F. N. 1970. A histochemical and fine structural study of early extracellular connective tissue in the chick embryo. *Anat Rec* 167: 425–438.

Odenthal, J., Haffter, P., Vogelsang, E., Brand, M., van Eeden, F. J., Furutani-Seiki, M., Granato, M., Hammerschmidt, M., Heisenberg, C. P., Jiang, Y. J., Kane, D. A., Kelsh, R. N., Mullins, M. C., Warga, R. M., Allende, M. L., Weinberg, E. S., and Nüsslein-Volhard, C. 1996. Mutations affecting the formation of the notochord in the zebrafish, *Danio rerio*. *Development* 123: 103–115.

Oettinger, H. F., Thal, G. J., Sasse, J. H., Holtzer, H., and Pacifici, M. 1985. Immunological analysis of chick notochord and cartilage matrix development with antisera to cartilage matrix macromolecules. *Dev Biol* 109: 63–71.

Ofer, L., Dean, M. N., Zaslansky, P., Kult, S., Shwartz, Y., Zaretsky, J., Griess-Fishheimer, S., Monsonego Ornan, E., Zelzer, E., and Shahar, R. 2019. A novel nonosteocytic regulatory mechanism of bone modeling. *PLoS Biol* 17(2): e3000140. doi: 10.1371/journal.pbio.3000140.

Ohisa, S., Inohaya, K., Takano, Y., and Kudo, A. 2010. sec24d encoding a component of COPII is essential for vertebra formation, revealed by the analysis of the medaka mutant, vbi. *Dev Biol* 342: 85–95.

Onai, T., Aramaki, T., Inomata, H., Hirai, T., and Kuratani, S. 2015. On the origin of vertebrate somites. *Zool Lett* 33: 1–10.

Oppenheimer, J. M. 1934. Experiments on early developing stages of *Fundulus*. *Proc Natl Acad Sci USA* 20: 536–538.

Oppenheimer, J. M. 1939. Transplantation experiments on developing teleosts (*Fundulus* and *Perca*). *J Exp Zool* 72: 409–437.

Oralová, V., Rosa, J. T., Larionova, D., Witten, P. E., and Huysseune, A. 2020. Multiple epithelia are required to develop teeth deep inside the pharynx. *Proc Natl Acad Sci USA* 117(21): 11503–11512.

Ota, K. G., Fujimoto, S., Oisi, Y., and Kuratani, S. 2011. Identification of vertebra-like elements and their possible differentiation from sclerotomes in the hagfish. *Nat Commun* 2: 373. doi: 10.1038/ncomms1355.

Ota, K. G., Oisi, Y., Fujimoto, S., and Kuratani, S. 2014. The origin of developmental mechanisms underlying vertebral elements: implications from hagfish evo-devo. *Zoology* 117: 77–80.

Overton, J., and Mapp, F. E. 1974. The fine structure of regenerating notochord in anuran tadpoles. *J Exp Zool* 187: 103–120.

Owen, R. 1847. Description of the atlas, axis and subvertebral wedge bones in the Plesiosaurus, with remarks on the homologies of those bones. *Ann Mag Nat Hist* 20(133): 217–225.

Owen, R. 1848. *The Archetype and Homologies of the Vertebrate Skeleton.* p. 203. Richard & John E. Taylor, London.

Owen, R. 1866. On the anatomy of vertebrates. *Fishes and Reptiles. Volume 1.* p. 650. Longmans, Green & Co., London.

Paavola, L. G., Wilson, D. B, and Center, E. H. M. 1980. Histochemistry of the developing notochord, perichordal sheath and vertebrae in Danforth's short-tail (Sd) and normal C57BL/6 mice. *J Embryol exp. Morph* 55: 227–245.

Pagnon-Minot, A., Malbouyres, M., Haftek-Terreau, Z., Kim, H. R., Sasaki, T., Thisse, C., Thisse, B., Ingham, P. W., Ruggiero, F., and Le Guellec, D. 2008. Collagen XV, a novel factor in zebrafish notochord differentiation and muscle development. *Dev Biol* 316: 21–35.

Pan, H., Yu, H., Ravi, V., Li, C., Lee, A. P., Lian, M. M., Tay, B.-H., Brenner, S., Wang, J., Yang, H., Zhang, G., and Venkatesh, B. 2016. The genome of the largest bony fish, ocean sunfish (*Mola mola*), provides insights into its fast growth rate. *GigaScience* 5: 36. doi: 10.1186/s13742-016-0144-3.

Park, J., Gebhardt, M., Golovchenko, S., Perez-Branguli, F., Hattori, T., Hartmann, C., Zhou, X., de Crombrugghe, B., Stock, M., Schneider, H., and von der Mark, K. 2015. Dual pathways to endochondral osteoblasts: a novel chondrocyte-derived osteoprogenitor cell identified in hypertrophic cartilage. *Biol Open* 4: 608–621.

Parmentier, E. 2016. Further insights into the metamorphosis process in a carapid fish. *J Zool* 298: 249–256.

Parsons, M. J., Pollard, S. M., Saúde, L., Feldman, B., Coutinho, P., Hirst, E. M. A., and Stemple, D. L. 2002. Zebrafish mutants identify an essential role for laminins in notochord formation. *Development* 129: 3137–3146.

Parton, R. G., and Simons, K. 2007. The multiple faces of caveolae. *Nat Rev Mol Cell Biol* 8: 185–194.

Pasteels, J. 1937. Etudes sur la gastrulation des Vertébrés méroblastiques. III. Oiseaux. IV. Conclusions générales. *Arch Biol* 48: 381–488.

Pasteels, J. 1958. Développement des agnathes. In *Traité de Zoologie* (P. Grassé, ed.), *Volume 13.* pp. 106–144. Masson et Cie, Saint-Germain, Paris.

Patterson, C. 1968. The caudal skeleton in Lower Triassic pholidophorid fishes. *Bull Brit Mus Kat Hist Geo* 16: 210–239.

Pears, J. B., Johanson, Z., Trinajstic, K., Dean, M. N., and Boisvert, C. A. 2020. Mineralization of the *Callorhinchus* vertebral column (Holocephali; Chondrichthyes). *Front Genet* 11: 571694. doi: 10.3389/fgene.2020.571694.

Peck, S. H., McKee, K. K., Tobias, J. W., Malhotra, N. R., Harfe, B. D., and Smith, L. J. 2017. Whole transcriptome analysis of notochord-derived cells during embryonic formation of the nucleus pulposus. *Sci Rep* 7: 10504. doi: 10.1038/s41598-017-10692-5.

Peignoux-Deville, J., Baud, C. A., Lallier, F., and Vidal, B. 1985. Perichondral ossification of vertebral arches from dogfish to man. *Prog Zool* 30: 65–68.

Peignoux-Deville, J., Lallier, F., and Vidal, B. 1982. Evidence for the presence of osseous tissue in dogfish vertebrae. *Cell Tissue Res* 222: 605–614.

Peña, C. E., Horvat, B. L., and Fisher, E. R. 1970. The ultrastructure of chordoma. *Am J Clin Pathol* 53: 544–551.

Peskin, B., Henke, K., Cumplido, N., Treaster, S., Harris, M. P., Bagnat, M., and Arratia, G. 2020. Notochordal signals establish phylogenetic identity of the teleost spine. *Curr Biol* 30: 1–10.

Peters, H. M. 1963. Eizahl und Gelegeentwicklung in der Gattung Tilapia (Cichlidae, Teleostei). *Int Revue Hydrobiol* 48: 547–576.

Peters, H., Wilm, B., Sakai, N., Imai, K., Maas, R., and Balling, R. 1999. Pax1 and Pax9 synergistically regulate vertebral column development. *Development* 126: 5399–5408.

Piiper, J. 1928. VI. On the evolution of the vertebral column in birds, illustrated by its development in *Larus* and *Struthio*. *Philos Trans R Soc London Ser B* 216: 285–351.

Pillai, M. K., and Nair, S. T. 2017. A true human tail in a neonate. *Sultan Qaboos Univ Med J* 17: e109–e111.

Placzek, M. 1995. The role of the notochord and floor plate in inductive interactions. *Curr Opin Genet Dev* 5: 499–506.

Pogoda, H-M., Riedl-Quinkertz, I., Löhr, H., Waxman, J. S., Dale, R. M., Topczewski, J., Schulte-Merker, S., and Hammerschmidt, M. 2018. Direct activation of chordoblasts by retinoic acid is required for segmented centra mineralization during zebrafish spine development. *Development* 145: dev159418. doi: 10.1242/dev.159418.

Pollard, S. M., Parsons, M. J., Kamei, M., Kettleborough, R. N., Thomas, K. A., Pham, V. N., Bae, M. K., Scott, A., Weinstein, B. M., and Stemple, D. L. 2006. Essential and overlapping roles for laminin alpha chains in notochord and blood vessel formation. *Dev Biol* 289: 64–76.

Poss, S. G., and H. T. Boschung. 1996. Lancelets (Cephalochordata: Branchiostomatidae): how many species are valid? *Israel J Zool* 42: S13–S66.

Pourquié, O., Coltey, M., Teillet, M., Ordahl, C., and Douarin, N. 1993. Control of dorsoventral patterning of somitic derivatives by notochord and floor plate. *Proc Natl Acad Sci USA* 90: 5242–5246.

Pownall, M. E., Strunk, M. E., and Emerson, C. P. Jr. 1996. Notochordal signals control the transcriptional cascade of myogenic bHLH genes in somites of quail embryos. *Development* 122: 1475–1488.

Prince, V. A., Joly, L., Ekker, M., and Ho, R. K. 1998a. Zebrafish hox genes: genomic organization and modified colinear expression patterns in the trunk. *Development* 125: 407–420.

Prince, V. A., Price, A. L., and Ho, R. K. 1998b. Hox gene expression reveals regionalization along the anteroposterior axis of the zebrafish notochord. *Dev Genes Evol* 208: 517–522.

Prinos, P., Joseph, S., Oh, K., Meyer, B. I., Gruss, P., and Lohnes, D. 2001. Multiple pathways governing Cdx1 expression during murine development. *Dev Biol* 239: 257–269. doi: 10.1006/dbio.2001.0446.

Pugin, E. M. 1973. Sur le comportement des troncons du tube neural et du la corde d'embryon de souris greffés à la place des organes homologues chez l'embryon de poulet. *CR Acad Sci* 276: 3477–3480.

Raff, E. C., Villinski, J. T., Turner, F. R., Donoghue, P. C. J., and Raff, R. A. 2006. Experimental taphonomy shows the feasibility of fossil embryos. *Proc Natl Acad Sci USA* 103: 5846–5851.

Reeves, W. M., Wu, Y., Harder, M. J., and Veeman, M. T. 2017. Functional and evolutionary insights from the *Ciona* notochord transcriptome. *Development* 144: 3375–3387.

Remak, R. 1855. *Untersuchungen über die Entwicklung der Wirbelthiere*. Berlin, Reimer, VI + p. 195.

Remane, A. 1936. Die Wirbelsäule und ihre Abkömmlinge. In *Handbuch der Vergleichenden Anatomie der Wirbeltiere. Volume 4* (L. Bolk, E. Göppert, E. Kallius, and W. Lubsch, eds), pp. 1–206. Urban and Schwarzenberg, Berlin, Wien.

Remane, A. 1952. *Die Grundlagen des natürlichen Systems, der vergleichenden Anatomie und der Phylogenetik*. Akademische Verlagsgesellschaft und Portig, K.-G., Leipzig.

Renn, J., Büttner, A., To, T. T., Chan, S. J. H., and Winkler, C. 2013. A col10a1:nlGFP transgenic line displays putative osteoblast precursors at the medaka notochordal sheath prior to mineralization. *Dev Biol* 381: 134–143.

Resende, T. P., Ferreira, M., Teillet, M. A., Tavares, A. T., Andrade, R. P., and Palmeirim, I. 2010. Sonic hedgehog in temporal control of somite formation. *Proc Natl Acad Sci USA* 107: 12907–12912.

Restović, I., Vukojević, K., Saraga-Babić, M., and Bočina, I. 2016. Ultrastructural features of the dogfish *Scyliorhinus canicula* (Pisces: Scyliorhinidae) notochordal cells and the notochordal sheath. *Ital J Zool* 83(3): 329–337.

Ribes, V., and Briscoe, J. 2009. Establishing and interpreting graded sonic hedgehog during vertebrate neural tube patterning: the role of negative feedback, *Cold Spring Harb Persp Biol* 1: a002014.

Richardson, S. M., Freenont, A. J., and Hoyland, J. A. 2014. Pathogenesis of intervertebral disk degeneration. In *The Intervertebral Disc* (I. M. Shapiro and M. V. Risbud, eds), pp. 177–200. Springer-Verlag, Wien.

Ridewood, W. G. 1921. On the calcification of the vertebral centra in sharks and rays. *Philos Trans R Soc London Ser B* 210: 312–407.

Ridley, M. 2004. *Evolution*. 3rd Edition, Blackwell. Oxford, UK.

Riedl, R. 1978. *Order in Living Organisms*. John Wiley and Sons, New York.

Risbud, M. V., Schaer, T. P., and Shapiro, I. M. 2010. Toward an understanding of the role of notochordal cells in the adult intervertebral disc: from discord to accord. *Dev Dyn* 239: 2141–2148.

Risbud, M. V., and Shapiro, I. M. 2011. Notochordal cells in the adult intervertebral disc: new perspective on an old question. *Crit Revs Eukary Gene Expression* 21: 29–41.

Roach, H. I. 1997. New aspects of enchondral ossification in the chick: chondrocyte apoptosis, bone formation by former chondrocytes, and acid phosphatase activity in the enchondral bone matrix. *J Bone Mineral Res* 12: 795–805.

Robinson, M. S. 2015. Forty-years of clathrin-coated vesicles. *Traffic* 16: 1210–1238.

Rockwell, H., Evans, F. G., and Pheasant, H. C. 1938. The comparative morphology of the vertebrate spinal column. Its form as related to function. *J Morph* 63: 87–117.

Romeo, S., and Hogendoorn, P. C. W. 2006. Brachyury and chordoma: the chondroid-chordoid dilemma resolved? *J Pathol* 209: 143–146.

Romer, A. S. 1951. *The Vertebrate Body*. W.B. Saunders Company, Philadelphia.

Romer A. S. 1956. *Osteology of the Reptile*. The University of Chicago Press, Chicago.

Romer, A. S. 1974. *Vertebrate Paleontology*. 3rd Edition. p. 468. University of Chicago Press, Chicago.

Rondelet, G. 1554–1555. *Libri de Piscibus Marinis, in Quibus Veræ Piscium Effigies Expressæ Sunt. Matthiam Bonhomme*, p. 583. Lugduni.

Roughley, P. J. 2004. Biology of intervertebral disc aging and degeneration: involvement of the extracellular matrix. *Spine* 29: 2691–2699.

Row, R. H., Tsotras, S. R., Goto, H., and Martin, B. L. 2016. The zebrafish tailbud contains two independent populations of midline progenitor cells. *Development* 143: 244–254. doi: 10.1242/dev.129015.

Rufai, A., Benjamin, M., and Ralphs, J. R. 1995. The development of fibrocartilage in the rat intervertebral disc. *Anat Embryol* 192: 53–62.

Ruggeri, A. 1972. Ultrastructural, histochemical and autoradiographic studies on the developing chick notochord. *Z Anat Entwickl* 138: 20–33.

Rugh, R. 1951. *The Frog: Its Reproduction and Development*. p. 336. The Blakiston Company, Philadelphia, PA.

Rychel, A. L., Smith, S. E., Shimamoto, H. T., and Swalla, B. J. 2006. Evolution and development of the chordates: collagen and pharyngeal cartilage. *Mol Biol Evol* 23: 541–549.

Sachdeva, S. W., Dietza, U. H., Oshimab, Y., Langa, M. R., Knapika, E. W., Hirakib, Y., and Shukunamib, C. 2000. Sequence analysis of zebrafish chondromodulin-1 and expression profile in the notochord and chondrogenic regions during cartilage morphogenesis. *Mech Dev* 105: 157–162.

Sagstad, A., Grotmol, S., Kryvi, H., Krossøy, C., Totland, G. K., Malde, K., Wang, S., Hansen, T., and Wargelius, A. 2011. Identification of vimentin- and elastin-like transcripts specifically expressed in developing notochord of Atlantic salmon (*Salmo salar* L.). *Cell Tissue Res* 346: 191–202.

Sallan, L., Giles, S., Sansom, R. S., Clarke, J. T., Johanson, Z. Sansom, I. J., and Janvier P. 2017. The 'tully monster' is not a vertebrate: characters, convergence and taphonomy in Palaeozoic problematic animals. *Palaenotology* 60: 149–157.

Sandell, L. J. 1994. In situ expression of collagen and proteoglycan genes in notochord and during skeletal development and growth. *Microsc Res Tech* 28: 470–482.

Sansom, R. S., Freedman, K., Gabbott, S. E., Aldridge, R. J., and Purnell, M. A. 2010. Taphonomy and affinity of an enigmatic Silurian vertebrate, *Jamoytius kerwoodi* white. *Palaeontology* 53: 1393–1140.

Sarmah, S., Barrallo-Gimeno, A., Melville, D. B., Topczewski, J., Solnica-Krezel, L., and Knapik, E. W. 2010. Sec24D-Dependent transport of extracellular matrix proteins is required for zebrafish skeletal morphogenesis. *PLOS One* 5: 1–14.

Satoh, N., Tagawa, K., Loew, C. J., Yu, J.-K., Kawashima, T., Ogasawara, M., Kirschner, M., Hisata, K., Su, Y. H., and Gerhart, J. 2014. On a possible evolutionary link of the stomochord of hemichordates to pharyngeal organs of chordates. *Genesis* 52: 925–934.

Satoh, N., Tagawa, K., and Takahashi, H. 2012. How was the notochord born? *Evol Dev* 14: 56–75.

Saúde, L., Woolley, K., Martin, P., Driever, W., and Stemple, D. L. 2000. Axis-inducing activities and cell fates of the zebrafish organizer. *Development* 127: 3407–3417.

Scaal, M. 2016. Early development of the vertebral column. *Sem Cell Dev Biol* 49: 83–91.

Scaal, M., and Wiegreffe, C. 2006. Somite compartments in anamniotes. *Anat Embryol* 211(1): S9–S19.

Scadeng, M., McKenzie, C., He, W., Bartsch, H., Dubowitz, D. J., Stec, D., and St. Leger, J. 2020. Morphology of the Amazonian teleost genus *Arapaima* using advanced 3D imaging. *Front Physiol* 11: 260. doi: 10.3389/fphys.2020.00260.

Schaeffer, B. 1967. Osteichthyan vertebrae. *J Linn Soc Zool* 47: 185–195.

Schaeffer, B., and Patterson, C. 1984. Jurassic Fishes from the Western United States, with Comments on Jurassic Fish Distribution. *American Museum Novitates* 2796: 1–86.

Schaffer, J. 1930. Die Stützgewebe. In *Handbuch der Mikroskopischen Anatomie des Menschen* (W. von Mallendorff, ed.), *Volume II.* Part 2, pp. 338–350. Springer-Verlag, Berlin.

Schauinsland, H. 1903. Beiträge zur Entwicklungsgeschichte und Anatomie der Wirbeltiere. I. II. III. In *Zoologica. Original-Abhandlungen aus dem Gesammtgebiete der Zoologie* (C. Chun, ed.), Bd 39. Verlag von Erwin Nägele, Stuttgard.

Schauinsland, H. 1906. Die Entwickelung der Wirbel-saule nebst Rippen und Brustbein. In *Handbuch der vergleichenden und experimentellen Entwickelungslehre der Wirbeltiere* (O. Hertwig, ed.), pp. 339–572. Gustav Fischer, Jena.

Schinko, I., Potter, I. C., Welsch, U., and Debbage, P. 1992. Structure and development of the notochord 'Elastica Externa' and nearby components of the elastic fibre system of agnathans. *Acta Zool* 73: 57–66.

Schmitz, R. J. 1995. Ultrastructure and function of cellular components of the intercentral joint in the percoid vertebral column. *J Morph* 226: 1–24.

Schmitz, R. J. 1998a. Comparative ultrastructure of the cellular components of the unconstricted notochord in the sturgeon and the lungfish. *J Morph* 236: 75–104.

Schmitz, R. J. 1998b. Immunohistochemical identification of the cytoskeletal elements in the notochord cells of bony fishes. *J Morph* 236: 105–116.

Schnell, N. K., Britz, R., and Johnson, G. D. 2010. New insights into the complex structure and ontogeny of the occipito-vertebral gap in barbeled dragonfishes (Stomiidae, Teleostei). *J Morph* 271: 1006–1022.

Schnell, N. K., and Bernstein, P. 2007. Vertebral centra lost-A synapomorphic feature within the Stomiid genera? *J Morph* 268: 1132–1132.

Schnell, N. K., and Johnson, G. D. 2017. Evolution of a functional head joint in deep-sea fishes (Stomiidae). *PLOS One* 1–11. doi: 10.1371/journal.pone.0170224.

Schoenwolf, G. C., Chandler, N. B., and Smith, J. L. 1985. Analysis of the origins and early fates of neural crest cells in caudal regions of avian embryos. *Dev Biol* 110: 467–479.

Schoenwolf, G. C., and Nichols, D. H. 1984. Histological and ultrastructural studies on the origin of caudal neural crest cells in mouse embryos. *J Comp Neurol* 222: 496–505.

Schroyens, W., Camon, J., and De Ridder, L. 1991. Resistance of the embryonic notochord to invasion by malignant neurogenic tumor cells in vitro. *Clin Exp Metastasis* 9: 403–409.

Schubert, M., Holland, L. Z., Stokes, M. D., and Holland, N. D. 2001. Three amphioxus Wnt genes (AmphiWnt3, AmphiWnt5, and AmphiWnt6) associated with the tail bud: the evolution of somitogenesis in chordates. *Dev Biol* 240: 262–273.

Schulte-Merker, S., Ho, R. K., Herrmann, B. G., and Nüsslein-Volhard, C. 1992. The protein product of the zebrafish homologue of the mouse T gene is expressed in nuclei of the germ-ring and the notochord of the early embryo. *Development* 116: 1021–1032.

Schultze, H.-P., and Arratia, G. 1986. Reevaluation of the caudal skeleton of actinopterygian fishes. I. *Lepisosteus* and *Amia*. *J Morph* 190: 215–241.

Schultze, H.-P., and Arratia, G. 1988. Reevaluation of the caudal skeleton of actinopterygian fishes. II. *Hiodon, Elops* and *Albula*. *J Morph* 195: 257–303.

Schultze, H-P., and Arratia, G. 2013. The caudal skeleton of basal teleosts, its conventions, and some of its major evolutionary novelties in a temporal dimension. In *Mesozoic Fishes 5 – Global Diversity and Evolution* (G. Arratia, H. P. Schultze, and M. V. H. Wilson, eds), pp. 187–246. Verlag Dr. Friedrich Pfeil, München, Germany. ISBN 978-3-89937-159-8.

Schultze, H.-P., and Cloutier, R. 1991. Computed tomography and magnetic resonance imaging studies of *Latimeria chalumnae*. *Env Biol Fishes* 32: 159–181.

Schwab, W., Galbiati, F., Volonte D., Hempel, U., Wenzel, K. W., Funk, R. H. W, Lisanti, M. P., and Kasper, M. 1999. Characterisation of caveolins from cartilage: expression of caveolin-1, -2 and -3 in chondrocytes and in alginate cell culture of the rat tibia. *Histochem Cell Biol* 112: 41–49.

Schwarz, W. 1961. Elektronenmikroskopische Untersuchungen an den Chordazellen von *Petromyzon*. *Z Zellforschung* 55: 597–609.

Scott, A., and Stemple, D. L 2004. Zebrafish notochordal basement membrane: signalling and structure. *Curr Topic Dev Biol* 65: 229–253.

Scott, H. W. 1934. The zoological relationships of the conodonts. *J Paleontol* 8: 448–455.

Šećerov, S. 1911. Über die Entstehung der Diplospondylie der Selachier. *Arb Zool Inst Wien* 19: 1–28.

Seeman, F., Peterson, D. R., Witten, P. E., Guo, B-S., Shantanagoudar, A. H., Ye, R. R., Zhang, G., and Au, D. W. T. 2015. Insight into the transgenerational effect of benzo[a]pyrene on bone formation in a teleost fish (*Oryzias latipes*). *Comp Biochem Physiol C* 178: 60–67.

Sefton, E. M., Piekarski, N., and Hanken, J. 2015. Dual embryonic origin and patterning of the pharyngeal skeleton in the axolotl (*Ambystoma mexicanum*). *Evol Dev* 17(3): 175–184.

Seleit, A., Gross, K., Onistschenko, J., Woelk, M., Autorino, C., and Centanin, L. 2021. Development and regeneration dynamics of the Medaka notochord. *Dev Biol* 463(1): 11–25.

Seligmann, H., Moravec, J. I. R., and Werner, Y. L. 2008. Morphological, functional and evolutionary aspects of tail autotomy and regeneration in the 'living fossil' *Sphenodon* (Reptilia: Rhynchocephalia). *Biol J Linn Soc London* 93: 721–743.

Selleck, M. A. J., and Stern, C. D. 1991. Fate mapping and cell lineage analysis of Hensen's node in the chick embryo. *Development* 112: 615–626.

Semper, C. 1875. Die Stammesverwandschaft der Wirbelthiere und Wirbellosen. *Arb Zool-Zool Inst Würzburg* 2: 25–76.

Semper, C. 1876–1877. Die Verwandschaftsbeziehungen der gegliederten Thiere. *Arb Zool-Zool Inst Würzburg* 3: 115–404.

Senevirathne, G., Baumgart, S., Shubin, N., Hanken, J., and Shubin, N. H. 2020. Ontogeny of the anuran urostyle and the developmental context of evolutionary novelty. *Proc Natl Acad Sci USA* 117: 3034–3044.

Senthinathan, B., Sousa, C., Tannahill, D., and Keynes, R. 2012. The generation of vertebral segmental patterning in the chick embryo. *J Anat* 220: 591–602.

Seufert, D. W., Hanken, J., and Klymkowsky, M. W. 1994. Type II collagen distribution during cranial development in *Xenopus laevis*. *Anat Embryol* 189: 81–89.

Shapiro, F. 1992, Vertebral development of the chick embryo during days 3–19 of incubation. *J Morph* 213: 317–333.

Shapiro, I. M., and Risbud, M. V. 2010. Transcriptional profiling of the nucleus pulposus: say yes to notochord. *Arthritis Res Therapy* 12: 117.

Shapiro, I. M., and Risbud, M. V. (Eds) 2014. *The Intervertebral Disc. Molecular and Structural Studies of the Disc in Health and Disease*. p. 446. Springer-Verlag, Wien, Heidelberg, New York, Dordrecht, London.

Shih, J., and Fraser, S. E. 1995. Distribution of tissue progenitors within the shield region of the zebrafish gastrula. *Development* 121: 2755–2765.

Shih, J., and Fraser, S. E. 1996. Characterizing the zebrafish organizer: microsurgical analysis at the early-shield stage. *Development* 122: 1313–1322.

Shu, D., Conway Morris, S., Zhang, Z.-F., Lie, J. N., Han, J., Chen, L., Zhang, Z. L., Yasui, K., and Yong, L. 2003. A new species of *Yunnanozoan* with implications for deuterostome evolution. *Science* 299: 1380–1384.

Shu, D.-G., Conway Morris, S., and Zhang, X.-L. 1996. A *Pikaia*-like chordate from the Lower Cambrian of China. *Nature* 384: 156–157.

Shu, D.-G., Luo, H.-L., Conway Morris, S., Zhang, X-L., Hu, S-X., Chen, L., Han, J., Zhu, M., Li, Y., and Chen, L.-Z. 1999. A Lower Cambrian vertebrate from south China. *Nature* 402: 42–46.

Shukunami, C., Iyama, K.-I., Inoue, H., and Hiraki, Y. 1999. Spatiotemporal pattern of the mouse chondromodulin-1 gene expression and its regulatory role in vascular invasion into cartilage during endochondral bone formation. *Int J Dev Biol* 43: 39–49.

Sinha, B., Köster, D., Ruez, R., Gonnord, P., Bastiani, M., Abankwa, D., Stan, R.V., Butler-Browne, G., Vedie, B., Johannes, L., Nobihuro, M., Parton, R. G., Raposo, G., Sens, P., Lamaze, C., and Nassoy, P. 2011. Cells respond to mechanical stress by rapid disassembly of caveolae. *Cell* 144: 402–413.

Sitia, R., and Braakman, I. 2003. Quality control in the endoplasmic reticulum protein factory. *Nature* 426: 891–894.

Sive, J. I., Baird, P., Jeziorsk, M., Watkins, A., Hoyland, J. A., and Freemont, A. J. 2002. Expression of chondrocyte markers by cells of normal and degenerate intervertebral discs. *Mol Pathol* 55: 91–97.

Slack, J. M. W. 2003. Phylotype and zootype. In *Keywords and Concepts in Evolutionary Developmental Biology*. (B. K. Hall and W. M. Olson, eds), pp. 319–318. Harvard University Press, Cambridge, MA.

Smith, J. C., and Watt, F. M. 1985. Biochemical specificity of *Xenopus* notochord. *Differentiation* 29: 109–115.

Smith, J. L. B. 1953. The second cœlacanth. *Nature* 171: 99–101.

Smith, L. J., Nerurkar, N. L., Choi, K. S., Harfe, B. D., and Elliott, D. M. 2011. Degeneration and regeneration of the intervertebral disc: lessons from development. *Disease Models Mech* 4: 31–41.

Smits, P., Dy, P., Mitra, S., and Lefebvre, V. 2004. Sox5 and Sox6 are needed to develop and maintain source, columnar, and hypertrophic chondrocytes in the cartilage growth plate. *J Cell Biol* 164: 747–758.

Smits, P, and Lefebvre, V. 2003. Sox5 and Sox6 are required for notochord extracellular matrix sheath formation, notochord cell survival and development of the nucleus pulposus of intervertebral discs. *Development* 130: 1135–1148.

Sordino, P., van der Hoeven, F., and Duboule, D. 1995. Hox gene expression in teleost fins and the origin of vertebrate digits. *Nature* 375: 678–681.

Spemann, H., and Mangold, H. 1924. Über Induktion von Embryonalanlagen durch Implantation artfremder Organisatoren. *Arch Mikrosk Anat Entwicklungs* 100: 599–638.

Spoorendonk, K. M., Peterson-Maduro, J., Renn, J., Trowe, T., Kranenbarg, S., Winkler, C., and Schulte-Merker, S. 2008. Retinoic acid and Cyp26b1 are critical regulators of osteogenesis in the axial skeleton. *Development* 135: 3765–3774.

Springer, V. G., and Johnson, G. D. 2000. Use and advantages of ethanol solution of alizarin red S dye for staining bone in fishes. *Copeia* 2000: 300–301.

Stach, T. 1999. The ontogeny of the notochord of *Branchiostoma lanceolatum*. *Acta Zool* 80: 25–33.

Stafford, D. Brunet, L. J., Khokha, M. K., Economides, A. N., and Harland, R. M. 2011. Cooperative activity of noggin and gremlin 1 in axial skeleton development. *Development* 138: 1005–1014.

Starck, D. 1979. *Vergleichende Anatomie der Wirbeltiere auf evolutionsbiologischer Grundlage. Bd. 2. Das Skeletsystem. Allgemeines, Skeletsubstanzen, Skelet der Wirbeltiere einschließlich Lokomotionstypen*. Springer Verlag, Berlin.

Stemple, D. L. 2005a. Structure and function of the notochord: an essential organ for chordate development. *Development* 132: 2503–2512.

Stemple, D. L. 2005b. The notochord. *Curr Biol* 14: R873–874.

Stern, C. D. 1990. Two distinct mechanisms of segmentation? *Sem Dev Biol* 1: 109–116.

Stern, C. D., Charité, J., Deschamps, J., Duboule, D., Durston, A. J., Kmita, M., Nicolas, J. F., Palmeirim, I., Smith, J. C. and Wolpert, L. 2006. Head-tail patterning of the vertebrate embryo: one, two or many unresolved problems? *Int J Dev Biol* 50: 3–15.

Stern, C. D., and Keynes, R. J. 1987. Interactions between somite cells: the formation and maintenance of segment boundaries in the chick embryo. *Development* 99: 261–272.

Stone, J. R., and Hall, B. K. 2004. Latent homologues for the neural crest as an evolutionary novelty. *Evol Dev* 6: 123–129.

Strudel, G. 1953. Influence morphogenese du tube nerveux et de la corde sur la différenciation de la colonne vertebrale. *Comptes Rendus des Seances de la Societe de Biologie et des ses Filiales* (Paris) 47: 132–133.

Strudel, G. 1955. L'action morphogène du tube nerveux et de la corde sur la différenciation des vertèbres et des muscles vertébraux chez l'embryon de poulet. *Arch Anat Microsc Morph Exp* 44: 209–235.

Strudel, G. 1962. Induction de cartilage in vitro par l'extrait de tube nerveux et de chorde de l'embryon de poulet. *Dev Biol* 4: 67–86.

Strudel, G. 1963. Autodifférenciation et induction de cartilage à partir de mésenchyme somitique de poulet cultivé in vitro. *J Embryol Exp Morph* 11: 399–412.

Strudel, G. 1967. Some aspects of organogenesis of the chick spinal column. In *Experimental Biology and Medicine. Morphological and Biochemical Aspects of Cytodifferentiation* (E. Hagen, W. Wechsler, and F. Zilliken, eds), *Volume 1.* pp. 183–198. S. Karger, Basel.

Strudel, G. 1971. Matériel extracellulaire et chondrogenèse vertébrale. *C R Hebd Seances Acad Sci* 272: 473–476.

Strudel, G. 1973. Relationship between the chick periaxial metachromatic extracellular material and vertebral chondrogenesis. In *Biology of the Fibroblast* (E. Kulonen and J. Pikkarainen, eds), pp. 93–102. Academic Press, New York.

Swalla, B. J., and Jeffery, W. R. 1996. Requirement of the Manx gene for expression of chordate features in a tailless ascidian larva. *Science* 274: 1205–1208.

Swalla, B. J., Makabe, K. W., Satoh, N., and Jeffery, W. R. 1993. Novel genes expressed differentially in ascidians with alternate modes of development. *Development* 119: 307–318.

Symmons, S. 1979. Notochordal and elastic components of the axial skeleton of fishes and their functions in locomotion. *J Zool London* 189: 157–206.

Takada, S., Stark, K. L., Shea, M. J., Vassileva, G., McMahon, J. A., and McMahon, A. P. 1994. Wnt-3a regulates somite and tailbud formation in the mouse embryo. *Genes Dev* 8: 174–189. doi: 10.1101/gad.8.2.174.

Takahashi, H., Hotta, K., Erives, A., Di Gregorio, A., Zeller, R. W., Levine, M., and Satoh, N. 1999. Brachyury downstream notochord differentiation in the ascidian embryo. *Genes Dev* 13: 1519–1523.

Talbot, W. S., Trevarrow, B., Halpern, M. E., Melby, A. E., Farr, G., Postlethwait, J. H., Jowett, T., Kimmel, C. B., and Kimelman, D. 1995. A homeobox gene essential for zebrafish notochord development. *Nature* 378: 150–157.

Tamplin, O. J. 2009. Development of the Mouse Notochord. PhD thesis. Department of Molecular Genetics and the Collaborative Program in Developmental Biology. University of Toronto, Toronto, Canada. p. 187.

Tan, J., He, W., Luo, G., and Wu, J. 2015. Involvement of impaired desmosome-related proteins in hypertrophic scar intraepidermal blister formation. *Burns* 41: 1517–1523.

Teillet, M. A., and Le Douarin, N. M. 1983. Consequences of neural tube and notochord excision on the development if the peripheral nervous system in the chick embryo. *Dev Biol* 98: 192–211.

Teillet, M., Watanabe, Y., Jeffs, P., Duprez, D., Lapointe, F., and Le Douarin, N. M. 1998a. Sonic hedgehog is required for survival of both myogenic and chondrogenic somitic lineages. *Development* 125: 2019–2030.

Teillet, M. A., Lapointe, F., and Le Douarin, N. M. 1998b. The relationships between notochord and floor plate in vertebrate development revisited. *Proc Natl Acad Sci USA* 95: 11733–11738.

Thompson, H., and Tucker, A. S. 2013. Dual origin of the epithelium of the mammalian middle Ear. *Science* 339: 1453–1456.

Thomson, A. 1918. Osteology. The Skeleton. In *Cunningham's Text-Book of Anatomy* (A. Robinson, ed.), 5th Edition, p. 1593. Willam Wood and Company, New York.

Thorogood, P. V., Bee, J., and von der Mark, K. 1986. Transient expression of collagen type II at epithelio-mesenchymal interfaces during morphogenesis of the cartilaginous neurocranium. *Dev Biol* 116: 497–509.

Toole, B. P. 1972. Hyaluronate turnover during chondrogenesis in the developing chick limb and axial skeleton. *Dev Biol* 29: 321–329.

Trapani, V., Bonaldo, P., and Corallo, D. 2017. Role of the ECM in notochord formation, function. *J Cell Sci* 130: 1–9. doi: 10.1242/jcs.175950.

Trelstad, R. L., Kang, A. H., Cohen, A. M., and Hay, E. D. 1973. Collagen synthesis in vitro by embryonic spinal cord epithelium. *Science* 179: 295–297.

Trout, J. J., Buckwalter, J. A., and Moore, K. C. 1982b. Ultrastructure of the human intervertebral-disk. 2. Cells of the nucleus pulposus. *Anat Rec* 204: 307–314.

Trout, J. J., Buckwalter, J. A., Moore, K. C., and Landas, S. K. 1982a. Ultrastructure of the human intervertebral-disk. 1. Changes in notochordal cells with age. *Tissue Cell* 14: 359–369.

Trueman, C. N. 2013. Chemical taphonomy of biomineralized tissue. *Palaeontology* 56: 475–486.

Tucker, A. S., and Slack, J. M. W. 1995. The *Xenopus laevis* tail-forming region. *Development* 121: 249–262.

Turner, S, Burrow, C. J., Schultze, H.-P., Blieck, A., Reif, W. E., Rexroad, C. B., Bultynck, P., and Nowlan, G. S. 2010. False teeth: conodont-vertebrate phylogenetic relationships revisited. *Geodiversitas* 32: 545–594.

Ueoka, C., Nadanaka, S., Seno, N., Khoo, K.-H., and Sugahara, K. 1999. Structural determination of novel tetra- and hexasaccharide sequences isolated from chondroitin sulfate H (oversulfated dermatan sulfate) of hagfish notochord. *Glycoconjugate J* 16: 291–305.

Vacaru, A. M., Unlu, G., Spitzner, M., Mione, M., Knapik, E. W., and Sadler, K. C. 2014. In vivo cell biology in zebrafish Providing insights into vertebrate development and disease. *J Cell Sci* 127: 485–495.

van Eeden, F. J. M., Granato, M., Schach, U., Brand, M., Furutani-Seiki, M., Haffter, P., Hammerschmidt, M., Heisenberg, C. P., Jiang, Y. J., Kane, D. A., Kelsh, R. N., Mullins, M. C., Odenthal, J., Warga, R. M., Allende, M. L., Weinberg, E. S., and Nüsslein-Volhard, C. 1996. Mutations affecting somite formation and patterning in the zebrafish, *Danio rerio*. *Development* 123: 153–164.

Vandewalle, P., Gluckmann, I., and Wagemans, F. 1998. A critical assessment of the alcian blue/alizarine double staining in fish larvae and fry. *Belg J Zool* 128: 93–95.

Vasan, N. S. 1987. Somite chondrogenesis: the role of the microenvironment. *Cell Differ* 21: 147–159.

Vasan, N. S., Lamb, K. M., and La Manna, O. 1986. Somite chondrogenesis in vitro: 2. Changes in the hyaluronic acid synthesis. *Cell Differ* 18: 91–99.

Vasan, N. S., and Miller, E. 1985. Somite chondrogenesis in vitro: differential inductions by modified matrix – a biochemical and morphological study. *Dev Growth Differ* 27: 405–418.

Verbout, A. J. 1976. A critical review of the 'neugliederung' concept in relation to the development of the vertebral column. *Acta Biotheor* 25: 219–258.

Vieira, L. G., Lima, F. C., Mendonca, S. H. S. T., Menezes, L. T., Hirano, L. Q. L., and Santos, A. L.Q. 2018. Ontogeny of the postcranial axial skeleton of *Melanosuchus niger* (Crocodylia, Alligatoridae). *Anat Rec* 301: 607–623. doi: 10.1002/ar.23722.

von Baer, K. E. 1828. *Über Entwicklungsgeschichte der Thiere: Beobachtung und Reflexion. Erster Theil*. Gebrüder Bornträger, Königsberg.

von Baer, K. E. 1837. *Über Entwicklungsgeschichte der Thiere: Beobachtungen und Reflexionen. Zweiter Theil*. Gebrüder Bornträger, Königsberg.

von der Mark, K. 1980. Immunological studies on collagen type transition in chondrogenesis. *Curr Topics Dev Biol* 14: 199–225.

von der Mark, H., von der Mark, K., and Gay, S. 1976a. Study of differential collagen synthesis during development of the chick embryo by immunofluorescence. I. Preparation of collagen type I and type II specific antibodies and their application to early stages of the chick embryo. *Dev Biol* 48: 237–249.

von der Mark, K., von der Mark, H., and Gay, S. 1976b. Study of differential collagen synthesis during development of the chick embryo by immunofluorescence. II. Localization of type I and type II collagen during long bone development. *Dev Biol* 53: 153–170.

von Ebner, V. 1888. Urwirbel und Neugliederung der Wirbelsäule. *Sitz Akad Wiss Wien* 111/97: 194–206.

von Ebner, V. 1896. Uber die Wirbel der Knochenfische und die Chorda dorsalis der Fische und Amphibien. *Sitz Akad Wiss Wien* 105: 123–161.

von Zittel, K. A. 1911. *Grundzüge Paläontologie (Paläozoologie). II. Abteilung Vertebrata*. 2nd Edition, p. 598. Von R. Oldenbourg, München.

Vujovic, S., Henderson, S., Presneau, N., Odell, S., Jacques, T. S., Tirabosco, R., Boshoff, C., and Flanagan, A. M. 2006. Brachyury, a crucial regulator of notochordal development, is a novel biomarker for chordomas. *J Pathol* 209: 157–165.

Wada, H. 2010. Origin and genetic evolution of the vertebrate skeleton. *Zool Sci* 27: 119–123.

Waddington, C. H. 1930. Developmental mechanics of chick and duck embryos. *Nature* 125: 924–925.

Waddington, C. H. 1932. Experiments on the development of chick and duck embryos cultivated in vitro. *Philos Trans R Soc London* 221: 179–230.

Waddington, C. H. 1937. Experiments on determination in the rabbit embryo. *Arch Biol* 48: 273–290.

Waddington, C. H., and Perry, M. M. 1962. The ultrastructure of the developing urodele notochord. *Proc R Soc London B* 156: 459–479.

Wagner, G. P. 1989a. The origin of morphological characters and the biological basis of homology. *Evolution* 43: 1157–1171.

Wagner, G. P. 1989b. The biological homology concept. *Annu Rev Ecol Syst* 20: 51–69.

Wainwright, S. A., Biggs, W. D., Currey, J. D., and Gosline, J. M. 1976. *Mechanical Design in Organisms*. Edward Arnold, London.

Wake, D. B. 1970. Aspects of vertebral evolution in the modern Amphibia. *Forma Funct* 3: 33–60.

Wake, D. B., and Lawson, R. 1973. Developmental and adult morphology of the vertebral column in the plethodontid salamander *Eurycea bislineata*, with comments on vertebral evolution in the amphibia. *J Morph* 139: 251–299.

Wake, H. M., and Wake, D. B. 2000. Developmental morphology of early vertebrogenesis in Caecilians (Amphibia: Gymnophiona): resegmentation and phylogenesis. *Zoology* 103: 68–88.

Walker, M. B., and Kimmel, C. B. 2007. A two-color acid-free cartilage and bone stain for zebrafish larvae. *Biotech Histochem* 82: 23–28.

Wang, S., Furmanek, T., Kryvi, H., Krossøy, C., Totland, G. K., Grotmol, S., and Wargelius, A. 2014. Transcriptome sequencing of Atlantic salmon (*Salmo salar* L.) notochord prior to development of the vertebrae provides clues to regulation of positional fate, chordoblast lineage and mineralisation. *BMC Genom* 15: 141. doi: 10.1186/1471-2164-15-141.

Wang, S., Kryvi, H., Grotmol, S., Wargelius, A., Krossøy, K., Epple, M., Neues, F., Furmanek, T., and Totland, G. K. 2013. Mineralization of the vertebral bodies in Atlantic salmon (*Salmo salar* L.) is initiated segmentally in the form of hydroxyapatite crystal accretions in the notochord sheath. *J Anat* 223: 159–170.

Ward, A. B., and Brainerd, E. L. 2007. Evolution of axial patterning in elongate fishes. *Biol J Linn Soc* 90(1): 97–116.

Ward, L., Pang, A. S. W., Evans, S. E., and Stern, C. D. 2018. The role of the notochord in amniote vertebral column segmentation. *Dev Biol* 439: 3–18.

Watterson, R. L., Fowler, I., and Fowler, B. J. 1954. The role of the neural tube and notochord in development of the axial skeleton of the chick. *Am J Anat (continued as Dev Dyn)* 95: 337–399.

Weinans, H., and Prendergast, P. J. 1996. Tissue adaptation as a dynamical process far from equilibrium. *Bone* 19: 143–149.

Weinstein, D. C., Ruiz I Altaba, A., Chen, W. S., Hoodless, P., Prezioso, V. R., Jessell, T. M., and Darnell, J. E. Jr. 1994. The winged-helix transcription factor HNF-3 is required for notochord development in the mouse embryo. *Cell* 78: 575–588.

Welsch, U. 1968. Über den Feinbau der Chorda dorsalis von *Branchiostoma lanceolatum*. *Z Zellforsch* 87: 69–81.

Welsch, U., and Storch, V. 1969. Zur Feinstruktur der Chorda dorsalis niederer Chordaten [*Dendrodoa grossularia* (v. Beneden) und *Oikopleura dioica* Fol]. *Z Zellforsch* 93: 547–559.

Welsch, U., and Storch, V. 1971. Fine structural and enzyme histochemical observations on notochord of *Ichthyophis glutinosus* and *Ichthyophis kohtaoensis* (Gymnophiona, Amphibia). *Z Zellforsch Mikros Anat* 117: 443–450.

Welsch, U., Chiba, A., and Honma, Y. 1998. The notochord. In *The Biology of Hagfishes* (J. M. Jørgensen, J. P. Lomholt, R. E. Weber, and H. Malte, eds), pp. 145–159. Springer-Science + Business Media, B.V. Dordrecht.

Werneburg, R. 2008. The notochord in brachiosaurs of the Permo-Carboniferous and possible sexual dimorphism. *Veröff Nat Mus Schl* 23: 63–86.

Werner, Y. L. 1967. Regeneration of the caudal axial skeleton in a Gekkonid lizard (*Hemidactlyus*) with particular reference to the 'latent' period. *Acta Zool* 48: 103–126.

West-Eberhard, M. J. 2003. *Developmental Plasticity and Evolution.* p. 794. Oxford University Press, New York.

Wettstein, O. 1931. Ordnung der klasse reptilia: rhynchocephalia. In *Handbuch der Zoologie: Sauropsida: Allgemeines, Reptilia, Aves* (W. Kükenthal and T. Krumbach, eds), p. 235. de Gruyter, Berlin.

White, E. I. 1946. *Jamoytius kerwoodi*, a new chordate from the Silurian of Lanarkshire [Scotland]. *Geol Mag* 83: 89–97.

Whitehouse, R. H. 1910. The caudal fin of the Teleostomi. *Proc Zool Soc London* 80(3): 590–627.

Wiley, E. O., Fuiten, A. M., Doosey, M. H., Lohman, B. K., Merkes, C., and Azuma, M. 2015. The caudal skeleton of the zebrafish, *Danio rerio*, from a phylogenetic perspective: a polyural interpretation of homologous structures. *Copeia* 103(4): 740–750.

Willems, B., Büttner, A., Huysseune, A., Renn, J., Witten, P. E., and Winkler, C. 2012. Conditional ablation of osteoblasts in medaka. *Dev Biol* 364: 128–137.

Willey, A. 1894. *Amphioxus and the Ancestry of the Vertebrates.* Macmillan, New York.

Williams, E. E. 1959. Gadow's arcualia and the development of tetrapod vertebrae. *Q Rev Biol* 34: 1–32.

Williams, J. L. 1942. The development of cervical vertebrae in the chick under normal and experimental conditions. *Am J Anat (continued as Dev Dyn)* 71: 153–177.

Williams, L. W. 1908. The later development of the notochord in mammals. *Am J Anat (continued as Dev Dyn)* 8: 251–284.

Williston, S. W. 1925. *The Osteology of the Reptiles.* p. 300. Harvard University Press, Cambridge, MA.

Winchester, L, and Bellairs, A. D'A. 1977. Aspects of vertebral development in lizards and snakes. *J Zool London* 181: 495–525.

Wintrich, T., Scaal, M., Böhmer, C., Schellhorn, R., Kogan, I., van der Reest, A., and Sander, P. M. 2020. Palaeontological evidence reveals convergent evolution of intervertebral joint types in amniotes. *Sci Rep* 10: 14106. doi: 10.1038/s41598-020-70751-2.

Witten, P. E., Fjelldal, P. G., Huysseune, A., McGurk, C., Obach, A., and Owen, M. A. G. 2019. Bone without minerals and its secondary mineralization in Atlantic salmon (*Salmo salar*): the recovery from phosphorus deficiency. *J Exp Biol* 222: jeb188763. doi: 10.1242/jeb.188763.

Witten, P. E., Gil-Martens, L., Hall, B. K., Huysseune, A., and Obach, A. 2005. Compressed vertebrae in Atlantic salmon (*Salmo salar*) evidence for metaplastic chondrogenesis as a skeletogenic response late in ontogeny. *Dis Aquat Org* 64: 237–246.

Witten, P. E., Gil-Martens, L., Huysseune, A., Takle, H., and Hjelde, K. 2009. Towards a classification and an understanding of developmental relationships of vertebral body malformations in Atlantic salmon (*Salmo salar* L.). *Aquaculture*, 295:6–14.

Witten, P. E., and Hall, B. K. 2015. Teleost skeletal plasticity: modulation, adaptation, and remodelling. *Copeia* 103: 727–739.

Witten, P.E., and Hall, B. K. 2021. The ancient, segmented, active and permanent notochord. In *Ancient Fishes and their Living Relatives: A Tribute to John G Maisey* (A. Pradel, J. S. S. Denton, and P. Janvier, eds), pp. 210–219. Verlag Dr. Friedrich Pfeil, München, Germany. ISBN 978-3-89937-269-4.

Witten, P. E., Harris, M. P., Huysseune, A., and Winkler, C. 2017. Small teleost fish provide new insights into human skeletal diseases. *Methods Cell Biol* 138: 321–346.

Witten, P. E., and Huysseune, A. 2009. A comparative view on mechanisms and functions of skeletal remodelling in teleost fish, with special emphasis on osteoclasts and their function. *Biol Rev Camb Philos Soc* 84: 315–346.

Witten, P. E., Huysseune, A., and Hall, B. K. 2010. A practical approach for the identification of the many cartilaginous tissues in teleost fish. *J Appl Ichthyol* 26: 257–262.

Witten, P. E., Obach, A., Huysseune, A., and Baeverfjord, G. 2006. Vertebrae fusion in Atlantic salmon (*Salmo salar*); development, aggravation and pathways of containment. *Aquaculture* 258: 164–172.

Wolpert, L. 1991. *The Triumph of the Embryo*. Oxford University Press, Oxford. 211 pp.

Wood, A., and Thorogood, P. 1994. Patterns of cell behaviour underlying somitogenesis and notochord formation in intact vertebrate embryos. *Dev Dyn* 201: 151–167.

Wopat, S., Bagwell, J., Sumigray, K. D., Dickson, A. L., Huitema, L. F. A., Poss, K. D., Schulte-Merker, S., and Bagnat, M. 2018. Spine patterning is guided by segmentation of the notochord sheath. *Cell Rep* 22: 2026–2038.

Wu, L. N. Y., Sauer, G. R., Genge, B. R., and Wuthier, R. E. 1989. Induction of mineral deposition by primary cultures of chicken growth plate chondrocytes in ascorbate-containing media. Evidence of an association between matrix vesicles and collagen. *J Biol Chem* 264: 21346–21355.

Yamada, T., Pfaff, S. L., Edlund, T., and Jessell, T. M. 1993. Control of cell pattern in the neural tube: motor neuron induction by diffusible factors from notochord and floor plate. *Cell* 73: 673–686.

Yamada, T., Placzek, M., Tanaka, H., Dodd, J. and Jessell, T. M. 1991. Control of cell pattern in the developing nervous system: polarizing activity of the floor plate and notochord. *Cell* 64: 635–647.

Yamamoto, M., Morita, R., Mizoguchi, T., Matsuo, H., Isoda, M., Ishitani, T., Chitnis, A. B., Matsumoto, K., Crump, J. G., Hozumi, K., Yonemura, S., Kawakami, K., and Itoh, M. 2010. Mib-Jag1-Notch signaling regulates patterning and structural roles of the notochord by controlling cell fate decisions. *Development* 137: 1527–1537.

Yamanaka, Y., Tamplin, O. J., Beckers, A., Gossler, A., and Rossant, J. 2007. Live imaging and genetic analysis of mouse notochord formation reveals regional morphogenetic mechanisms. *Dev Cell* 13: 884–896.

Yan, Y. L., Hatta, K., Riggleman, B., and Postlethwait, J. H. 1995. Expression of a type-II collagen gene in the zebrafish embryonic axis. *Dev Dyn* 203: 363–376.

Yang, L., Tsang, K. Y., Tang, H. C., Chan, D., and Cheah, K. S. E. 2014. Hypertrophic chondrocytes can become osteoblasts and osteocytes in endochondral bone formation. *Proc Natl Acad Sci USA* 111: 12097–12102.

Yang, X. R., Ng, D., Alcorta, D. A., Liebsch, N. J., Sheridan, E., Li, S., Goldstein, A. M., Parry, D. M., and Kelley, M. J. 2009. Brachyury gene duplication confers susceptibility to familial chordomas. *Nat Genet* 41: 1176–1178.

Yu, T., Graf, M., Renn, J., Schartl, M., Larionova, D., Huysseune, A., Witten, P. E., and Winkler, C. 2017. A vertebrate specific and essential role for sp7/osterix in osteogenesis revealed by gene knock-out in the teleost medaka. *Development* 144: 265–271.

Zachos, F. E., and Hoßfeld, U. 1996. Adolf Remane (1898–1976) and his views on systematics, homology and the Modern Synthesis. *Theory Biosci* 124: 335–348.

Zakin, L., Chang, E. Y., Plouhinec, J. L. and De Robertis, E. M. 2010. Crossveinless-2 is required for the relocalization of Chordin protein within the vertebral field in mouse embryos. *Dev Biol* 347: 204–215.

Zangerl, R. 1981. Chondrichthyes I. Paleozoic Elasmobranchii. In *Handbook of Paleoichthyology* (H.-P. Schultze, ed.), *Volume 3A*. p. 113. Verlag Dr. Friedrich Pfeil, Munich.

Zhang, G. 2009. An evo-devo view on the origin of the backbone: evolutionary development of the vertebrae. *Inter Comp Biol* 49(2): 178–186.

Zhang, G. J., and Cohn, M. J. 2006. Hagfish and lancelet fibrillar collagens reveal that type II collagen-based cartilage evolved in stem vertebrates. *Proc Natl Acad Sci USA* 103: 16829–16833.

Zhang, G. J., Miyamoto, M. M., and Cohn, M. J. 2006. Lamprey type II collagen and Sox9 reveal an ancient origin of the vertebrate collagenous skeleton. *Proc Natl Acad Sci USA* 103: 3180–3185.

Zhao, Q., Eberspaecher, H., Lefebvre, V., and de Crombrugghe, B. 1997. Parallel expression of Sox9 and Col2a1 in cells undergoing chondrogenesis. *Dev Dyn* 209: 377–386.

Zykoff, W. 1893. Uber das Verhältnis des Knorpels zur Chorda bei *Siredon pisciformis*. *Bull Soc Imp Nat Moscou* 7: 30–36.

Index

Note: **Bold** page numbers refer to tables and *italic* page numbers refer to figures

Evolutionary Cell Biology

Series Editors
Brian K. Hall – *Dalhousie University, Halifax, Nova Scotia, Canada*
Sally A. Moody – *George Washington University, Washington DC, USA*

EDITORIAL BOARD
Michael Hadfield – *University of Hawaii, Honolulu, USA*
Kim Cooper – *University of California, San Diego, USA*
Mark Martindale – *University of Florida, Gainesville, USA*
David M. Gardiner – *University of California, Irvine, USA*
Shigeru Kuratani – *Kobe University, Japan*
Nori Satoh – *Okinawa Institute of Science and Technology, Japan*
Sally Leys – *University of Alberta, Canada*

SCIENCE PUBLISHER
Charles R. Crumly – *CRC Press/Taylor & Francis Group*

PUBLISHED TITLES

The Notochord
Development, Evolution and
Contributions to the Vertebral Column
By P. Eckhard Witten and Brian K. Hall

The Evolution of Multicellularity
Edited by Matthew D. Herron, Peter L. Conlin, and William C. Ratcliff

Evolution of Neurosensory Cells and Systems
Gene Regulation and Cellular Networks and Processes
Edited by Bernd Fritzsch and Karen L. Elliott

Evolutionary Cell Processes in Primates
Two Volume Set
Edited by M. Kathleen Pitirri and Joan T. Richtsmeier

Evolutionary Cell Processes in Primates
Bone, Brains, and Muscle, Volume I
Edited by M. Kathleen Pitirri and Joan T. Richtsmeier

Evolutionary Cell Processes in Primates
Genes, Skin, Energetics, Breathing, and Feeding, Volume II
Edited by M. Kathleen Pitirri and Joan T. Richtsmeier

For more information about this series, please visit: www.crcpress.com/Evolutionary-Cell-Biology/book-series/CRCEVOCELBIO